The Laws of Nature

Legal Thought
Across Disciplines
Published in Cooperation with
The University of Akron School of Law

The Laws of Nature

★　　★　　★　　★　　★

Reflections on the Evolution of Ecosystem Management Law & Policy

Edited by Kalyani Robbins

University of Akron Press
Akron, Ohio

All rights reserved • First Edition 2013 • Manufactured in the United States of America. All inquiries and permission requests should be addressed to the Publisher, the University of Akron Press, Akron, Ohio 44325–1703.

17   16   15   14   13   5   4   3   2   1

LIBRARY OF CONGRESS CATALOGING-IN-PUBLICATION DATA
The laws of nature : reflections on the evolution of ecosystem management law and policy / [edited by] Kalyani Robbins.
     p. cm.   ISBN 978-1-935603-63-4 (pbk.)
     1. Ecosystem management—Law and legislation—United States. 2. Conservation of natural resources—Law and legislation—United States. 3. Biodiversity conservation—Law and legislation—United States. 4. Environmental policy—United States. I. Robbins, Kalyani, 1970–
     KF5505.L39 2012   346.7304'4—DC23
                              2012034307

The paper used in this publication meets the minimum requirements of ANSI/NISO Z39.48–1992 (Permanence of Paper). ∞

The Laws of Nature was designed and typeset by Amy Freels, with assistance from Lauren McAndrews. The typeface, Stone Print, was designed by Sumner Stone in 1991. *The Laws of Nature* was printed on sixty-pound natural and bound by BookMasters of Ashland, Ohio.

Cover photo: Grizzly bear sow and cubs, Yellowstone National Park, by Kim Keating, used with permission.

For Skyler and Maxfield

Contents

Contributors

Robert Adler, James I. Farr Chair in Law and Professor of Law, University of Utah S.J. Quinney College of Law. B.A. Johns Hopkins University; J.D. Georgetown University.

James W. Boyd, Senior Fellow and Co-Director, Center for the Management of Ecological Wealth, Resources for the Future. B.A. University of Michigan; Ph.D. Wharton Business School, University of Pennsylvania.

Susan G. Clark, Joseph F. Cullman 3rd Adjunct Professor of Wildlife Ecology and Policy Sciences, School of Forestry & Environmental Studies and Fellow, Institution for Social and Policy Studies, Yale University. B.S. Northeastern Oklahoma State College; M.S. University of Wyoming; Ph.D. University of Wisconsin.

David Cherney, Research Affiliate at the University of Colorado at Boulder's Center for Science and Technology Policy Research and Research Associate with the Northern Rockies Conservation Cooperative. B.A. Claremont McKenna College; M.A. Yale University; Ph.D. University of Colorado.

Jamison E. Colburn, Joseph H. Goldstein Faculty Scholar and Professor of Law, Penn State Dickinson School of Law. B.A. State University of New York, Plattsburgh; J.D. Rutgers University; LL.M. Harvard University; J.S.D. Columbia University.

Travis Greenwalt, Senior Economist at Cardno ENTRIX. B.S. and M.B.A. University of Montana.

Robert B. Keiter, Wallace Stegner Professor of Law, University Distinguished Professor, and Director of Wallace Stegner Center for Land, Resources, and the

Environment, University of Utah S.J. Quinney College of Law. B.A. Washington University; J.D. Northwestern University.

Judith A. Layzer, Associate Professor, Head, Environmental Policy and Planning, Massachusetts Institute of Technology. B.A. University of Michigan; Ph.D. Massachusetts Institute of Technology.

Deborah A. McGrath, Associate Professor & Chair of Biology, Sewanee: University of the South. B.A. University of Wisconsin; M.S. and Ph.D. University of Florida.

Martin Nie, Professor, Natural Resource Policy; Chair, Department of Society & Conservation, University of Montana. B.A. University of Nebraska; Ph.D. Northern Arizona University.

Sara O'Brien, Director, Conservation Planning, Defenders of Wildlife. B.A. Grinnell College; M.S. University of Arizona.

Kalyani Robbins, Associate Professor of Law, University of Akron School of Law. B.A. University of California at Berkeley; J.D. Stanford University Law School; LL.M. Lewis & Clark Law School.

Daniel Rohlf, Associate Professor of Law and Of Counsel, Earthrise Law Center, Lewis & Clark Law School. B.S. Colorado College; J.D. Stanford University Law School.

J.B. Ruhl, David Daniels Allen Distinguished Chair in Law, Vanderbilt University Law School. B.A. and J.D. University of Virginia; LL.M. George Washington University; Ph.D. Southern Illinois University.

Lynn Scarlett, Visiting Scholar and Co-Director, Center for the Management of Ecological Wealth, Resources for the Future. B.A. and M.A. University of California, Santa Barbara.

Sara Vickerman, Senior Director, Biodiversity Partnerships, Defenders of Wildlife. A.A. Fullerton Junior College; B.S. California State University at Fullerton; M.S. Southern Oregon University.

I

Understanding and Evaluating
Ecosystem Management Thus Far

1

An Ecosystem Management Primer
History, Perceptions, and Modern Definition
Kalyani Robbins, The University of Akron

"When we try to pick out anything by itself, we find it hitched to everything else in the Universe."
—*John Muir*

Ecosystem management is still a relatively new field of study—then Forest Service Chief F. Dale Robertson coined the term just two decades ago in 1992[1]—so its membership is still fairly small. But the issues are too important, too potentially life-altering, to leave to a handful of experts to worry about. This book is for everyone: law students, college and graduate students, experts, and weekend readers alike. Because it is for everyone, it is essential that it begin at the beginning.

Much like we have shortened biological diversity into the now common term 'biodiversity,' the term 'ecosystem' is the short (and now more common) way of saying ecological system.[2] Systems in general exist on multiple scales, so it is likewise the case that the term 'ecosystem' applies to discrete natural units such as a lake or a valley, as well as vast regions in which the interconnectedness of nature has been observed.[3] Indeed, when multiple systems interact, that is itself a system, and so on, giving rise to a complex and nearly infinite concept. The spatial definition of an ecosystem is any unit of nature, at any scale, in which the biotic organisms and abiotic environment interact in a manner that results in an ongoing and dynamic biotic structure.[4] However, some adhere to a more 'process-based' view, in which an ecosystem is defined by the processes through which it functions, such as "productivity, energy flow among trophic levels, decomposition, and nutrient cycling."[5] Regardless of the ecosystem understanding one prefers, there is no question that ecosystems provide humans with many essential services, some of which are even subject to economic valuation via a replacement-cost analysis.[6]

The phrase 'ecosystem management' already gives away quite a bit, if we simply look at the combination of terms. The term 'ecosystem' evokes nature. An ecosystem is the most fundamental unit in nature, and the relationships it embodies are essential to understanding our natural world. Arthur Tansley, a pioneer of the science of ecology, coined the term 'ecosystem' in 1935. Tansley stated: "Though the organisms may claim our prime interest, when we are trying to think fundamentally, we cannot separate them from their special environments, with which they form one physical system."[7] 'Management,' on the other hand, suggests human control. It is a very unnatural word, the opposite of letting nature take its course. Indeed, in spite of the fact that the Clinton administration introduced the ecosystem management concept in an effort to incorporate scientific principles into the management of the national forests (recognizing that ecosystems were the focus for scientists),[8] the initial effort involved such excessive top-down government control that it met with great resistance.[9] The concept later evolved into one involving greater shared decision making at multiple levels,[10] though management is still management, a human domination over nature. As such, the term 'ecosystem management,' without more, already gives away the inherent tension between nature and humanity—a tension that spawns both the need for, and the problems with, ecosystem management.

This chapter will first take the reader on a journey through the history of ecosystem management, providing a summary of how it has grown and developed over the past two decades. This will only naturally lead to the next part of the chapter, which focuses on the present understanding of how ecosystem management is to be defined and applied, as well as the variety in perceptions of this modern understanding. Finally, it will serve as an introduction to the remainder of the book, previewing the various contributions collected here, offered by some of the leading scholars in the field of ecosystem management.

I. THE LIFE AND TIMES OF ECOSYSTEM MANAGEMENT

In spite of the development of ecosystem-orientation in the 1930s, the next half-century remained focused on narrowly targeted single-jurisdiction management of land and natural resources. The lack of a more holistic approach capable of respecting the intricate web of ecosystem relationships accelerated the damage we caused to the natural environment. By the 1970s and 1980s the scientific community had begun to emphasize the need for a broader landscape-based approach to not only understanding, but also regulating, the natural environment. This coincided with the culmination of decades of ecological research that had disproved the previous theory—based on a notion of 'equilibrium'—basically, that ecosystems were stable and self-regulating fully-enclosed entities. What we were discovering instead was that ecosystems were in fact dynamic and interactive with external forces, including humans. And disturbances—such as fires, hurricanes, floods, and drought—previously viewed as potentially harmful, were found to be incredibly valuable players in the evolution of ecosystems and their relationships with one another.[11] We needed to move toward a management approach that could take everything into account—ecological, social, economic, and climate realities—rather than cordoning off a particular tract of land for focused management.

Decades of controlling disturbances and expecting already-fragmented ecosystems to manage on their own, even to benefit from a lack of further human interference, led to fragile ecosystems unable to withstand potentially unavoidable disturbance, much like a coddled child forced to enter the real world. This "command-and-control approach implicitly assumes that the problem is well-bounded, clearly defined, relatively simple, and generally linear with respect to cause and effect."[12] The reality, as we were discovering, is a far more complex, interactive, unpredictable world beyond our complete grasp. Not only was it harmful to attempt to control disturbances, but it was unwise to expect nature preserves to take care of themselves if we simply prevented further human interference with them, given that we had already done the greatest misdeed: turning them into islands forced to devour themselves due to lack of interaction with other ecosystems.[13]

By the 1980s it had become clear to many environmentalists, ecologists, and conservation biologists that our policy decisions for land and resource management needed to take greater care to heed the decades-old advice of Aldo Leopold, that everything is dependent on everything else and no part can be sacrificed without great risk to the whole.[14] Environmental problems cannot be addressed individually in a vacuum; rather, the entire field must be viewed holistically and in a comprehensive manner. Political boundaries and property lines mean nothing to the natural world and as such make for terrible management scales. It finally dawned on administrators, in response to substantial pressure from environmental and scientific stakeholders, that land and resource management should ideally take place on a landscape scale. People began to understand the complexity of the situation, realizing that in order to "understand realistically complex ecological systems, it is necessary to study how the components affect and are affected by the larger, more complicated systems within which they are located."[15] That said, what is ideal, or even finally understood as ideal, is not always what actually takes place,[16] which is why we are here, over two decades later, still talking about this problem.

It is tough to turn back from a direction already traveled for some time, and we faced the two somewhat-related problems of too many cooks and too many items on the menu. First, as to the excessive menu, we had a very long-standing, firmly entrenched, multiple-use framework for managing natural resources. Agencies at both the state and federal levels had cut their teeth on the primary goal of sustainable commodity extraction and commercial development. To the extent that we restrained ourselves at all, it was only about allowing our economic use to continue into the future. It is not realistic to simply add to such multiple-use goals the new goal of ecological integrity and hope to make everyone happy at once. Throw in the numerous cooks—over a large-scale ecosystem there may be several jurisdictions and numerous private land owners—and it is nearly impossible to manage on a landscape scale. The U.S. Forest Service and Bureau of Land Management (both focused on commodity production and recreation), the Fish and Wildlife Service and National Park Service (arguably a bit more concerned with conservation), the Bureau of Reclamation and Army Corps of Engineers (managing federal irrigation and flood-control projects), state agencies, municipalities, and numerous private land owners would all have to somehow work together to coordinate their various mandates and needs with the needs of the overall ecosystem. Such comprehensive cooperation is the truth of our natural landscape but the impossibility of our political landscape.

In response to this massive-scale problem, concerned environmentalists and scientists began to speak of the concept of 'ecosystem management,' in which land and resource regulation would focus on interactions within and among ecosystems and adapt to changes in either scientific information or ecosystem functioning.[17] The goal of this new methodology was ecological restoration, but the approach included comprehensive consideration of social and economic functioning as well as ecosystem functioning (to the extent that these are even separate considerations; many argue that humans, with all our constructs, are an integral part of the ecosystems we inhabit).

Even before ecosystem management had been formally proposed or adopted, there were already a few examples of the (as-yet-untitled) approach that helped with the concept's development. The earliest examples of such a multi-jurisdiction effort to save a large-scale ecosystem date back to the 1970s—the multistate restorations of the Great Lakes and Chesapeake Bay, both of which involved an ecosystem-based approach.[18] Two of the most famous examples took place in the late 1980s as a result of concern for the habitat of two vulnerable species: the Greater Yellowstone grizzly bear and the northern spotted owl.

The first highly popularized program of ecosystem-focused management was for the Greater Yellowstone Ecosystem (GYE).[19] The GYE spans over eighteen million acres of land, overlapping the states of Montana, Wyoming, and Idaho. In addition to housing critical habitat for the grizzly bear, whooping crane, bald eagle, peregrine falcon, and trumpeter swan, the GYE is home to one of the last free-roaming bison herds and the world's largest herds of elk.[20] The GYE's "complex patchwork of management and ownership"[21] includes two national parks (Yellowstone and Grand Teton); three national wildlife refuges; land held by the Bureau of Land Management, states, and private owners; and overlaps six national forests—twenty-eight distinct political units in all. There are about six million acres of National Park Service and National Forest Service wilderness lands, as well as another six million acres of National Forest Service multiple-use lands. Depending on how the ecological boundaries of the GYE are defined, only about 7 to 30 percent is state or privately owned land, but this is nonetheless land of significant value, encompassing critical wildlife migration zones such as river valleys and other low-elevation areas.[22]

This diffuse set of stakeholders without shared goals had thus far resulted in ecologically harmful circumstances, such as habitat fragmentation, disruption of ecological processes, and an increase in human-wildlife confrontations.[23] The patchwork of habitat and human activity led to these problems, so environmentalists began to push for more integrated land management throughout the area. After their late 1970s discovery that the grizzly bear was foraging throughout the area, and far beyond the borders of Yellowstone National Park, biologists Frank and John Craighead coined the term 'Greater Yellowstone Ecosystem' and began the movement toward a unified bear-management scheme throughout the GYE.[24] Following their lead, the various environmental groups in the area came together to create an umbrella group called the Greater Yellowstone Coalition, in order to advocate for a more comprehensive ecosystem-based management strategy for the GYE.[25]

By the mid-1980s the Park Service and Forest Service still had not adequately coordinated their management of the region, leading to harsh criticism in a congressional hearing, which spurred the agencies to bring back a defunct inter-agency partnership from the 1960s, called the Greater Yellowstone Coordinating Committee (GYCC). Several years later, in 1990, the GYCC issued a draft vision document recommending ecosystem management and suggesting an interest in keeping the area largely wild. The vision document stated:

> ... the overall mood of the GY[E] will be one of naturalness, a combination of ecological processes operating with little restraint and humans moderating their activities so that they become a reasonable part of, rather than encumbrances upon, those processes ... the overarching goal is to conserve the sense of naturalness and maintain ecosystem integrity in the GY[E] through respect for ecological and geological processes and features that cross administrative boundaries.[26]

Naturally, this looked great, if perhaps a bit unrealistically optimistic, to environmentalists, but it inflamed local politicians and economic groups, who saw it as a threat to private property rights and local economies. Negotiations began and the final document cut back dramatically on what had been achieved in the draft. Rather than 19 million acres under diverse ownership, it was now a mere 11.7 million acres of national forest and national park lands. The stated goal of focusing on ecological integrity to preserve a natural state was removed; so much was removed that the length of the document itself was only about a sixth of the draft. In this new form the vision statement fell flat and failed to get any attention.[27]

The tale of the northern spotted owl, while perhaps more infamous than that of the GYE, actually fared quite a bit better in the end. We discovered that spotted owl populations were in decline in the 1970s when Oregon State University graduate student Eric Forsman proposed that the extensive cutting of old-growth forests was threatening the spotted owl with extinction. The state of Oregon listed the spotted owl as 'threatened' in 1975, but the U.S. Fish and Wildlife Service concluded that a listing under the federal Endangered Species Act was not warranted. Still, it was clear that the species was at least vulnerable, so federal land managers did adopt minimal protective measures to avoid the need for listing. Still, because of the power of the region's timber industry, there was little impact from these measures. The owl's condition only worsened, and as the science demonstrated this, the pressure from environmentalists rose to meet the economic pressure. The Pacific Northwest became a battle zone over the now nationally infamous

spotted owl. Bumper stickers carried phrases like "Kill an owl, save a logger."
Judges and their families required police protection. The intensity of the old-
growth-forest battle grew through the 1980s, finally culminating in 1988 in the
federal courts of Portland and Seattle, which essentially shut down the logging
of federally owned old-growth forests. The first Bush administration failed to
solve the problem before leaving office, so it was passed on to President Clinton.[28]

Shortly after his inauguration, Clinton invited all the major stakeholders in
the Pacific Northwest's old-growth forests to a summit, after which he arranged
for a team of experts to develop a forest management plan for the region that
would pull it out of the mess it had been in for so long. This resulted in the 1994
Northwest Forest Management Plan. Although the plan did not engage stakeholders to the extent generally envisioned for ecosystem management, it is otherwise a nice early example of the methodology. It was large-ecosystem-scaled,
bounded according to the spotted owl's range rather than political lines, and
included federal, state, and private lands. It considered other species besides the
owl, such as salmon, in recognition of the interconnectedness within an ecosystem. It utilized cutting-edge scientific information to create a network of interconnected reserves to facilitate migration of old-growth-dependent species and
embraced a return to normal disturbance regimes. It suggested ten different
adaptive management areas to give land managers laboratories for new interventions. It even took into account socioeconomic issues, such as job training
to help former timber workers move into new fields of work.[29] Such considerations are essential to a successful ecosystem management plan.

II. WHAT IS ECOSYSTEM MANAGEMENT?

A. Defining Ecosystem Management

With the movement from the early application of ecosystem management
principles to the formalization of ecosystem management in the 1990s, definitions became more concrete, even if still somewhat ambiguous.

> Ecosystem management is management driven by explicit goals, executed by
> policies, protocols, and practices, and made adaptable by monitoring and
> research based on our best understanding of the ecological interactions and
> processes necessary to sustain ecosystem composition, structure, and function.[30]

As Nagle and Ruhl point out, "this only begs the question: What are the
goals, policies, protocols, and practices of ecosystem management?"[31] Of course,
much of this cannot be answered by science, as goals and policies are determined

at political levels, a common problem for science-based policy that has been raised by many scholars.[32] As such, this definition is arguably where the scientific community throws the ball into the regulatory community's court, awaiting a response to the policy questions before determining such things as protocols and practices.

As definitions go, it may be easiest to think of ecosystem management in terms of what it does, generally speaking, saving the specifics for further discussion. Much like we have the process-based option for understanding the ecosystem itself, this is the process-based approach to understanding ecosystem management. Arguably the best definition of ecosystem management ever put forward came from R. Edward Grumbine, whose 1994 article fleshed out the concept with brilliant coherence. He began with a relatively simple definition: "Ecosystem management integrates scientific knowledge of ecological relationships within a complex sociopolitical and values framework toward the general goal of protecting native ecosystem integrity over the long term."[33]

What made Grumbine's article so important was not so much his own substantive contribution to the question of how to define ecosystem management, albeit quite valuable, but rather the fact that he took it upon himself to synthesize all of the then-existing scholarship on ecosystem management in search of common themes and goals. He found ten common themes, which are useful to this chapter's goal of providing a basic understanding of ecosystem management: 1) "Hierarchical Context," which is another way of describing the systems perspective, where such systems include multiple levels or scales, such as "genes, species, populations, ecosystems, [and] landscapes;" 2) "Ecological Boundaries," which is another way of saying that political boundaries do not apply; 3) "Ecological Integrity," which requires the protection of native diversity and processes, including disturbance regimes; 4) "Data Collection," which is considered a necessary component of ecosystem-wide planning; 5) "Monitoring," with which we maintain a continuous loop of feedback on the successes and failures of management actions to use as a basis for setting policy; 6) "Adaptive Management," which "focuses on management as a learning process or continuous experiment where incorporating the results of previous actions allows managers to remain flexible and adapt to uncertainty," and remains into the twenty-first century the most analyzed aspect of ecosystem management; 7) "Interagency Cooperation," which becomes necessary if we are to manage based on ecological boundaries rather than political ones; 8) "Organizational Change," which is the notion that moving to an ecosystem management approach will necessitate a restructuring of land management agencies and the manner in which they operate; 9) "Humans Embedded In Nature," or the idea that ecosystems include human beings, who interact with them at a level that must be taken into account in assessing an ecosystem's functioning; and 10) "Values," specifically human values, which unavoidably play a dominant role in determining the goals of ecosystem management, regardless of what we can learn from science.[34]

Grumbine drew his ecosystem management definition from these themes and further noted that most scholars shared the overarching goal of sustaining ecological integrity, most commonly focusing on the following specific goals for ecosystem management:

1. Maintain viable populations of all native species in situ.
2. Represent, within protected areas, all native ecosystem types across their natural range of variation.
3. Maintain evolutionary and ecological processes (i.e., disturbance regimes, hydrological processes, nutrient cycles, etc.).
4. Manage over periods of time long enough to maintain the evolutionary potential of species and ecosystems.
5. Accommodate human use and occupancy within these constraints.[35]

From this list of goals one can see just how difficult a task this is—indeed, potentially internally inconsistent, depending upon the size of human population at issue. Grumbine points out the greatest obstacle of all, which is the need to reconcile "the new goal of protecting ecological integrity and the old standard of providing goods and services for humans."[36] Of course, this leads to the question: Whose goal? Arguably this is simply a framing of the scientific community's goal as 'new' and the goals of our broader society and voting constituents as 'old.' If ecological integrity is indeed to become our new goal, this is only attainable with the very reconciliation Grumbine describes.

This is where ecosystem services may come in, which are detailed in far greater depth in chapters five and twelve. Consumptive value of land and natural resources is arguably a national tradition, but thankfully there is evidence of significant value to humans in the maintenance of healthy-functioning ecosystems. Ecosystem services are the benefits—many of which we depend on for life—derived from natural ecosystems. "Ecosystems, if properly protected and maintained, provide a wide array of valuable services to humans, ranging from the purification of water to the sequestration of carbon to the provision of pollinating insects essential to agricultural crop production."[37] Our work in discovering the range of ecosystem services and evaluating our ability to survive without them has only just begun, and may well pave the road to different attitudes toward conservation in the future.

B. Implementing Ecosystem Management

In implementing ecosystem management, arguably the most core universally expected element is adaptive management, in which land and resource managers treat their management actions themselves as a research study, always prepared to alter them according to the feedback received. Nearly every ecosystem management scholar considers this the essence of ecosystem management, a completely indispensable component.[38] Of course, this creates the question of what sorts of data are of interest,[39] and our response to that data is of course purely a policy question, so it becomes extremely important to consider who is in charge of adaptive management, as it necessarily entails a great deal of power. Thankfully, that power can be somewhat limited via detailed advance directives for responding to a range of potential management outcomes. Perhaps the greater risk is one of lacking the necessary funding to follow up with adaptive management programs once begun, which is a potentially catastrophic situation.[40]

A study by the National Academy of Sciences' National Research Council, asked to advise on agency planning for the Klamath River Basin,

> recommended using adaptive management and outlined its eight essential steps: (1) define the problem; (2) determine management goals and objectives; (3) determine the resource baseline; (4) develop conceptual models; (5) select future restoration actions; (6) implement management actions; (7) monitor ecosystem response; and (8) evaluate restoration efforts and proposals for remedial actions.[41]

Indeed, much of what defines adaptive management overlaps significantly with our understanding of ecosystem management, so the two go hand-in-hand. Ecosystem management, as dependent as it is on adaptive management, is a constantly evolving process.

Implementing ecosystem management also requires the employment of a diverse group of experts, given the variability of concerns to be taken into account. Such interdisciplinary teams must be trained to communicate effectively with one another, in addition to being informed regarding the specific nature of the ecosystem management projects they are to work on together. Further, a system must be in place to receive input from a range of local interests throughout implementation.

Finally, as a practical matter, ecosystem management can require, or at least benefit from, the use of modeling techniques.[42] Management planning requires a significant quantity of data, much of which will be collected after implementation has begun via adaptive management techniques. However, given that we must begin somewhere, the substantial data gaps can be filled via modeling, in which predictions and probabilities are formed into a hypothetical image of the future.[43] Of course, the use of modeling data can also be controversial, and certainly should be applied with care to minimize the risk of error.

C. *The Trouble with Ecosystem Management*

Given that ecosystem management is so widely considered the ideal approach to land and resource management, why do we continue to flounder in our effort to meaningfully implement it? Robert Lackey suggested that the problems facing ecosystem management have five general characteristics:

1. fundamental public and private values and priorities are in dispute, resulting in partially or wholly mutually exclusive decision alternatives;
2. there is substantial and intense political pressure to make rapid and significant changes in public policy in spite of disputes over values and priorities and the presence of mutually exclusive decision alternatives;
3. public and private stakes are high, with substantial costs and substantial risks of adverse effects (some also irreversible ecologically) to some groups regardless of which option is selected (think of the Endangered Species Act);
4. technical facts, ecological and sociological, are highly uncertain (after all, how certain are we over the long term consequences of farming nearly all of the tall grass prairie?);
5. ecosystem policy problems are meshed in a large framework assuring that policy decisions will have effects outside the scope of the problem (think about the "taking" issue: which "rights" take precedence in public policy?).[44]

The problems Lackey identified fourteen years ago are the same we continue to face today. Moving forward, it is imperative that we find ways of working together, both by clarifying the need for ecosystem management and by addressing some of the concerns of those who stand in its way. It is our hope that this book will take us a step further in the right direction.

III. THE STRUCTURE OF THIS BOOK

Where this book breaks ground is not with the concept of ecosystem management itself—as we have seen in this chapter, the matter has been bounced around for at least two decades. Rather, what I have endeavored to do is to bring together some of the leading scholars (from a range of disciplines) who have put thought into ecosystem management policy and present their input on the state of ecosystem management thus far and going forward. My concern was that ecosystem management had hit a wall. The concept was the result of incredible breakthroughs in our understanding of the natural world, but was not compatible with our existing routine. How, I wondered, will we ever make this happen?

This book is divided into four parts. Part one reviews and evaluates the work we have already done to design and implement an ecosystem management approach. Part two provides us with some valuable theoretical insights, which can support a deeper understanding of ecosystem management concepts. Part three considers how we might work with existing federal statutes to move toward a more systemic, landscape-scale approach to managing land and natural resources. In part four, we take a variety of creative approaches to future policy-making.

A. Understanding and Evaluating Ecosystem Management Thus Far

Once this chapter provides the reader, especially the novice, with some basic background on the history and meaning of ecosystem management, we move through several critical analyses of that background. In chapter two, Judith Layzer draws on her own research and systematic assessments of several landscape-scale ecosystem management projects. Although Layzer finds that ecosystem-based management provides great benefits, both ecologically and educationally, she takes a scalpel to it in an effort to keep only what works best. In so doing, she discovers that one of the common elements of ecosystem management may be doing a disservice to the overall goal of restoring ecosystems.

In chapter three, Dan Rohlf focuses on the integration of law, science, and policy in the management and restoration of ecosystems. Utilizing the more process-based understanding of ecosystem management, Rohlf points out that part of its allure comes from our ability to project our own goals onto the concept, which naturally results in a positive assessment. In reality, however, we live with rather significant constraints—culturally, economically, and politically—that must be taken into account in our efforts to apply the scientific principles of ecosystem management. We must be especially careful in the context of adaptive management, a key component of ecosystem management, so that we use it as a tool for learning as we go, but not as an opportunity to postpone difficult choices.

Finally, in chapter four, Martin Nie offers up the book's strongest criticism of our attempt at ecosystem management, a project he sees as already on the outs. Nie walks us through the dark side of adaptive management, collaboration, and landscape-scale restoration—three of the hallmarks of ecosystem management—noting that the same obstacles we faced two decades ago continue to prevent us from effective use of the methodology. Some examples of such "obstacles include disparate agency missions and planning processes, shifting political priorities, problematic budgets and an assortment of other legal, organizational, and political challenges."[45] Thankfully, Nie offers several suggestions as to how we might move out of the rut in which we find ourselves.

B. Letting Theory Inform Practice

Part two provides valuable theoretical insight to support our effort to grasp our relationship with ecosystem management policy. In chapter five, J.B. Ruhl discusses the relationship between ecosystem services theory and ecosystem management. How do the needs for ecological integrity and human prosperity relate to one another? Ruhl takes a comprehensive approach to the analysis by sifting through Grumbine's ten themes of ecosystem management and considering the relationship each has with our interest in ecosystem services. Ruhl then applies this analysis to a case study to determine whether his conclusions work in a practical context. While Ruhl finds great potential value in ecosystem services theory to support the goals of ecosystem management, he cautions that there are also risks involved with this economic perspective.

In chapter six, Susan Clark and David Cherney explore the tension between two competing ecosystem management paradigms for implementation. They consider the Scientific Management outlook, which views policy-making as an expert-driven technical exercise, and contrast it with the Adaptive Governance standpoint, promoting shared control by diverse stakeholders. Clark and Cherney ground their analysis in case material from one of the most famous early efforts at landscape-scale management, the Greater Yellowstone Ecosystem, in order to demonstrate the advantages and pitfalls of both management theories. They conclude with recommendations based on these observations.

C. Making Better Use of Existing Federal Law

The chapters in part three take a look at the federal statutes that predate the emergence of ecosystem management principles and suggest how they might be applied in light of our current scientific understandings. In chapter seven, Jamison Colburn tackles the National Environmental Policy Act (NEPA) from a philosophical perspective, providing insight into how we might adjust our NEPA routines to respect systemic ideals, particularly with regard to determining the spatial and temporal scales on which to focus.

In chapter eight, Robert Adler focuses on the restoration goal of ecosystem management, mining through a vast array of federal statutes for provisions that might be useful in achieving it. Adler organizes this material into four categories of applicability to restoration, creating a valuable road map to ecological restoration, and then wraps up with a discussion of the challenges in using this legislative material for this goal.

Finally, in chapter nine, Lynn Scarlett and James Boyd analyze how existing federal statutes can be leveraged to support the two emerging trends (from ecosystem management theory) of landscape-scale conservation and growing interest in ecosystem services.

D. *Finding the Right Tools Going Forward*

The fourth and final section of this book looks the future squarely in the face, recognizes the gaps in our existing regulatory structure, and begins the brainstorming process for creative approaches that may have some chance of improving our lot. Chapter ten is Robert Keiter's somewhat frightening discussion of the relationship between climate change and wildlife conservation, noting how dramatically our approach must change in the interest of climate adaptation. The good news; ecosystem management methodology, such as landscape-scale planning and adaptive management, are absolutely essential to adapting wildlife to a rapidly changing climate.

We see perhaps our most detailed policy planning in chapter eleven, in which Sara O'Brien and Sara Vickerman argue in favor of a national network of conservation lands, which would help give shape to our thus far limited efforts at an ecosystem management approach to land and natural resources. O'Brien and Vickerman go on to describe the necessary policies (both new policies and new spins on existing ones) to make such a nationally connected system work, allowing for better multilevel collaboration. Their proposed system would be designed in light of conservation principles, emphasizing landscape connectivity and ecosystem resilience to anthropogenic climate change.

Finally, in chapter twelve, Deborah McGrath and Travis Greenwalt explain the processes for economic valuation of ecosystem services, as well as how programs setting up payment mechanisms for such services (PES programs) can create financial incentives to protect and provide them. McGrath and Greenwalt ultimately propose that such programs may offer a valuable contribution to the improvement of ecosystem management.

In the end we must, as a society, imagine defending our choices to future generations. Often, when we look at the damage our ancestors caused, we give them some moral credit for not knowing what they were doing. What this book makes quite clear is that our generation has no such excuse. We have the benefit of a strong academic understanding of the issues and brilliant efforts to carve out practical plans to implement our scientific knowledge. We ignore this material at our peril.

NOTES

1. *See* James M. Guldin & T. Bently Wigley, *Intensive Management—Can the South Really Live Without It?*, Trans. 63rd No. Am. Wildl. & Natu. Resour. Conf. 362 (1998), *available at* http://www.srs.fs.usda.gov/pubs/ja/ja_guldin003.pdf.

2. John Copeland Nagle & J.B. Ruhl, The Law of Biodiversity and Ecosystem Management 318 (2nd ed. 2006).

3. John M. Blair et al., *Ecosystems as Functional Units in Nature*, 14 Nat. Res. & Env. 150 (2000).

4. Eugene P. Odum, Basic Ecology (1983).

5. Blair, *supra* note 4.

6. *See* James Salzman, *Valuing Ecosystem Services*, 24 Ecology L.Q. 887 (1997); Edward Farnworth et al., *The Value of Natural Ecosystems: An Economic and Ecological Framework*, 8 Envtl. Conserv. 275 (1981).

7. Arthur G. Tansley, *The Use and Abuse of Vegetational Terms and Concepts*, 16 Ecology 284–307 (responding to the contemporary focus on organisms in the field of ecology).

8. The Forest Service chief stated that the new methodology would "blend the needs of people and environmental values in such a way that the National Forests and Grasslands represent diverse, healthy, productive, and sustainable ecosystems." *See* Guldin & Wigley, *supra* note 1.

9. Gary K. Meffe et al., Ecosystem Management, at 4 (2002).

10. *See id.*

11. *See* David A Wardle et al., *Ecosystem Properties and Forest Decline in Contrasting Long-Term Chronosequences*, 305 Science 509 (2004).

12. Crawford S. Holling & Gary K. Meffe, *Command and Control and the Pathology of Natural Resource Management*, 10 Conserv. Biology 328, 329 (1996).

13. *See* Daniel B. Botkin, Discordant Harmonies (1992).

14. Aldo Leopold, A Sand County Almanac and Sketches Here and There 35 (1949).

15. James H Brown et al., *Complex Species Interactions and the Dynamics of Ecological Systems: Long-Term Experiments*, 293 Science 643 (2001).

16. *See* Thomas R. Stanley, Jr., *Ecosystem Management and the Arrogance of Humanism*, 9 Conserv. Biology 255 (1995) (arguing that the "problem is not how to maintain current levels of resource output while also maintaining ecosystem integrity; the problem is how to control population growth and constrain resource consumption.").

17. *See* Christensen et al., *The Report of the Ecological Society of America Committee on the Scientific Basis for Ecosystem Management*, 6 Ecological Applications 665–91 (1996); Interagency Ecosystem Management Task Force (IEMTF), *The Ecosystem Approach: Healthy Ecosystems and Sustainable Economies* Vol. 1 (1995); R. Edward Grumbine, *What is Ecosystem Management?*, 8 Conserv. Biology 27 (1994).

18. Howard R. Ernst, Chesapeake Bay Blues (2003); D. Scott Slocombe, Lessons From Experience With Ecosystem-Based Management, 40 Landscape and Urban Planning 31–39 (1998).

19. Judith Layzer, Ecosystem-Based Management and Restoration, The Oxford Handbook of U.S. Environmental Policy (forthcoming 2012).

20. *See* Robert B. Keiter, *An Introduction to the Ecosystem Management Debate, in* The Greater Yellowstone Ecosystem: Redefining America's Wilderness Heritage (R.B. Keiter and M.S. Boyce ed., 1991).

21. Bruce Goldstein, *The Struggle Over Ecosystem Management at Yellowstone*, 42 Bioscience 183–87 (1991).

22. *Id.*

23. D.A. Glick & T.W. Clark, *Overcoming Boundaries: The Greater Yellowstone Ecosystem, in* Stewardship Across Boundaries (R.L. Knight & P.B. Landres ed., 1991).

24. *See* Layzer, *supra* note 20.

25. Allan K. Fitzsimmons, Defending Illusions (1999).

26. *See id.* for quotation.

27. Robert B. Keiter, Keeping Faith With Nature (2003).

28. *See* Judith A. Layzer, *Jobs vs. the Environment: Saving the Northern Spotted Owl, in* The Environmental Case (3d ed. 2012); Steven L. Yaffee, Wisdom of the Spotted Owl (1994).

29. *See* Layzer, *supra* note 29.

30. Norman L. Christensen, *The Report of the Ecological Society of America Committee on the Scientific Basis for Ecosystem Management*, 6 Ecological Applications 665 (1996).

31. Nagle & Ruhl, *supra* note 2 at 335.

32. *See, e.g.,* Kalyani Robbins, *Strength in Numbers: Setting Quantitative Criteria for Listing Species under the Endangered Species Act*, 27 UCLA J. Envtl L. & Pol. 1, 15 (2009); Katherine Renshaw, *Leaving the Fox to Guard the Henhouse: Bringing Accountability to Consultation Under the Endangered Species Act*, 32 Colum. J. Envtl. L. 161, 174–75 (2007); Carden, *supra* note 77 at 202; Cary Coglianese & Gary E. Marchant, *Shifting Sands: The Limits of Science in Setting Risk Standards*, 152 U. Pa. L. Rev. 1255, 1257–58 (2004); Wendy E. Wagner, *The Science Charade in Toxic Risk Regulation*, 95 Colum. L. Rev. 1613, 1628 (1995).

33. R. Edward Grumbine, *What is Ecosystem Management?*, 8 Conserv. Biology 27 (1994).

34. *Id.*

35. *Id.*

36. *Id.*

37. Barton H. Thompson, Jr., *Ecosystem Services & Natural Capital: Reconceiving Environmental Management*, 17 N.Y.U. Env. L.J. 460 (2008).

38. *See, e.g.*, Ronald D. Brunner & Tim W. Clark, *A Practice-Based Approach to Ecosystem Management*, 11 Conserv. Biology 48 (1997); Paul L. Ringold et al., *Adaptive Management Design for Ecosystem Management*, 6 Ecological Applications 745 (1996); Anne E. Heissenbuttel, *Ecosystem Management—Principles for Practical Application*, 6 Ecological Applications 730 (1996).

39. For somewhat different approaches to this question, *see* Karen V. Root et al., *A Multispecies Approach to Ecological Valuation and Conservation*, 17 Conserv. Biology 196 (2003), and Kenneth F.D. Hughey et al., *Integrating Economics into Priority Setting and Evaluation in Conservation Management*, 17 Conserv. Biology 93 (2003).

40. *See* D. James Baker, *What Do Ecosystem Management and the Current Budget Mean for Federally Supported Environmental Research?*, 6 Ecological Applications 712 (1996).

41. J.B. Ruhl, *The Disconnect Between Environmental Assessment and Adaptive Management*, 36 Trends 1 (July/Aug. 2005).

42. *See* Robert Costanza, *Ecological Economics: Reintegrating the Study of Humans and Nature*, 6 Ecological Economics 978 (1996).

43. For an example of this process, *see* Erik Nelson et al., *Modeling Multiple Ecosystem Services, Biodiversity Conservation, Commodity Production, and Tradeoffs at Landscape Scales*, 7 Frontiers in Ecology and Environment 4, 4–10 (2009).

44. Robert T. Lackey, *Ecosystem Management: Paradigms and Prattle, People and Prizes*, 16 Renewable Resources Journal 1, 8–13 (1998).

45. Nie abstract.

2 Ecosystem-Based Management

An Empirical Assessment

Judith A. Layzer, Massachusetts Institute of Technology

Since the 1980s, ecosystem-based management (EBM) experiments have been undertaken around the world. Although detailed descriptions of individual projects abound,[1] systematic assessments of their results remain sparse. In the 2000s, however, scholars began investigating the actual performance of EBM projects. The results of that research suggest that EBM has yielded some important environmental benefits, including the acquisition of ecologically valuable land and the protection of marine reserves.[2] EBM has also generated improvements in scientists' understanding of large-scale ecosystems.[3] But those scientific advances have not necessarily translated into the kinds of political and policy changes that proponents of EBM had hoped. Instead, the pursuit of multiple goals simultaneously—often the result of stakeholder collaboration—has led to solutions that are environmentally risky and unlikely to bring about sustainable and resilient landscapes.

I. THEORETICAL BENEFITS OF EBM

Scholars and practitioners have offered a variety of definitions of EBM, but most agree on three key elements: landscape-scale planning, collaboration with stakeholders, and flexible, adaptive management. In combination, proponents argue, those elements should lead to more comprehensive management of larger spatial scales on a longer time frame than conventional management, and should therefore lead to more sustainable and resilient human-ecological systems (see table 1).[4]

A. Landscape-Scale Planning

First and foremost, EBM involves holistic planning at a landscape scale—that is, planning across a mosaic of ecosystems.[5] Holistic planning is essential because, as forest ecologists David Perry and Michael Amaranthus explain, "[t]he critical role of landscapes and regions in buffering the spread of disturbances, providing pathways of movement for organisms, altering climate, and mediating key processes such as the hydrologic cycle means that the fate of any one piece of ground is intimately linked to its larger spatial context."[6] Although there is no single, accepted scientific definition of an ecosystem,[7] there are distinctive landscapes, such as watersheds, that are widely recognized as meaningful. The essential points are that planning, and therefore management, should (1) be organized around the problem(s) to be solved, not political units or property lines; (2) address concerns across a full range of spatial scales; and (3) focus on the relationships among landscape elements, rather than simply attending to each element in isolation.[8]

Table 1. Ecosystem-Based Management Versus Traditional Management

Attribute	Traditional Management	Ecosystem-Based Management
Underlying View of Nature	A collection of resources to be controlled	Complex, dynamic, interrelated, and inherently unpredictable systems
Relevant Science	Equilibrium perspective: succession leads to stable climax communities; reductionist methods; goal is predictability	Flux-of-nature perspective: disturbance is normal; holism; embrace of uncertainty and surprise
Goal(s) of Management	Maximum sustainable yield of commodities	Sustainable ecosystems, ecological integrity OR Balance between commodity production, amenities, and ecological integrity
Decision Making	Centralized, top-down, expert-driven	Decentralized, participatory, collaborative
Implementation/ Solutions	Prescriptive, uniform, piecemeal, technology-based; emphasis on control and remediation of damage	Incentive-based or voluntary, locally tailored, holistic, and performance-based; emphasis on prevention
	Management that is rigid and aims for control	Management that is experimental, adaptive

In theory, landscape-scale planning yields several environmental benefits. It requires scientists to develop integrative assessments that illuminate the relationships among ecosystem components, as well as the ecological structures and functions that are critical to a system's long-term resilience; it also requires them to consider multiple scales and the dynamic character of ecosystems. Such assessments should raise policy makers' and stakeholders' awareness and knowledge of critical ecological processes. As a result, they should be more likely to design solutions that are holistic and comprehensive—and therefore more effective at conserving biological diversity—than uniform, national-level policies.[9]

In addition, landscape-scale planning requires coordination among the numerous entities with jurisdiction over a given region. Such coordination, in turn, ought to alleviate the problems that arise when federal and state agencies within a single ecosystem pursue inconsistent policies. It should also avert the death by a thousand cuts that occurs when localities make decisions that disregard spillovers across jurisdictions.[10]

B. Stakeholder Collaboration

Second, EBM involves collaboration among stakeholders—an attribute that gained prominence as scholars and practitioners contemplated the challenges of implementation.[11] In particular, proponents contend that stakeholders ought to select the desired states of the landscape and formulate the means to achieve those states. In most collaborative planning processes, participants deliberate with the aim of reaching consensus, generally defined as willingness by all to accept the decisions of the group. When properly structured, consensus-based problem solving identifies solutions that promise gains for all participants—although no single group is likely to get everything it wants.

In theory, engaging stakeholders in a collaborative process of defining the goals, objectives, and outputs of EBM produces several environmental benefits. Over time, repeated interactions among stakeholders are likely to increase their knowledge and understanding of one another's interests, eventually fostering trust among the participants.[12] Trust, in turn, generates creative interactions, which can yield innovative solutions.[13] As political scientists Philip Brick and Edward Weber explain, "[i]nstead of a system premised on hierarchy, collaboratives devolve significant authority to citizens, with an emphasis on voluntary participation and compliance, unleashing untapped potential for innovation latent in any regulated environment."[14]

Proponents of collaboration also expect that it will yield solutions that are more effective at solving environmental problems because the process incorporates more and better information, and does so more thoroughly, than top-down approaches. Collaboration ought to engage scientists more productively than adversarial approaches because in a deliberative forum, reasoning rather than tactics is paramount.[15] Moreover, unlike decision making by narrowly trained experts, collaborative processes incorporate local knowledge, which is based on extended, close observation of how an ecosystem behaves.[16] In the process, collaboration filters out the biases and broadens the perspectives of experts.[17]

Involving all interested parties can ease implementation, since everyone who might obstruct the implementation of a decision will have participated in formulating the solution.[18] As forestry scientist John Gordon and environmental writer Jane Coppock point out, "[t]he inclusiveness of the process broadens the base of support, making it harder for die-hard opponents to overturn agreements as soon as they see a political advantage."[19] By contrast, locals tend to perceive mandates issued by federal officials as unfair and illegitimate, and hence resent and resist them.[20] In short, according to marine scientists Heather Leslie and Karen McLeod, "meaningful engagement with stakeholders is needed to create management initiatives that are credible, enforceable, and realistic."[21]

C. Flexible, Adaptive Implementation

The third element of EBM is implementation that is flexible and adaptive. A flexible implementation strategy is one that employs information, incentives, performance standards, and voluntarism, rather than prescriptive rules and deterrence.[22] Such flexibility is essential because next-generation environmental problems are fundamentally different from those tackled in the 1970s: whereas centralized rules may have been appropriate for problems caused by large factories, they are inadequate for dealing with suburban sprawl, agricultural runoff, and other problems caused by myriad individual decisions.[23] Adaptive management entails designing interventions to illuminate ecosystem responses and adjusting management to reflect new scientific knowledge. In its ideal form, adaptive management begins with the establishment of baseline conditions and the identification of gaps in knowledge about a system; next, scientists devise management interventions as experiments that test hypotheses about the behavior of the system and monitor the results of those interventions; finally, managers modify their practices in response to information gleaned from monitoring.[24]

Flexible, adaptive implementation promises at least two major environmental benefits. In theory, flexibility fosters a sense of stewardship among regulated entities, increasing the likelihood they will take protective measures that exceed minimal legal requirements.[25] By contrast, according to critics of the status quo, traditional regulatory approaches appear unreasonably burdensome and arbitrary, so they provoke resistance or efforts to circumvent the rules. Those who do comply are likely to engage in the minimum legally required behavior change.[26] According to its proponents, adaptive management promotes continuous learning, which is essential because both ecological and social systems are complex, dynamic, and inherently unpredictable.[27] Adjustment in response to new information can, in theory, result in management practices that are more reflective of the best scientific understanding of ecosystem processes and functions.

II. CRITIQUES OF EBM

Despite great enthusiasm about the potential benefits of EBM, critics fear the concept is too ambiguous to bring about genuine environmental protection, and that absent a shift in values, EBM will yield more of the same while breeding complacency.[28] Some critics have suggested that existing statutory frameworks that give precedence to commodity production or species-level obligations might impede ecosystem-based approaches.[29] Others have worried that institutional factors—particularly long-standing agency missions and standard operating procedures—will obstruct EBM initiatives.[30] Still others have argued that flexible implementation allows evasion of protective measures by recalcitrant managers and stakeholders,[31] while adaptive management—although desirable in theory—is likely to encounter resistance in practice.[32]

But the gravest concerns about EBM focus on stakeholder collaboration. Critics argue that collaboration aimed at consensus yields lowest-common-denominator solutions rather than environmentally protective ones. According to this logic, collaboration undermines efforts to depart dramatically from the status quo because, in an effort to attain consensus, planners exclude or marginalize those with 'extreme' views, skirt contentious issues, focus on the attributes of the ecosystem that are easiest to control, and avoid considering solutions that impose costs on participating stakeholders.[33] Some skeptics charge that collaboration actually exacerbates the power imbalance between environmental and development interests, and therefore generates *worse* outcomes than the traditional regulatory approach.[34]

The uncertain role of stakeholder collaboration is reflected in the extent to which proponents believe that restoring ecological integrity and biological diversity should constrain human activity—that is, whether they adopt an ecocentric or anthropocentric point of view. Some proponents of EBM, many of them scientists, emphasize restoring and sustaining healthy ecosystems and moderating human behavior to accommodate natural constraints.[35] For example, environmental-policy analyst Christopher Wood argues that "[t]o embrace the ecosystem management concept is to accept that ecological factors such as maintaining biological diversity, ecological integrity, and resource productivity dictate strict limits on social and economic uses of the land."[36] Similarly, environmental educator Edward Grumbine contends that "[e]cosystem management integrates scientific knowledge of ecological relationships within a complex sociopolitical and values framework toward the general goal of protecting native ecosystem integrity over the long term."[37] And ecologist Hal Salwasser argues "[t]he aim of ecosystem management on national forests should be to sustain healthy land first, then to provide people with the variety of benefits and options they need and want, consistent with basic land stewardship."[38]

By contrast, many others have proffered a view of sustainability in which social, economic, and ecological benefits are pursued simultaneously. For example, in 1996 the Keystone Center defined ecosystem management as "a collaborative process that strives to reconcile the promotion of economic opportunities and livable communities with the conservation of ecological integrity and biological diversity."[39] For planners Robert Szaro, William Sexton, and Charles Malone "the mandate [of EBM] should be to protect environmental quality while also producing the resources that people need. Therefore, ecosystem management cannot simply be a matter of choosing one over the other."[40] And, according to the Ecosystem-Based Management Tools Network, a web-based alliance of EBM researchers and practitioners seeking to promote EBM for coastal and marine environments, EBM "is concerned with the ecological integrity of coastal-marine systems and the sustainability of both human and ecological elements."[41]

President Clinton's Interagency Ecosystem Management Task Force issued what is perhaps the consummate statement of the multiple-goals perspective. According to the task force:

> The goal of the ecosystem approach is to restore and sustain the health, productivity, and biological diversity of ecosystems and the overall quality of life through a natural resource management approach that is fully integrated with social and economic goals. This is essential to maintain the air we breathe, the water we drink, the food we eat, and to sustain natural resources for future populations. The ecosystem approach recognizes the interrelationship between natural systems and healthy, sustainable economies. It is a common sense way for public and private managers to carry out their mandates with greater efficiency.[42]

Cortner and Moote acknowledge the tension that pervades such definitions of EBM. "While ecosystem management explicitly recognizes that social goals and objectives play a central role in framing management direction," they note, "it also presumes that humans will decide to make protection of ecological processes their overriding social objective."[43] Similarly, the U.S. Government Accountability Office (GAO) observes that "[p]roponents of ecosystem management believe that coordinating human activities across large geographic areas to maintain or restore healthy ecosystems ... would, among other things, better address declining ecological conditions and ensure the sustainable long-term use of natural resources, including the production of natural resource commodities."[44] The GAO also recognizes, however, that "[i]n the absence of a clear statement of federal priorities for sustaining and restoring ecosystem and the minimum level of ecosystem health needed to do so, ecosystem management has come to represent different things to different people."[45]

In sum, critics worry that EBM is likely to yield insufficiently protective outcomes because it empowers development interests and disadvantages environmentalists. In particular, those with an ecocentric perspective suspect that stakeholder collaboration will generate a combination of social, economic, and environmental goals, rather than making the resilience of natural systems paramount. Such a result, they fear, will inevitably undermine efforts at ecological restoration.

III. EBM IN PRACTICE

Despite its ambiguities and disagreements over its effectiveness, EBM has become the dominant approach to natural resource management worldwide. By 1999, federal officials in the U.S. had stopped referring to ecosystem management, retreating in the face of vitriolic reactions from commodity interests and western wise-use advocates.[46] But the concepts that underpin EBM have persisted, and initiatives continued under different names. Moreover, even as the term fell out of favor among U.S. land managers, enthusiasm grew for applying EBM principles to the management of marine ecosystems.[47] California's 1999 Marine Life Management Act required EBM for managing all marine wildlife in the state's waters. In 2003 and 2004, two prestigious scientific panels—the Pew Oceans Commission[48] and the U.S. Commission on Ocean Policy[49]—recommended taking an ecosystem-based approach to managing marine systems. In fact, by the mid-2000s EBM had become the dominant paradigm for managing natural resources around the world, and several international organizations—

including the United Nations,[50] the World Conservation Union,[51] and the Millennium Ecosystem Assessment[52]—had developed case studies and principles for successful implementation of EBM.[53]

Despite widespread and sustained enthusiasm for EBM, systematic assessments of its efficacy have been few and far between. In part this is because, until recently, few initiatives had existed long enough for evaluators to assess their substantive benefits. Of those few, their complexity and heterogeneity made rigorous evaluation particularly challenging. Since the mid-2000s, however, scholars have sought to document the results of EBM and, to the extent possible, analyze the reasons for its success or failure.

A. A Comparative Assessment

During the 2000s, I sought to conduct a systematic assessment of the results of EBM and the mechanisms by which those results were produced. To that end, I chose four prominent cases of EBM: two terrestrial and two aquatic-system initiatives, all in rapidly urbanizing regions of the United States. The Balcones Canyonlands Conservation Plan (BCCP) in Austin, Texas and the San Diego Multiple Species Conservation Plan (MSCP) were two of the country's most prominent habitat conservation plans. The Everglades Restoration in south Florida and the California Bay-Delta Program (CALFED) were two major aquatic-system restoration projects.

Upon further investigation, it became clear that although they differed in their particulars, all four cases were generating results that were only minimally environmentally beneficial. In hopes of clarifying which attributes of EBM were responsible for these disappointing outcomes, I identified three additional cases that were similar in terms of locations and the problems being addressed, but seemed to be producing more substantial environmental benefits (see table 2).[54] These cases were Sonoran Desert Conservation Plan (SDCP) in Pima County, Arizona, the Kissimmee River Restoration in central Florida, and California's Mono Basin Restoration.

A detailed analysis of these seven cases yielded some surprising results. All of the initiatives examined generated concrete policies and practices that appeared likely, over time, to produce some environmental benefits. Each prompted the creation of a deeper and more holistic understanding of how specific ecosystems work, which in turn fostered a more widespread recognition among policy makers and stakeholders of the relationships among the landscape's ecological elements and functions. Without exception, the programs furnished participants with rationales for raising large sums of money that were used to acquire ecologically valuable land or undertake activities aimed at restoring ecological functions. And each empowered environmentally oriented personnel within agencies and jurisdictions, some of whom tried to institutionalize more environmentally beneficial practices. Only those that did not rely on collaborative planning, however, yielded policies and practices that appeared likely to conserve and restore biological diversity and, therefore, ecological resilience.

Table 2. Case Selection

	Minimal Environmental Benefit	Substantial Environmental Benefit
Terrestrial Ecosystem	Austin (Texas) BCCP	Pima County (Ariz.) SDCP
	San Diego (Calif.) MSCP	
Aquatic Ecosystem	Comprehensive Everglades Restoration (Fla.)	Kissimmee River Restoration (Fla.)
	California Bay-Delta Program (Calif.)	Mono Basin Restoration (Calif.)

A comparison among the seven cases suggests that a landscape-scale focus was an important catalyst for the adoption of more protective policies and practices. In each of the cases, scientists described a defining moment when they realized that what happens in one part of a system affects the other parts; for the first time, they saw the system as a whole, not just a set of parts. In addition, scientists identified key drivers of ecological damage and documented the mechanisms by which that damage was occurring. And they recommended measures for conserving key species and the ecological processes they depend on. Furthermore, in every case, trying to address problems at a landscape scale prompted planners to adopt more comprehensive approaches to environmental problem solving and led to new forms of coordination among disparate agencies and jurisdictions.

The beneficial effects of collaborating with stakeholders were more elusive, however. In the four cases where policy makers deferred to stakeholders to set goals, the policies and practices that emerged appeared unlikely to conserve or restore ecological health because, to gain consensus, planners skirted tradeoffs and opted instead for solutions that promised something for everyone. The resulting plans typically featured management-intensive approaches with little buffering. As a result, they imposed the risk of failure on the natural system.

There are several explanations for this result. First, although collaboration did enhance trust, there is little evidence that stakeholders' interests were genuinely transformed, or that participants generated innovative solutions. Instead, consensus-oriented groups tended to marginalize advocates who espoused 'extreme' views. Even with a carefully selected stakeholder group, negotiations often resembled bargaining more than deliberation, particularly as plans became more specific. And stakeholder groups tended to avoid the most difficult issues or mask differences by using vague language—a decision that ultimately haunted implementation.

Stakeholder collaboration also did not ensure that the best information would prevail. The four cases that involved collaboration featured scientific enterprises that were difficult to penetrate, so local knowledge was often ignored. Nor did collaboration put an end to bickering among stakeholders over science. (It is noteworthy that the plans that did *not* rely on collaboration were actually more recognizably grounded in science than those that did.) The evidence also failed to support the notion that collaboration ensures durable implementation. Instead, implementation exposed many of the differences papered over during the collaborative planning processes, as stakeholders sought to prevent or modify projects that threatened their interests.

A commitment to flexible, adaptive implementation did not compensate for the failings of these four environmentally risky plans and, in fact, sometimes exacerbated them. Adaptive management did not translate into a willingness to alter policies in the face of new information, partly because minimalist plans actually provided little room for adjustment, but also because management and monitoring were insufficiently funded, and research by scientists did not translate automatically into management changes. Flexible implementation allowed managers with missions that were incompatible with ecological restoration to resume user-friendly practices when political conditions shifted.

By contrast, when policy makers—elected officials, administrators, or judges—endorsed an environmentally protective goal and used regulatory leverage to prevent development interests from undermining that objective, as they did in the three comparison cases, the resulting policies and practices were more likely than their counterparts to conserve or restore ecological integrity. In these cases, a willingness among political leaders to make ecological health the preeminent aim changed the balance of power and altered perceptions of what was politically feasible. When restoring ecological health was the paramount goal, planners were more likely to approve, and managers to implement, approaches that relied less on energy-intensive manipulation and more on enhancing the ability of natural processes to sustain themselves—even if doing so imposed costs on some stakeholders.

B. Analyses of Individual Cases

The insights generated by this comparative analysis should be taken as cautionary rather than disparaging. In particular, it is important to note that the four cases of genuine EBM were more complex, both geographically and organizationally, than the comparison cases. That said, the findings are consistent with the results of other research into particular examples of EBM. For example, detailed examinations of the Chesapeake Bay Program (CBP) have yielded comparable results. Established in 1983, the CBP is a collaborative effort among the states of Maryland, Virginia, and Pennsylvania; the federal government; and Washington, D.C. that aims to restore the resilience of the nation's largest and most productive estuary. It relies primarily on nonregulatory mechanisms—such as educational programs, incentives, and stewardship initiatives—to reduce nutrient runoff from agricultural lands and suburban lawns, curb overharvesting of the bay's resources, and arrest suburban sprawl. However, on January 5, 2010, despite more than twenty-five years of work, the CBP missed another in a series of cleanup deadlines: efforts to reduce pollution of the bay had fallen more than 40 percent short of their goals, and—with nearly $6 billion spent—monitoring data suggested that the impacts of relentless growth were overwhelming pollution-control efforts.[55]

Most observers attribute the bay program's failures to its reliance on collaborative planning and flexible implementation.[56] Historically, the program has operated as a multistate cooperative, with the Environmental Protection Agency (EPA) in a supporting role and the states following different paths depending on their political culture and proximity to the bay. While the state of Maryland has typically adopted more stringent measures, particularly with respect to land-use planning, Virginia and Pennsylvania have taken advantage of the program's flexibility to adopt minimally restrictive land-use policies and practices. Reliance by all on nonregulatory approaches to encourage adoption of agricultural best-management practices has yielded little in the way of results. In fact, the single most effective measures taken in the watershed to date have been sewage treatment plant upgrades and the bans on phosphate detergent adopted by the bay states in the 1980s and 1990s.[57]

Similarly, in the Great Lakes, a multibillion-dollar international effort to restore ecological integrity has been under way for decades. Ecosystem-based thinking emerged in the Great Lakes in the mid-1970s, during discussions between the United States and Canada about programs overseen by the International Joint Commission (IJC) and the Great Lakes Fishery Commission (GLFC). At this time, several entities identified the need for an "integrative framework" for managing this 300,000-square-mile basin.[58] As political scientist Barry Rabe explains, by the late 1990s "[a] flurry of collective activity at multiple levels of government [had] occurred in the basin over the [preceding] quarter-century, much of it linked by the common goal of comprehensive basin protection."[59] Nevertheless, the effort "remain[ed] vulnerable to uneven and ever-changing levels of commitment from its respective jurisdictions.[60]

There have been signs of progress in the Great Lakes: phosphorus inputs have declined; the walleye has recovered dramatically, as a result of fishing limits and phosphorus controls; and the population of the burrowing mayfly, historically the dominant benthic invertebrate in the lakes, rebounded between 1990 and 2001.[61] The available data, however, suggest that wetland-dependent birds have been static or declining, nonnative fish dominate prey fish in most areas, and native freshwater mussel communities have been decimated by invasive zebra mussels.[62] Moreover, progress to date is likely to be negated by increasing population and urban sprawl, the effects of which will be exacerbated by global warming.[63]

In May 2004, in an effort to rejuvenate the undertaking President George W. Bush issued an executive order that recognized the Great Lakes as a "national treasure" and created a federal Great Lakes Interagency Task Force. The task force, along with several other groups, in turn convened what became known as the Great Lakes Regional Collaboration (GLRC) in December 2004. A year later, the regional collaborative released the "GLRC Strategy to Restore and Protect the Great Lakes." Observers were skeptical about the prospects for implementing the plan, however; the team estimated that restoring the Great Lakes would cost upwards of $20 billion over five years—mostly for upgrades to sewage systems and restoration of wetlands and streams—about ten times what the United States was spending on the Great Lakes when the report was released.

A third case of landscape-scale collaboration, the Platte River Restoration, yielded similarly mediocre results. Like the Chesapeake Bay and the Great Lakes, the Platte River watershed is a complex ecosystem. It is dominated by a wide, shallow river that flows from Colorado across the Great Plains to the Missouri River. As legal scholar John Echeverria notes, it is also "one of the most heavily developed rivers in the world."[64] In the mid-1990s, as part of a broader effort to shield the Endangered Species Act from legislative reforms, Interior Secretary

Bruce Babbitt endorsed the Central Platte River Basin Endangered Species Recovery Implementation Program—a commitment by stakeholders in Colorado, Nebraska, and Wyoming to collaborate on a basin-wide program to protect and restore Platte River habitat. (A similar process had been tried, and failed, a decade earlier.)

In an (admittedly impressionistic) account of his experiences with collaborative watershed planning for the Platte River, Echeverria argues that from the outset, planning was "heavily weighted in favor of parochial economic interests."[65] According to Echeverria, the inhospitable political context in which the process began largely determined the extent to which genuinely protective measures were possible. He argues that water users and political leaders in the river-basin states embraced the Platte program because it gave them more say than a purely federal program would have, and created opportunities to argue for taxpayer help in paying project mitigation costs. Echeverria concludes that the negotiated solution would almost certainly be a failure, both in absolute terms and relative to what could have been accomplished through the exercise of regulatory muscle.

Some other, less prominent initiatives have also fallen short of restoring ecological integrity because planners have been unwilling to tackle thorny issues. For example, Arizona's collaborative Upper San Pedro Partnership (USPP), formed in 1998, failed to prevent the San Pedro River from running dry for the first time in 2005. Although the partnership has generated and disseminated an impressive amount of information to stakeholders, and local governments have taken several steps to reduce water consumption in the region, there has been no consensus on the ultimate issue: growth control. According to environmental planners George Saliba and Katharine Jacobs, "[p]erhaps the largest criticism leveled against the USPP is its inability to make difficult decisions regarding growth.... The politics and economics of growth in Arizona make this conversation very difficult."[66]

C. Studies of Marine EBM

Scholars investigating marine EBM have also discerned pallid results. In a review of EBM in fisheries management around the world, fisheries scientist Tony Pitcher and his colleagues found that while many countries had articulated EBM principles, few had actually taken the steps to achieve effective implementation. Only a handful of countries in the developed world were clearly moving toward EBM, and even many European countries received dismal ratings. The authors concluded that "[w]hilst the late nineties . . . saw the blossoming of

'Oceans' approaches aimed at developing and applying EBM principles to multiple sectors in multi-stakeholder processes, the gradual pace of these reforms
and their perceived expense has meant that few have been implemented."[67]
    Similarly, ecologists Katie Arkema, Sarah Abramson, and Bryan Dewsbury[68]
investigated forty-nine management plans for eight large marine and coastal
ecosystems to assess the extent to which managers were actually practicing EBM.
They found that implementation of EBM principles was lagging. In particular,
management objectives included more detailed human than ecological management criteria; for example, many of the plans were dominated by objectives that
focused on promoting commercial and recreational uses such as maintaining
public access and rebuilding depleted fisheries.

D. Studies of Collaborative Planning and Adaptive Management

    Other studies support claims about the propensity of collaborative planning
and adaptive management, in particular, to yield environmentally risk-tolerant
solutions. For example, in her detailed analysis of the Quincy Library Group
(QLG), political scientist Sarah Pralle found that the decision by key activists to
work collaboratively in a local forum led planners to redefine the central problem
facing the region as forest fires, rather than as excessive logging. Doing so
defused conflict and allowed for a solution that gave something to everyone
without necessarily addressing the root cause of the region's environmental
degradation. Pralle noted that the focus on process disarmed environmental
challengers, who found it difficult to combat the "overwhelmingly positive characterization"[69] of local, collaborative decision making. She observed that "[i]n
a world of polarized interest groups and partisan gridlock, policymakers may
be more than willing to settle for outward signs of consensus rather than true
political compromises."[70]
    Similarly, in her study of two projects within the broader Everglades resto-
ration in South Florida, political scientist Kathryn Frank found that collaboration was better at resolving conflict than at problem solving, which tended to be
a subordinate objective. She concluded that:

> Since collaborative processes did not significantly change power relations, col-
> laborative outputs and the political capital upon which they depended were
> largely transient. Collaboration produced a delicate balancing act of aligned
> interests in keeping with the rhetoric of win-win and sustainability. Collab-
> orative recommendations appeared highly integrated, yet under the surface
> there were strategic motivations and shallow commitments. The agreements
> began to unravel when system dynamics of technical shortcomings changed

the conditions upon which the agreements depended. Combine this with the long-range dominance of economic interests, and the result was poorer implementation performance for environmental plan features.[71]

More generally, scholars have ascertained that collaborative problem-solving appears to increase human and social capital.[72] But they have struggled to document a causal link between social capital and improved environmental outcomes.[73]

Although scholars continue to advocate adaptive management, most acknowledge that in practice it has fallen short of expectations. Forest Service researchers W.H. Moir and W.M. Block[74] argue that the information-feedback system is the weak link in adaptive management. They note that land managers devote most of their time and energy to the planning stage of problem solving. Then, "[t]o placate or get the support of skeptical and adversarial elements of the public, the planners propose AM [adaptive management]. The doubters (including scientists) are told that the activity will be modified or stopped when it becomes clear that there is significant divergence from the trajectory towards stated goals."[75] But monitoring programs are not designed to capture slower, longer-term ecosystem responses to management intervention.

Carl Walters, one of the original proponents of adaptive management, concedes that of the upwards of a hundred case studies where attempts were made to apply adaptive management to natural resource management, most have been failures in the sense that no experimental management was ever implemented, and there have been serious problems with monitoring programs in the handful of cases where an experimental plan was implemented.[76] Walters attributes the failure of adaptive management to a lack of resources for large-scale monitoring, an unwillingness by decision makers to embrace uncertainty, and a dearth of leadership. Craig Allen and Lance Gunderson proffer a more expansive list of explanations for failures in the design and execution of adaptive management, but the message is similar: adaptive management generally has not lived up to expectations.[77]

E. Examples of Success

Observers have identified examples of EBM that have incontrovertibly resulted in environmental improvements, including the Malpai Borderlands Group on the New Mexico-Arizona border[78] and the Applegate Partnership in Oregon.[79] Such projects tend to feature the characteristics posited by political scientist Elinor Ostrom[80] and her colleagues as essential to effective local, collaborative management of common pool resources: appropriators believe they will be harmed if they do not adopt rules to govern use of the resource; are affected in similar ways by the proposed rules; value the continued use of this common property resources (discount rates are low); face low information, transformation, and enforcement costs; share generalized norms of reciprocity and trust; and constitute a relatively small and stable group. In addition, the target resource is in sufficiently good shape that efforts to protect it will confer benefits, there are valid and reliable indicators of system health, the flow of resources is relatively predictable, and the system is sufficiently small to allow knowledge of external boundaries and internal microenvironments.[81] Unfortunately, such conditions hold in a dwindling number of places and are particularly rare at larger scales.

Furthermore, in many cases evaluators who discern positive results have relied on the testimony of participants, rather than on actual evidence of ecological improvement. For example, planning scholar Steven Yaffee has followed 105 partnerships throughout the U.S.[82] According to surveys of those initiatives, a majority have not only produced better relationships and greater awareness of the ecosystem, but also improved scientific understanding, ecological restoration, increased native species populations, and led to improvements in overall ecosystem integrity.[83] Similarly, in a survey by the U.S. GAO of seven collaborative initiatives, participants claimed they had improved natural resource conditions—although none had collected any data to show their actual impact at a landscape scale.[84] Alternatively, analysts hold EBM to process standards, but not to outcome standards. For example, zoologist Heather Tallis and her colleagues proffer two case studies of 'successful' EBM—one in Raja Ampat, Indonesia, and the other in Washington's Puget Sound, Washington—despite the fact that neither can yet demonstrate actual ecological results.[85]

IV. CONCLUSIONS

Overall, the evidence suggests that the theoretical benefits of EBM, although widely touted, do not always materialize in practice. In cases where EBM has been fully implemented, landscape-scale planning has yielded discernible benefits, but the effects of collaboration with stakeholders and flexible, adaptive implementation are more ambiguous. This is not entirely surprising: proponents of EBM often gloss over the potential tradeoffs among environmental, economic, and social considerations, particularly in the short run. They assume that a preoccupation with long-term ecological sustainability will somehow emerge from a collaborative process. For this to happen, however, participants must be convinced that healthy, functioning ecosystems are essential to human well-being; they must embrace a land ethic and eschew a short-term economic point of view.

Such a transformation is unlikely under any circumstances but, counterintuitively, may be facilitated by the exercise of political leadership and regulatory leverage. That is why, as legal scholar Bradley Karkkainen observes, the federal government plays a critical role in EBM. He argues that productive collaboration is most likely when the most powerful actors have their backs against the wall— usually as a result of a stringent federal law that is likely to be enforced. Mobilization and litigation by environmental advocates can also generate the kind of "pervasive, persistent, and profound uncertainty, and the associated recognition of mutual dependence" necessary to bring about fundamental shifts in the balance of power.[86]

NOTES

1. *See, e.g.*, Tim W. Clark, Elizabeth Dawn Amato, Donald G. Whittemore & Ann H. Harvey, *Policy and Programs for Ecosystem Management in the Greater Yellowstone Ecosystem: An Analysis*, 5 Conserv. Biology 3, 412–22 (1991); Mary Doyle & Cynthia A. Drew, Large-Scale Ecosystem Restoration (2008); Robert B. Keiter, Keeping Faith With Nature (2003); Jean-Yves Pirot, Peter-John Meynell & Danny Elder, Ecosystem Management (2000); Barry G. Rabe, *Sustainability in a Regional Context: The Case of the Great Lakes Basin, in* Toward Sustainable Communities (Daniel A. Mazmanian & Michael E. Kraft eds., 1999); U.N. Convention on Biological Diversity, Ecosystem Approach: Further Elaboration, Guidelines for Implementation and Relationship With Sustainable Forest Management, Report of the Expert Meeting on the Ecosystem Approach, U.N. Doc. UNEP/CBD/SBSTTA/9/INF/4 (Sept. 29, 2003); U.N. Development Programme, U.N. Environment Programme, World Bank & World Resources Institute, A Guide to World Resources 2002–2004—Decisions for the Earth—Balance, Voice, and Power (2003).

2. *See, e.g.*, Judith A. Layzer, Natural Experiments (2008); Mary Ruckelshaus, Terrie Klinger, Nancy Knowlton & Douglas P. DeMaster, *Marine Ecosystem-Based Management in Practice: Scientific and Governance Challenges*, 58 Biosci. 1, 53–63 (2008).

3. Doyle & Drew, *supra* note 1; Layzer, *supra* note 2; Heather Tallis et al., *The Many Faces of Ecosystem-Based Management: Making the Process Work Today in Real Places*, 34 Marine Pol'y 340–48 (2010).

4. Howard Browman et al., *Perspectives on Ecosystem-Based Approaches to the Management of Marine Resources*, 274 Marine Ecology Progress Series, 269–303 (2004); Peter F. Brussard, J. Michael Reed & C. Richard Tracy, *Ecosystem Management: What Is It Really?*, 40 Landscape & Urb. Plan. 1, 9–20 (1998); Norman L. Christensen et al., *The Report of the Ecological Society of America Committee on the Scientific Basis for Ecosystem Management*, 6 Ecological Applications 3, 665–91 (1996); Jamie Rappaport Clark, *The Ecosystem Approach From a Practical Point of View*, 13 Conserv. Biology 3, 679–81 (1999); U.S. Gov't Accountability Office, GAO/RCED-94-111, Ecosystem Management: Additional Actions Needed to Adequately Test a Promising Approach (1994); John Gordon & Jane Coppock, *Ecosystem Management and Economic Development, in* Thinking Ecologically (Marian R. Chertow & Daniel C. Esty eds., 1997); Jerry

F. Franklin, *Ecosystem Management: An Overview, in* Ecosystem Management (Alan Haney & Mark S. Boyce eds., 1997); R. Edward Grumbine, *What is Ecosystem Management?*, 8 Conserv. Biology 1, 27–38 (1994); Robert B. Keiter, *Ecosystems and the Law: Toward an Integrated Approach*, 8 Ecological Applications 2, 332–41 (1998); Keiter, *supra* note 1; Averil Lamont, *Policy Characterization of Ecosystem Management*, 113 Envtl. Monitoring & Assessment 1–3, 5–18 (2006); Kai N. Lee, Compass and Gyroscope (1993); Gary K. Meffe, Larry A. Nielsen, Richard L. Knight & Dennis A. Schenborn, Ecosystem Management (2002); Mary G. Wallace, Hanna J. Cortner, Margaret A. Moote & Sabrina Burke, *Moving Toward Ecosystem Management: Examining a Change in Philosophy for Resource Management*, 3 J. of Pol. Ecology 1–36 (1996); Steven L. Yaffee, *Three Faces of Ecosystem Management*, 13 Conserv. Biology 4, 713–25 (1999); Steven L. Yaffee, *Ecosystem Management in Policy and Practice, in* Ecosystem Management (Gary K. Meffe, Larry A. Nielsen, Richard L. Knight & Dennis A. Schenborn eds., 2002) [hereinafter *Ecosystem Management*].

5. Richard T.T. Forman, Land Mosaics (1995).

6. David A. Perry & Michael P. Amaranthus, *Disturbance, Recovery, and Stability, in* Creating a Forestry for the 21st Century 49 (Kathryn A. Kohm & Jerry F. Franklin eds., 1997).

7. Robert C. Szaro, William T. Sexton & Charles R. Malone, *The Emergence of Ecosystem Management as a Tool for Meeting People's Needs and Sustaining Ecosystems*, 40 Landscape & Urb. Plan. 1–3, 1–7 (1998).

8. Brussard, *supra* note 4.

9. Christensen et al., *supra* note 4; Gary K. Meffe & C. Ronald Carroll, Principles of Conservation Biology (1994); Dennis D. Murphy, *Case Study, in* Bioregional Assessments (K. Norman Johnson, Frederick Swanson, Margaret Herring & Sarah Greene eds., 1999).

10. Timothy Beatley & Kristy Manning, The Ecology of Place (1997).

11. Timothy P. Duane, *Community Participation in Ecosystem Management*, 24 Ecology L.Q. 24, 771–97 (1997).

12. Robert M. Axelrod, The Evolution of Cooperation (1984); John S. Dryzek, Discursive Democracy (1990); Judith E. Innes & David E. Booher, *Consensus Building and Complex Adaptive Systems: A Framework for Evaluating Collaborative Planning*, 65 J. of the Am. Plan. Ass'n 4, 412–23 (1999); Lawrence Susskind & Jeffrey Cruikshank, Breaking the Impasse (1987).

13. Dryzek, *supra* note 12; Judith E. Innes, *Planning Through Consensus Building*, 62 J. of the Am. Plan. Ass'n 4, 460–72 (1996); Julia M. Wondollek & Steven L. Yaffee, Making Collaboration Work (2000).

14. Philip Brick & Edward P. Weber, *Will the Rain Follow the Plow? Unearthing a New Environmental Movement, in* Across the Great Divide 18 (Philip Brick, Donald Snow & Sarah Van De Wetering eds., 2001).

15. C.J. Andrews, Humble Analysis (2002); Connie P. Ozawa, Recasting Science (1991).

16. Fikret Berkes, Sacred Ecology (1999); Ronald D. Brunner et al., Adaptive Governance (2005); Frank Fischer, Citizens, Experts, and the Environment (2000).

17. Brick & Weber, *supra* note 14; Susskind & Cruikshank, *supra* note 12.

18. Dana Blumenthal & Jean-Luc Jannink, *A Classification of Collaborative Management Methods*, 4 Conserv. Ecology 2 (2000), *available at* http://www.ecologyandsociety.org/vol4/iss2/art13/; Meffe et al., *supra* note 4.

19. Gordon & Coppock, *supra* note 4, at 44.

20. Susskind & Cruikshank, *supra* note 12.

21. Heather M. Leslie & Karen L. McLeod, *Confronting the Challenges of Implementing Marine Ecosystem-Based Management*, 5 Frontiers in Ecology & the Env't 10, 542 (2007).

22. Daniel J. Fiorino, *Flexibility, in* Environmental Governance Reconsidered (Robert F. Durant, Daniel J. Fiorino & Rosemary O'Leary eds., 2004).

23. Daniel C. Esty & Marian R. Chertow, *Thinking Ecologically: An Introduction, in* Thinking Ecologically (Marian R. Chertow & Daniel C. Esty eds., 1997); Mary Graham, The Morning After Earth Day (1999); Daniel A. Mazmanian & Michael E. Kraft, *The Three Epochs of the Environmental Movement, in* Toward Sustainable Communities (Daniel A. Mazmanian & Michael E. Kraft eds., 1999).

24. International Institute for Applied Systems Analysis, Adaptive Environmental Assessment and Management (C.S. Holling ed., 1978).

25. Fiorino, *supra* note 22.

26. Fiorino, *supra* note 22; Jody Freeman, *Collaborative Governance in the Administrative State*, 45 UCLA L. Rev. 1, 1–98 (1997).

27. Hanna J. Cortner & Margaret A. Moote, The Politics of Ecosystem Management (1999); C.S. Holling, *What Barriers? What Bridges?, in* Barriers and Bridges to the Renewal of Ecosystems and Institutions (Lance H. Gunderson, C.S. Holling & Stephen S. Light eds., 1995); C.S. Holling, *Surprise for Science, Resilience for Ecosystems, and Incentives for People*, 6 Ecological Applications 3, 733–35 (1996); Bradley C. Karkkainen, *Collaborative Ecosystem Governance: Scale, Complexity, and Dynamism*, 21 Va. Envtl. L.J. 189–243 (2002); Lee, *supra* note 4.

28. Robert T. Lackey, *Seven Pillars of Ecosystem Management*, 40 Landscape & Urb. Plan. 1–3, 21–30 (1998); Donald Ludwig, Ray Hilborn & Carl Walters, *Uncertainty, Resource Exploitation, and Conservation: Lessons From History*, 260 Sci. 17, 36 (1993); Thomas R. Stanley, Jr., *Ecosystem Management and the Arrogance of Humanism*, 9 Conserv. Biology 3, 255–62 (1995).

29. Keiter, *supra* note 4; A. Dan Tarlock, *Slouching Toward Eden: The Eco-pragmatic Challenges of Ecosystem Revival*, 87 Minn. L. Rev. 1173–208 (2003).

30. Cortner & Moote, *supra* note 27; Keiter, *supra* note 4.

31. Theodore J. Lowi, *Frontyard Propaganda*, 24 Boston Rev. 5, 17–18 (1999); Rena I. Steinzor, *The Corruption of Civic Environmentalism*, 30 Envtl. L. Rep. 10909–21 (2000).

32. Barry L. Johnson, *Introduction to the Special Feature: Adaptive Management—Scientifically Sound, Socially Challenged?*, 3 Conserv. Ecology 1 (1999), *available at* http://www.ecologyandsociety.org/vol3/iss1/art10; George H. Stankey, Bernard T. Bormann, Clare Ryan, Bruce Shindler, Victoria Sturtevant, Roger N. Clark & Charles Philpot, *Adaptive Management and the Northwest Forest Plan*, 101 J. of Forestry 1, 40–46 (2003); Carl Walters, Challenges in Adaptive Management of Riparian and Coastal Ecosystems, 1 Conserv. Ecology 2, 1 (1997), *available at* http://www.consecol.org/vol1/iss2/art1.

33. Thomas C. Beierle & Jerry Cayford, Democracy in Practice (2002); Cary Coglianese, *Is Consensus an Appropriate Basis for Regulatory Policy?, in* Environmental Contracts (Eric W. Orts & Kurt Deketelaere eds., 2001); Robyn Eckersley, *Environmental Pragmatism, Ecocentrism, and Deliberative Democracy: Between Problem-Solving and Fundamental Critique, in* Democracy and the Claims of Nature (Ben A. Minteer & Bob Pepperman Taylor eds., 2002); M.

Nils Peterson, Markus J. Peterson & Tarla Rai Peterson, *Conservation and the Myths of Consensus*, 19 Conserv. Biology 3, 762–67 (2005); Stanley, Jr., *supra* note 28.

34. Douglas J. Amy, *Environmental Dispute Resolution: The Promise and the Pitfalls*, in Environmental Policy in the 1990s (Norman J. Vig & Michael E. Kraft eds., 1990); George Cameron Coggins, *Of Californicators, Quislings, and Crazies: Some Perils of Devolved Collaboration*, in Across the Great Divide (Philip Brick, Donald Snow & Sarah Van de Wetering eds., 2001); Michael McCloskey, *The Skeptic: Collaboration Has Its Limits*, High Country News, May 13, 1996; Andy Stahl, *Ownership, Accountability, and Collaboration*, in Across the Great Divide (Philip Brick, Donald Snow & Sarah Van de Wetering eds., 2001); Steinzor, *supra* note 31.

35. J. Baird Callicott, *Harmony Between Man and Land: Aldo Leopold and the Foundations of Ecosystem Management*, 98 J. of Forestry 5, 4–13 (2000); Christensen et al., *supra* note 4; Grumbine, *supra* note 4; Lamont, *supra* note 4; Reed F. Noss & J. Michael Scott, *Ecosystem Protection and Restoration: The Core of Ecosystem Management*, in Ecosystem Management (Mark S. Boyce & Alan Hanley eds., 1997); Christopher A. Wood, *Ecosystem Management: Achieving the New Land Ethic*, 12 Renewable Resources J. 1, 6–12 (1994).

36. Wood, *supra* note 35, at 7.

37. Grumbine, *supra* note 4, at 31.

38. Hal Salwasser, *Ecosystem Management: A New Perspective for National Forest and Grasslands*, in Ecosystem Management 90 (Jennifer Aley, William R. Burch, Beth Conover & Donald Field eds., 1998).

39. Allan K. Fitzsimmons, Defending Illusions 6 (1999).

40. Szaro, Sexton & Malone, *supra* note 7, at 3.

41. About Ecosystem-Based Management (EBM), *available at* http://www.ebmtool.org/about_ebm.html (last visited Dec. 8, 2011).

42. 1 Interagency Ecosystem Management Task Force, The Ecosystem Approach: Healthy Ecosystems *and* Sustainable Economies, 3 (1995).

43. Cortner & Moote, *supra* note 27, at 42.

44. U.S. Gov't Accountability Office, *supra* note 4, at 4.

45. U.S. Gov't Accountability Office, *supra* note 4, at 38.

46. While environmentalists feared that EBM would yield insufficiently protective solutions, conservatives argued that EBM was a vehicle of nature-worshipping environmentalists to elevate protection of ecosystems above all else. *See, e.g.,* Fitzsimmons, *supra* note 39.

47. Browman et al., *supra* note 4; Karen L. McLeod et al., Scientific Consensus Statement on Marine Ecosystem-Based Management (2005), *available at* http://www.compassonline.org/sites/all/files/document_files/EBM_Consensus_Statement_v12.pdf; Andrew A. Rosenberg & Karen L. McLeod, *Implementing Ecosystem-Based Approaches to Management for the Conservation of Ecosystem Services*, in Politics and Socioeconomics of Ecosystem-Based Management of Marine Resources, 300 Marine Ecology Progress Series, 270–74 (Howard I. Browman & Konstantinos I. Stergio eds., 2005); Ruckelshaus et al., *supra* note 2; U.N. Environment Programme (UNEP) & Global Programme of Action for the Protection of the Marine Environment from Land-Based Activities (GPA), Ecosystem-Based Management: Markers for Assessing Progress (2006), *available at* http://gpa.unep.org/documents/ecosystem-based _management _english.pdf.

48. Pew Oceans Commission, America's Living Oceans (2003).

49. U.S. Commission on Ocean Policy, An Ocean Blueprint for the 21st Century (2004).

50. U.N. Convention on Biological Diversity, *supra* note 1; U.N. Development Programme, U.N. Environment Programme, World Bank & World Resources Institute, *supra* note 1.

51. Pirot et al., *supra* note 1.

52. Millennium Ecosystem Assessment, Ecosystems & Human Well-Being (2003).

53. Elsewhere, EBM goes by other names, including Integrated Coastal Management, Integrated Water Resources Management, and Integrated River Basin Management.

54. For details on both the case selection and the empirical analysis, *see* Layzer, *supra* note 2, at 32–40.

55. David A. Fahrenthold, *Efforts to Clean Up Bay Off to Another Fresh Start; US Regional Officials Seem Optimistic Despite History of Missed Goals*, Washington Post, January 6, 2010.

56. Howard R. Ernst, Chesapeake Bay Blues (2003); Tom Horton & William M. Eichbaum, Turning the Tide (2003); Judith A. Layzer, *Ecosystem-Based Management in the Chesapeake Bay, in* The Environmental Case (3d ed. 2012).

57. Scientific and Technical Advisory Committee, Chesapeake Futures (Donald F. Boesch & Jack Greer eds., 2003), *available at* http://www.mdsg.umd.edu/issues/watersheds/ chesapeake_futures/index.php; D'Vera Cohn, *Bans on Phosphates Said to Aid Bay Cleanup*, Washington Post, January 23, 1989.

58. George Francis, *Ecosystem Management*, 33 Nat. Resources J. 315–45 (1993).

59. Rabe, *supra* note 1, at 249.

60. Rabe, *supra* note 1, at 251.

61. Harvey Shear, *The Great Lakes, an Ecosystem Rehabilitated, but Still Under Threat*, 113 Envtl. Monitoring & Assessment 1–3, 199–225 (2006).

62. *Id.*

63. *Id.*

64. John D. Echeverria, *No Success Like Failure: The Platte River Collaborative Watershed Planning Process*, 25 Wm & Mary Envtl. L. & Pol'y Rev. 559, 562 (2001).

65. *Id.* at 560.

66. George Saliba & Katharine L. Jacobs, *Saving the San Pedro River*, 50 Env't 6, 41 (2008).

67. Tony J. Pitcher, Daniela Kalikoski, Katherine Short, Divya Varkey & Ganapathiraju Pramod, *An Evaluation of Progress in Implementing Ecosystem-Based Management in Fisheries in 33 Countries*, 33 Marine Policy 231 (2009).

68. Katie K. Arkema, Sarah C. Abramson & Bryan M. Dewsbury, *Marine Ecosystem-Based Management: From Characterization to Implementation*, 4 Frontiers in Ecology & the Env't 10, 525–32 (2006).

69. Sarah B. Pralle, Branching Out, Digging In 202 (2006).

70. *Id.*

71. Kathryn Irene Frank, The Role of Collaboration in Everglades Restoration, 238 (Aug. 21 2009) (unpublished Ph.D. dissertation, Georgia Institute of Technology).

72. Beierle & Cayford, *supra* note 33; Thomas I. Gunton, J.C. Day & Peter W. Williams, *Evaluating Collaborative Planning: The British Columbia Experience*, 31 Env'ts 3, 1–11 (2003); Judith E. Innes, Judith Gruber, Michael Neuman & Robert Thompson, California Policy Seminar, Coordinating Growth and Environmental Management Through Consensus Building (1994); William D. Leach, Neil W. Pelkey & Paul A. Sabatier, *Stakeholder Partnerships As Collaborative Management: Evaluation Criteria Applied to Watershed Management in California and Washington*, 21 J. of Pol'y Analysis & Mgmt. 4, 645–70 (2002); Mark Lubell, *Do Watershed Partnerships Enhance Beliefs Conducive to Collective Action?*, in Swimming Upstream (Paul Sabatier, Will Focht, Mark Lubell, Zev Trachtenberg, Arnold Vedlitz & Marty Matlock eds., 2005); Edward P. Weber, Bringing Society Back In (2003).

73. Leach et al., *supra* note 72; William D. Leach & Paul A. Sabatier, *Are Trust and Social Capital the Keys to Success?*, in Swimming Upstream (Paul Sabatier, Will Focht, Mark Lubell, Zev Trachtenberg, Arnold Vedlitz & Marty Matlock eds., 2005); Mark Lubell, *Collaborative Environmental Institutions: All Talk and No Action*, 23 J. of Pol'y Analysis & Mgmt. 3, 549–73 (2004); Leigh Raymond, *Cooperation Without Trust: Overcoming Collective Action Barriers to Endangered Species Protection*, 34 Pol'y Stud. J. 1, 37–57 (2006).

74. William H. Moir & William M. Block, *Adaptive Management on Public Lands in the United States: Commitment or Rhetoric?*, 28 Envtl. Mgmt 2, 141–48 (2001).

75. *Id.* at 142.

76. Carl J. Walters, *Is Adaptive Management Helping to Solve Fisheries Problems?*, 36 Ambio: A J. of the Human Env't 4, 304–07 (2007).

77. Craig R. Allen & Lance H. Gunderson, *Pathology and Failure in the Design and Implementation of Adaptive Management*, 92 J. of Envtl. Mgmt. 5, 1379–84 (2011).

78. Nathan F. Sayre, Working Wilderness (2005).

79. Weber, *supra* note 72.

80. Elinor Ostrom, Governing the Commons (1990).

81. Elinor Ostrom, *Reformulating the Commons, in* Protecting the Commons (Joanna Burger, Elinor Ostrom, Richard B. Norgaard, David Policansky & Bernard D. Goldstein eds., 2001).

82. *Ecosystem Management, supra* note 4.

83. *Ecosystem Management, supra* note 4, at 91.

84. U.S. Gov't Accountability Office, GAO-08-262, Opportunities Exist to Enhance Federal Participation in Collaborative Efforts to Reduce Conflicts and Improve Natural Resource Conditions (2008)

85. Tallis et al., *supra* note 3.

86. Joel Rogers & Joshua Cohen, *Power and Reason, in* Deepening Democracy 252 (Archon Fung & Erik Olin Wright eds., 2003).

3 Integrating Law, Policy, and Science in Managing and Restoring Ecosystems

Daniel J. Rohlf, Lewis & Clark Law School

The term 'ecosystem management' reminds me of an episode of my favorite television program as a kid, the classic *Star Trek* series. In one episode, Dr. McCoy seemingly finds a long-lost love on a remote planet, looking even younger and lovelier than he remembers her. However, the young lady turns out to be a salt-sucking monster which is able to appear as a beautiful paramour from its victim's past. While the good doctor of course survives his encounter with the creature to boldly go through many more *Trek* episodes, other hapless members of the space colony and *Enterprise* crew are not so lucky; enticed by the monster's ability to make its targets see what they wish to see, they meet an unfortunate end.[1]

Like the fatal mistake made by the fictional twenty-fifth century spacefarers, it is tempting to view the concept of ecosystem management as embodying everything that one could wish for in managing natural resources. Indeed, it is not difficult to find calls for applying ecosystem management from scientists, agencies, environmental advocates, landowners, and, in at least one instance, even Congress.[2] But just as Dr. McCoy did not pause to think about why his still-young former girlfriend turned up years later on a remote planet, the disparate array of interests that express enthusiasm for ecosystem management may not understand it fully, but nonetheless see in it elements that appear to dovetail with their own interests. Wilcove and Blair clearly had this contradiction in mind when they noted that "[t]he fact that no one really knows what ecosystem management really means has not diminished enthusiasm for the concept. To the contrary, the vagueness of the term ensures that people can make of it what they want."[3]

This 'eye of the beholder' quality helps explain why, despite widespread support for the concept, ecosystem management has rarely proven to be the elusive strategy that finally provides the common ground to minimize controversies over land and resource management decisions. Indeed, in perhaps the concept's most high-profile application, a vast ecosystem management strategy years in the making for the Upper Columbia River Basin was panned by most stakeholders and eventually faded away before it was ever implemented.[4]

Why does ecosystem management have an attractive exterior but often the disposition of a monster? Part of the answer stems from the shape-shifting nature of its key components. 'Management' is of course an extremely broad term, so the elements of an ecosystem management strategy viewed as crucial by one group in a given setting may be precisely the steps that another stakeholder bitterly opposes. And the boundaries of a given ecosystem are far from self-evident, even to scientists. As a result, an ecosystem management strategy that sounds appealing in a plan's executive summary may prove to be highly contentious or otherwise problematic when those responsible for implementing the plan attempt to define specific management parameters and make the trade-offs that inevitably come with a strategy's on-the-ground application.

Ecosystem management can also have a false allure, stemming from the term's juxtaposition of scientific scope and human decision making. Many see this combination as a promise of technical, dispassionate answers to contentious questions that often accompany natural resources management. An important innovation of ecosystem management was the recognition that effective natural resource management demands taking into account the ecological characteristics and limitations of a given ecosystem in defining management options. However, this recognition often leads to the mistaken conclusion that science can guide us to the 'correct' way to manage in a given place or under a particular set of circumstances. Rather than discussing and perhaps even fighting about how to protect and manage biodiversity and other resources, to some ecosystem management suggests we can simply ask experts to outline the 'scientific' way to proceed (and as Professor Oliver Houck noted, the "best science" not surprisingly usually turns out to be the information that happens to support the position of the interest group advocating its application in a given situation).[5]

This is far from an accurate view of ecosystem management. Even the ecological limitations of a given ecosystem do not necessarily set hard-and-fast limits on management decisions. In 1979 for example, Congress passed—and President Carter reluctantly signed—revisions to the Endangered Species Act (ESA) that exempted from the statute the infamous Tellico Dam.[6] Political leaders took this step even though they were fully aware that the legislation over-ruled a landmark Supreme Court ruling halting completion of the dam to protect endangered snail darters, a reversal that scientists at the time believed would consign the fish to extinction.[7] Even given this scientific understanding of the consequences of their decision, leaders nonetheless determined that it was in the public interest to proceed with completion of the dam. In this light, ecosystem management simply means making resource management decisions by considering possible options in an ecological as well as policy context. Choices made in this fashion can result in stringent protections for wilderness or wild-life—or they could favor a dam over the continued existence of an endangered species. Of course, in many settings this practically guarantees that not all stake-holders and interests will be happy with ecosystem management decisions.

Recognizing that science should inform—but cannot determine the outcome of—value-based natural resource management decisions, ecologist Robert Lackey designated the following as one of his "pillars" of ecosystem man-agement; "[s]cientific information is important for effective ecosystem manage-ment, but it is only one element in a decision-making process that is fundamen-tally one of public or private choice."[8]

In light of both *Star Trek's* cautionary tale and Lackey's insight about the role of science in making management choices, it is important to avoid seeing ecosystem management as the *answer* to vexing questions of how to use and protect land, biodiversity, and other natural resources—it really just provides a *better way to ask* these tough questions. Therefore, rather than conceiving of ecosystem management as a substantive prescription for managing land and natural resources, we should instead view this concept as an imperative for func-tionally integrating science with law and policy in order to more effectively make decisions. Ecosystem management does not tell us how to manage; rather, it tells us what we need to consider when making decisions about how to use and protect natural resources.

As such, effective ecosystem management must consciously link cutting-edge technical knowledge with the legal and policy intricacies of public decision making. Recent years have shown significant progress in ecosystem science and in the law and policies that affect important ecosystem resources and services. Science has made great strides in understanding how ecosystems work and how human actions affect them, and there is an ever-growing body of legal mandates and policy initiatives governing management and restoration of a wide variety of ecosystems and their components. For example, recent ecological research in Yellowstone has revealed the cascade of negative consequences in an ecosystem due to the loss of a keystone species such as wolves, as well as demonstrated the surprising restoration success that reintroducing such a species can bring. On the legal side of the ledger, Congress has added to early federal mandates to manage and protect ecosystems—such as the ESA's 1973 goal of conserving the ecosystems upon which threatened and endangered species depend—through statutes such as the 2000 Water Resources Development Act's authorization of a multibillion dollar effort to restore the Everglades ecosystem.

Unfortunately, however, the scientific community as well as managers and policy makers have historically failed to prioritize work toward integrating ecosystem science with the laws and policies that determine what happens on the ground. As a consequence, ecosystem management itself has, in general, been more of a hopeful label than a deliberate effort to connect science, law, and policy. Indeed, even the concept of ecosystem management itself has lost favor to some extent over the past decade, particularly among federal land managers. In 2011 for example, the U.S. Forest Service, once a world leader in describing and touting the benefits of ecosystem management, continued to emphasize the importance of ecosystems and ecosystem services but effectively abandoned references to ecosystem management in favor of an amorphous "all lands approach" to land management.[9] Ecosystem management has languished in significant part due to continuing disconnections between its scientific and regulatory elements.

A commitment to strengthening the fundamental precept of ecosystem management—its interdisciplinary nature—is thus essential to move this concept toward increased prominence and influence in land and natural resources management. Science alone cannot provide the basis for decisions that involve value-based trade-offs between resource use and protection within an ecosystem, but the best technical information available is an essential ingredient to making such choices. Successful ecosystem management thus depends to a significant degree on our ability to understand the interactions between science, law, and policy, as well as on engaging decision makers, regulators, the business community, and the public in a dialog about the choices and trade-offs involved in managing land and natural resources.

In this chapter, I suggest a three-part framework for meeting the challenge of integrating management decisions with ecological knowledge. First, greater focus on standards could enable ecosystem management to transcend buzz-words and define more meaningful management benchmarks. Next, designing sound 'science process' can help managers achieve established standards and avoid—or at least fairly resolve—disputes that arise in managing toward identi-fied goals. Finally, emphasis on workable plan design and commitment to metic-ulous—yet flexible—implementation is crucial for actually achieving these goals on the ground.

I. THE STANDARDS CHALLENGE

Law professors are fond of telling students that their classes are designed to train them to "think like a lawyer." While we take three years in law school to accomplish this feat, in my experience attorneys actually have a lot in common with three-year-olds. "What does that mean?" and "Why?" are staples of con-versations with toddlers, but as any parent knows these seemingly simple ques-tions often turn out to be quite complex. They are also more or less the same questions that attorneys often spend endless hours analyzing and arguing about. These queries should also be the starting point for developing the standards for applying ecosystem management.

A standard sets forth a goal, measuring stick, threshold, or benchmark that governs conduct or a set of decisions. Subjecting ecosystem management to the questions a lawyer (or three-year-old) would ask reveals two key standards that must be established for this concept to work in a given situation. First, ecosystem management suggests that we consider ecological boundaries and processes in managing resources, which may require looking beyond human-created admin-istrative lines and jurisdictions and instead focus on an ecosystem. But what does that mean? What exactly is an 'ecosystem'? Successful ecosystem manage-ment requires a clear and explicit definition of the geographical area or resources that are the target of a management scheme. Second, why are we interested in managing that ecosystem and its resources? What goals do we seek to achieve through our management efforts? What management practices will allow us to meet those goals? Workable ecosystem management needs explicit—and pref-erably measurable—assertions of exactly what management is intended to accomplish.

Establishing the first standard by defining the target area or resource to manage may seem like a relatively straightforward task, but a closer look reveals its inherent challenges. While 'ecosystem' is a scientific term, asking ecologists to define it produces a result that may surprise many people. When lawmakers posed precisely that question, the resulting report by the Congressional Research Service (CRS) described a wide divergence of views among scientists over how to define an ecosystem—and even regarding whether drawing lines on a map separating one ecosystem from another is possible at all.[10] In assessing whether ecosystem boundaries should include the range of one particular species or a suite of species, or whether an ecosystem description should also account for physical processes such as soil nutrients and weather patterns, the report ironically reached a conclusion that would make a lawyer proud—"it depends." Even further, CRS noted that watersheds or biomes may provide alternative biologically-based boundaries for management focus.

What does it mean when even a scientific report concludes that "it depends" when attempting to define an ecosystem? The answer is twofold. First, simply declaring that we will employ ecosystem management in a given situation just kicks the can down the road; it really says very little of use unless we also define much more precisely how we are defining the 'ecosystem' we intend to manage. For example, the Northwest Forest Plan defines its target management area by the range of northern spotted owls in the contiguous United States, with the significant caveat that its management requirements generally only apply to federal land within the owls' range.[11] On the other hand, the Comprehensive Everglades Restoration Plan (CERP) applies to a sixteen county area of south Florida and includes an aggressive campaign to purchase hundreds of thousands of acres of private land to facilitate changes in land and water management.[12]

This leads to the second (and most important) part of the meaning of "it depends." Determining the boundaries of an ecosystem to target for management—or whether management boundaries will instead follow watershed boundaries, biomes, or the range of a particular species—requires making decisions that are not just scientific. These decisions clearly must be *informed* by science, but they also turn on value judgments and other nonscientific factors. Managers chose to develop forest management regulations over the range of northern spotted owls because a federal judge had halted logging on national forests throughout the owls' range, and the birds had recently been protected under the ESA. But spotted owls can fly over private land and the ESA's strongest protections apply only to federal agencies, so the Northwest Forest Plan applies almost exclusively to federal forests. In contrast, restoring sheet flows of water to the Everglades requires replumbing (again) most of south Florida, and a broad array of interests was able to persuade Congress to allocate billions of dollars to the task. As a result, CERP has had a major effect on state, private, and federal lands over a vast area. Defining the relevant 'ecosystems' to manage for spotted owls, as well as the area to consider when restoring the Everglades, thus involved choices that integrated a) scientific information about the target resources; b) value judgments about a resource's importance and the public's willingness to pay to protect it; and c) a variety of other legal and policy considerations such as the strength of federal power over the owner of the specific area of land in question.

To apply ecosystem management more broadly and more successfully, it is important that we appreciate the need for this integration. Policy makers, attorneys, and judges often do not understand that ecologists cannot simply draw lines around ecosystems. Many scientists bristle at the suggestion that policy experts and lawyers should participate in defining an ecosystem or other resource to protect and manage. These misunderstandings can significantly slow progress in identifying important resources to manage.

There are similar misunderstandings involving decisions that generate much of the controversy surrounding ecosystem management, namely formulating the management standards governing an ecosystem or other natural resource. Many saw ecosystem management as focusing on protecting either species within a given ecosystem or qualities such as ecosystem 'function' or ecosystem 'integrity,' particularly in its early years. For example, an influential science journal article by Edward Grumbine entitled "What is Ecosystem Management" concluded that most scientific literature identified protection of "ecological integrity" as a key goal of ecosystem management; integrity includes protection of viable populations of all native species at a level that maintains their evolutionary potential, perpetuating ecological processes such as natural disturbance regimes and nutrient cycles, and ensuring that human occupancy and uses of ecosystem resources remains within these constraints.[13] These goals align quite closely with those of most environmental organizations, which were quick to endorse ecosystem management, often portraying its focus on ecological integrity as representing the "best science" for resource management decisions.

However, a closer reading of Grumbine's article reveals that the author himself has a more nuanced view of ecosystem management's goals. After recounting the above elements of ecological integrity, Grumbine acknowledged that they are "value statements derived from current scientific knowledge that aim to reduce (and eventually eliminate) the biodiversity crisis."[14] Other scientists took issue with what they termed 'biodiversity buzzwords' such as integrity. But Robert Lackey best articulated the need for ecosystem management standards to reflect an integration of policy and science, value decisions informed by—not made by—technical experts, "[e]cosystem management should maintain ecosystems in the appropriate condition to achieve desired social benefits; the desired social benefits are defined by society, not scientists. . . . Ecosystem management may or may not result in emphasis on biological diversity as a desired social benefit." Like defining the ecosystem or resource we wish to manage, determining the goals of management demands much more than just 'good science.'

Both scientists and advocates are often uncomfortable with this interdisciplinary, process-oriented definition of ecosystem management. Scientists see it as undercutting their role in decision making, relegating technical information to merely a factor for consideration rather than a driving force in determining target resources and their management standards. Advocates for a particular management strategy, often interests favoring increased protection for biodiversity, express frustration that viewing ecosystem management as merely a tool for achieving "desired social benefits" means that protection for species and ecosystem services will inevitably lose out to goals driven by short-term financial gain and political expediency.

There are two responses to these concerns. First, it may indeed be frustrating to some to acknowledge that science does not in fact drive standards decisions, and that goals such as biodiversity conservation are not the default choices for 'science-based' management. However, in practice these facts are inescapable. For example, if instructed to manage on an ecosystem basis, the U.S. Forest Service will have to identify the land and resources that are and are not included within an 'ecosystem,' a decision that scientists expressly acknowledge cannot be made solely based on scientific factors. Similarly, even management plans aimed at protecting ecological processes or a particular species must determine the level of security to achieve for these resources by making trade-offs involving economic and other social considerations. It is no coincidence that the Forest Service initially considered eight options for managing old-growth forests under the Northwest Forest Plan—and ultimately chose option nine. The decision was the culmination of a massive struggle to find a way to satisfy legal mandates to protect northern spotted owls while at the same time avoiding a meltdown of the Northwest's timber industry. It ultimately represented a policy-based compromise between the degree of protection to afford to owls and their habitat on one hand and the timber industry and associated employment on the other. Since the ESA does not prescribe a specific level of species security to define 'recovery' of a listed species, federal agencies made this call themselves in crafting the Northwest Forest Plan's management standards.

The Northwest Forest Plan example also helps explain the second response to those frustrated by the law-policy-science integration model of ecosystem management. The U.S. Fish and Wildlife Service (FWS) rarely justifies the plan's substantive management standards by explaining how these benchmarks are consistent with the goals of Congress as expressed in the ESA. Nor does FWS attempt to explain why it went along with a level of security for the birds that still allows logging within remaining owl habitat. The Northwest Forest Plan calls for an 80 percent chance that northern spotted owls will remain widely distributed for a hundred years, but its management standards would have had to protect considerably more habitat if this security level had instead been a 95 percent chance of wide distribution—a figure often used by federal agencies in other contexts.[15] Instead of explaining the rationale behind their choice of a management standard by pointing to the need to balance owl conservation and timber harvest, FWS almost always defends the plan—particularly in court—as simply the result of the "best science" related to spotted owls and their habitat needs. In reality, however, while science did indeed *inform* the standards decisions for the Northwest Forest Plan, it was in fact only one of several considerations in the identification of the desired social benefits of forest management throughout the range of northern spotted owls—as the Forest Service candidly acknowledged in the Northwest Forest Plan's Record of Decision.[16] But by using the popular notion that ecosystem management decisions are science-driven, FWS was able to make its value judgments and trade-offs related to owl recovery behind closed doors. On the other hand, had it explicitly acknowledged that ecosystem management requires standards that integrate science with legal and policy decisions, FWS would have had to be more forthcoming about the choices it made—and perhaps allow greater public participation in making them.

This final point deserves particular emphasis. Fully integrating science with law and policy to make standards decisions for ecosystem management also requires both transparency by the standard-setter, as well as participation by stakeholders and the public in the standard-setting process in order for the entity establishing standards to determine the desired public benefits in managing an ecosystem's resources. If legislators establish such standards, the political process affords (at least theoretically) the public an opportunity to make its wishes known to decision makers. However, lawmakers typically delegate authority to establish ecosystem management standards to administrative agencies. But when such agencies purport to make standards decisions solely based on science, they are essentially cutting stakeholders and the public out of the process of making value-based judgments—informed by science—about how to protect and use natural resources.

Given this integration model for ecosystem management, adequate standards share two important qualities. First, they are as specific as possible, both in describing the target ecosystem (or watershed, biome, or species), and in setting forth management prescriptions. Some interests may or may not be comfortable with the Northwest Forest Plan's standard to manage old growth forests in order to provide an 80 percent chance of wide distribution of spotted owls over a hundred years, but this standard at least allows for any interested party to make a science-based assessment of whether federal agencies' activities are consistent with this benchmark. In contrast, a standard to manage for "ecosystem integrity" would, without a much more specific definition of this term, provide little actual guidance for on-the-ground decision making. Ironically, agencies often prefer the latter type of standard because it maximizes their management discretion. When combined with the widespread notion that standards decisions are based on—rather than informed by—science, this discretion becomes almost unreviewable.

Additionally, sound ecosystem management standards are established through a transparent decision-making process that provides a mechanism for stakeholder and public input into the process of making the policy and value-based choices inherent in the standards. Though Dr. Lackey argues that ecosystem management does not necessarily focus on biodiversity restoration and conservation, a significant majority of the public may wish to make this goal a priority in managing, for example, the Yellowstone ecosystem. Such public input would provide a strong push for setting standards for managing the area that explicitly favor needs of wildlife over other potential resource uses. In fact, our growing knowledge about the value of ecosystem services should in many cases give stakeholders and the public a strong argument to favor protection of natural processes and biodiversity when defining desired social benefits in managing many ecosystems and resources.

II. SCIENCE PROCESS

Putting standards to use in actually managing ecosystems and related resources requires a series of scientific and technical determinations of exactly what measures and protections are necessary to meet those benchmarks. So-called 'good science' is of course important to carrying out these assessments, but it is important to recognize that even the processes by which we make determinations that are primarily scientific in nature also raise a host of important issues at the intersection of policy, law, and science.

An example can best illustrate the many interdisciplinary issues inherent in deciding how to apply scientific tools to determine the management measures and prescriptions necessary to meet a set of identified standards. Suppose that an ecosystem restoration strategy calls for modifying a land management agency's timber harvest practices over a specific area in order to restore historic populations of a suite of native species. Agency managers must turn to scientists and silvicultural experts to identify specific forest management practices and other necessary steps these experts that they believe support their group's perspectives. However, in such situations it is common for various stakeholders and interest groups to hold very different views on what these management practices should be; these groups often back up their opinions with technical information and even scientific experts that they believe support their perspectives. Such situations present managers with a set of difficult 'science process' choices, that is, decisions about how to use science to make important determinations. Which scientists should the agency rely upon to carry out the scientific analyses needed to identify the proper course of action to meet the plan's standards—the agency's own scientists, outside scientists who are recognized experts, scientists favored by stakeholder or interest groups, or some combination of these? What data should these experts use? In the absence of direct observational data, is it acceptable for scientists to employ assumptions, models, or best professional judgment to resolve uncertainties? How much uncertainty—or margin of error—is acceptable in deciding how to manage important resources? Should managers employ quality control measures to ensure the reliability of the scientific determinations that form the basis for their decisions, such as independent peer review? And if disputes arise over scientific determinations, how are they resolved?

Answers to these 'science process' questions often have a huge influence over the substance of management prescriptions. Choosing which experts, data, or methods of analysis to use can itself sometimes determine the outcome of the resulting 'scientific' determinations, and these choices almost always have at least some level of impact on the resulting management prescriptions and actions. However, it is not possible to make such choices based on just science. Hence, much like establishing standards, decisions involving science process have significant policy and even legal components.

These nontechnical considerations—and thus the need to make policy or legal judgments—arise in many areas of science process. For example, a land

management agency tasked with making specific ecosystem management decisions must decide whether its experts' key scientific findings, which will determine the course of on-the-ground management, should be submitted to independent peer review. Such review is likely to provide managers and the public
with greater assurance of the accuracy of agency scientists' conclusions, which
in turn determine specific management actions and prescriptions. On the other
hand, peer review often requires considerable time and expense, eating up the
agency's limited resources and delaying important and perhaps time-sensitive
decisions. Deciding on an appropriate course of action given such trade-offs is
not a scientific decision, though scientific knowledge (i.e. are the scientific questions at issue sufficiently novel that outside perspectives would be particularly
useful?) can *inform* this decision.

Given their pervasive influence, it is not surprising that science process decisions also sometimes serve as a sort of 'back door' mechanism to make value or
policy-based ecosystem management decisions that were left ambiguous during
the standard-setting process, or even to effectively modify standards decisions
themselves. For instance, various runs of salmon and steelhead in the Columbia
River Basin were listed as threatened and endangered under the ESA, due in part
to adverse impacts on these fish stemming from operations of the Northwest's
federal hydropower system, a series of massive dams on the Columbia and Snake
Rivers that supply a significant percentage of the region's electric power generation. The task of determining whether to make major operational changes
to the system in order to protect salmonids—and the political pressures aligned
against modifications carrying a potential price tag of hundreds of millions of
dollars—fell to the National Marine Fisheries Service (NMFS). Notwithstanding
the listings, hydrosystem managers proposed to make only a few minor changes
in dam operations. NMFS assessed this proposal's impacts on salmon using a
computer-based model that compared salmon populations in the past with those
likely to result from the planned operations. Based on its modeling, the agency
concluded that few changes were needed to adequately protect the runs.
However, a federal court disagreed with the agency's conclusion. The court noted
that NMFS's model used years with the lowest salmon returns on record as a
baseline for comparison, making only modest population improvements appear
significant and thus ensuring that the proposal calling for only minor adjustments to the status quo would pass muster. The court therefore concluded that
the agency's analysis process "focuses more upon system capability than upon
the needs of the species" while other data showed that "the situation literally

cries out for a major overhaul."[17] The decision ultimately prompted NMFS to require significant changes in hydropower operations to benefit salmon.

It is vital to both recognize the influence of science process decisions in ecosystem management, as well as guard against their misuse. In viewing ecosystem management as an imperative to better integrate science with law and policy, we therefore must make science process considerations an important—and explicit—part of designing management regimes. But just as choices about standards must be made on a case by case basis in light of society's goals for a particular ecosystem or resource, it is impossible to specify specific science process elements that should be a part of any particular ecosystem management strategy. However, identifying general characteristics of effective science process can provide guidance for shaping the elements of science process that will work best in a given situation.

A key goal of science process obviously must be accuracy. In order for management measures to realize a plan's goals, the science used to determine those measures must be as accurate as possible. A variety of science process mechanisms can help ensure accuracy, such as submitting key findings to independent peer review, employing science advisory panels, seeking advice from professional societies, and using high quality data and analytical models. Science process can also attempt to minimize errors through tools such as establishing maximum levels of uncertainty for using modeling data or other predictive tools.

In many ways, the second indicator of effective science process is at odds with the first. For making on-the-ground decisions to achieve standards for ecosystem management—often with limited availability of time, resources, and even data—efficiency in reaching necessary scientific determinations is crucial. In many cases, the need for efficiency may preclude use of tools such as independent peer review or studies to gather additional data. Tools such as standardized methods for data gathering and analysis, sharing data, and setting priorities to limit use of expensive and time-consuming steps such as peer review can also allow managers to act quickly and use fewer resources, while still making determinations that are as accurate as possible. However, an unavoidable aspect of science process is working to achieve a workable compromise between ensuring accuracy and maintaining efficiency in obtaining the data and making the scientific determinations needed to implement ecosystem management.

Scientists have always valued transparency in assessing scientific validity because it allows other scientists to validate a given result by evaluating the analytical methodology used to reach the conclusion, as well as by attempting to replicate the result. Similarly, transparency in science process allows public and stakeholder scrutiny of the policy and legal judgments undergirding the way an entity reached important scientific determinations that influence or determine management. For example, if an agency employs a scientific advisory board to assist it in making science calls but acts against the board's advice in making a particularly controversial determination, the agency should disclose its action as well as its reasons for disregarding the board's views. This openness guards against the temptation to use science process as a tool to revisit or reverse policy judgments made through the standard-setting process. It also reminds both the agency and the public that choices related to applying concepts of science in a given situation have important policy components that can have a significant impact on scientific results.

Finally, given that ecosystem management often affects important resources over large scales, it is vital to devise an ecosystem management scheme's science process in a manner that allows for public confidence in the resulting scientific determinations. Explicit attention to the accuracy, efficiency, and transparency can go a long way toward providing this confidence. However, additional steps can enhance confidence. Providing opportunities for public input into scientific assessment processes, as well as providing meaningful responses to this input, can help reassure the public that an agency's scientific determinations are not the result of closed, 'black box' processes, but instead allow for consideration of all available data and provide clearly explained scientific results.

In recent years, high-profile ecosystem management efforts have begun to explicitly include provisions for science process. For example, after several controversies over funding for salmon conservation projects in the Columbia River basin as part of the Northwest Power Act's mandate to restore salmon and steelhead, Congress amended the Act to establish an independent scientific review panel.[18] The panel, drawn from a list of qualified scientists suggested by the National Academy of Sciences and appointed based on very specific measures—to avoid conflicts of interest—reviews proposals for using money from a fund reserved for salmon conservation measures. The panel's comments must be disclosed to the public, and any decision by the agency overseeing these expenditures which is inconsistent with the panel's recommendations must be accompanied by a written explanation. At an administrative level, CERP has extensive science process requirements for making and evaluating key scientific determinations that guide specific measures toward restoration of the Everglades ecosystem.[19]

III. IMPLEMENTATION

Implementation is the final and perhaps most obvious element of the intersection between science, law, and policy in managing ecosystems. Ironically, however, implementation is commonly overlooked as a separate and deliberate consideration in the management process. Instead, many people simply take implementation as a given, assuming that responsible managers will translate the provisions of a plan to action on the ground. Unfortunately, such assumptions often prove to be too generous. At the same time, however, it may be possible to avoid or at least minimize failures of implementation by more effectively accounting for interdisciplinary issues in developing ecosystem management strategies.

Implementation issues are both simple and deceptively complex. At its core, of course, implementation poses a basic threat to ecosystem management when considered in the negative: even a management strategy with brilliantly designed standards and effective science processes to carry out those standards will be ineffective if it is not faithfully implemented. But adding a layer of irony as well as complexity, a management strategy may also be ineffective if it is meticulously implemented exactly as specified. If circumstances change or new information reveals that initial assumptions underlying a plan were incorrect, carrying out the measures identified in a management plan may in fact *not* succeed in meeting the plan's goals.

Finally, implementation considerations should also inform decisions about both standards and science process. A regulatory scheme may include sophisticated standards and a meticulously-designed science process, but if these standards and processes are too complicated or too expensive to successfully implement over time in the real world, they will merely be paper tigers.

The above implementation strategies are not conceptually difficult, but they require qualities sometimes in short supply: careful planning, attention to detail, follow-through, willingness to change or make adjustments, and self evaluation. Many efforts to implement ecosystem management have foundered due to a lack of one or more of these qualities.

One of federal land management agencies' most significant attempts at applying ecosystem management over a large scale provides a cautionary case study of implementation failures. The Interior Columbia Basin Ecosystem Management Plan (ICBEMP), a joint effort by the U.S. Forest Service and Bureau of Land Management (BLM) initiated in 1994, aimed to coordinate management of federal land in the northwest United States, as well as consider federal land management decisions in a broader ecological context in connection with adjoining and interspersed nonfederal land. The project sought to achieve a litany of goals ranging from protecting habitat for self-sustaining populations of plant and animal species and restoring damaged ecosystems to providing for natural resources production to support the long-term needs of rural communities. However, after nearly a decade of scientific analyses, planning, and environmental impact analyses under the National Environmental Policy Act (NEPA), the project essentially fell apart. While some of its 'interim' protections for old growth and aquatic species continue to influence land management decisions, ICBEMP as a whole never became the comprehensive ecosystem management framework originally intended by the Forest Service and BLM.

The project's extraordinarily vast scale proved to be a key challenge to effective implementation. ICBEMP targeted more than 63 million acres of federal land, with a scientific assessment area of 144 million acres covering portions of seven states and affecting 100 counties and 22 tribal governments.[20] Federal agencies intended ICBEMP to amend 62 individual federal land management plans, including those of 23 national forests and nine BLM districts over the covered region. Though the initiative's science teams assembled an impressive array of ecological, economic, and other information, the agencies could not effectively grapple with the sheer magnitude of the administrative task of managing in a coordinated manner such a large number of administrative units covering such a huge area.

Cost, both in terms of time and money, also became a significant barrier to ICBEMP implementation. The agencies' original timeline called for completion of comprehensive baseline scientific studies eighteen months into the project, but delays pushed the finishing date out to nearly five years. The final management strategy called for aggressive ecosystem restoration actions such as forest thinning projects to counter years of fire suppression and logging of mature forests. However, implementing these measures carried a hefty price tag, estimated at $67 million per year.[21] By 2000, lawmakers were slashing rather than increasing land management agencies' budgets, rendering unrealistic the plan's expensive restoration strategy.

Changing politics also changed federal agencies' outlook on ecosystem management and restoration. Whereas the Clinton Administration was willing to consider innovative approaches to land management, using science as a key decision-making tool and constraining resource production within limitations that also protected wildlife populations and ecosystem services, the new decision makers under George W. Bush favored increased commodity production on public lands and upward trends in uses likely to be curtailed under ICBEMP such as motorized recreation. ICBEMP's centralized management of a vast area also did not mesh with the new administration's de-emphasis of federal control and the primacy of decision making by states and rural communities. The project thus quickly became a political liability and was all but abandoned not long after the 2000 federal election.

Finally, when ICBEMP ran into trouble there were virtually no stakeholders or interest groups willing to come to its rescue. The project's multiple goals, which included ecosystem restoration and species protection on one hand and an emphasis on sustaining rural communities on the other, were designed to appeal almost everyone. However, this also ensured that environmental advocates were suspicious of the extent to which land managers would end up favoring commodity production goals during actual implementation, and those primarily interested in economic issues likewise were concerned that environmental restrictions would ultimately result in the plan's failure to sustain rural economies. Therefore, almost no one mourned ICBEMP's demise.

Such implementation challenges are not unique to such as ICBEMP. The Everglades restoration effort mandated by Congress, CERP, has also fallen far short of expectations and well behind schedule. A report by the National Academy of Sciences cites a number of factors to explain CERP's implementation problems, including a complex and contentious process for planning specific restoration measures, federal funding levels far below the level originally envisioned, and a lack of clear priorities.[22] Many other ecosystem management strategies suffer from similar problems.

What lessons can we learn from these examples that might improve implementation for future ecosystem management efforts?

First, there is a clear relationship between the standards and science process elements of ecosystem management and the likelihood that a plan will be implemented. Careful attention to formulating well-defined, scientifically meaningful standards can avoid management ambiguities similar to those that dampened enthusiasm for ICBEMP and have made it difficult to set priorities for Everglades restoration. While clear standards will draw opposition from interests that disagree with the value choices the standards represent, well-defined benchmarks will also produce supporters who can help ensure that implementation stays on track. It is also important to carefully consider efficiency in designing science process. While ICBEMP produced valuable new data on ecological conditions and relationships over a vast land area, the unexpected length of time necessary to obtain and analyze this information significantly slowed momentum of the overall plan, which in turn allowed political opposition to intensify and ultimately doom the process. Delays and cost overruns are among the most common causes of implementation failure, so a science process that is unduly cumbersome or costly increases the risk a plan will fail.

Additionally, the reach of ecosystem management can sometimes exceed its political and administrative grasp. ICBEMP was visionary in scope; it had very clearly defined boundaries, and its science team compiled an impressive amount of data on the ecological, economic, and even social conditions across all land ownerships. However, the sheer size of the ecosystem it proposed to manage posed significant administrative challenges by attempting to coordinate management of many administrative units across two agencies. On top of this task, the plan's huge coverage affected the constituencies of a large number of western politicians, many of whom were wary of more centralized control of resource management decisions and displeased by increased emphasis on species conservation. These elected officials held the purse strings for necessary funding, and thus had considerable leverage over the plan's implementation. The Northwest Forest Plan faced similar challenges, but it was driven primarily by the pointed legal mandates of the ESA's protections for northern spotted owls and thus much more insulated from changes in political winds (interestingly, efforts to protect listed species are also a key driver of restoration efforts in the Everglades, but implementation there has suffered due in part to conflicts between the needs of different listed species). ICBEMP, in contrast, was grounded in the considerably more general and malleable provisions of federal land management statutes. Therefore, implementation success may be inversely proportional to a plan's administrative complexity and geographical scope, and directly related to the strength of the legal requirements upon which it is based.

Finally, availability of resources—both financial and human—will always have a great deal of influence on implementation of ecosystem management regimes. Congress was well aware of this limitation when it allowed for incidental take of listed species—unintentional death or injury to protected species due to an otherwise lawful activity—upon agency approval of conservation plans (typically called habitat conservation plans) submitted by a nonfederal entity; the ESA establishes assured funding for such plans as a condition for plan approval.[23] Such long-term funding assurances obviously increase the likelihood of plan implementation over time. However, for ecosystem management initiatives carried out or funded in whole or part by federal agencies, it is difficult to escape year-to-year funding decisions. Accordingly, maximizing a plan's political support over time is important to securing annual congressional appropriations for both personnel and project expenditures needed to carry out an ecosystem management strategy. The internal administrative structure and reward systems of federal agencies or other entities charged with day-to-day implementation of ecosystem management is also crucial. Managers' ability to translate agreed-upon standards and science processes to on-the-ground management decisions depends upon having the expertise, information, and resources to actually apply these standards and science processes—and support from their superiors for doing so. The latter may depend in part on whether an organization or agency has a culture and internal reward structure (formal or informal) that provides incentives for its personnel to manage ecosystem resources according to the specified strategy, even if it is politically unpopular. If, on the other hand, scientists are pressured to alter their findings to conform to a decision path that is expedient rather than consistent with the specified standards and science process, implementation will suffer. Therefore, implementation success is also likely to be enhanced by legal provisions and personnel policies that protect employee scientists and managers from reprisal for resisting such coercion.

Even if all of the elements for effective implementation discussed above fall into place, changes in circumstances, unanticipated ecological, economic, or social responses to management actions, or new information may call into question whether an ecosystem management strategy can meet its goals even if managers follow the plan to the letter. A strategy for making mid-course adjustments to management actions is thus an essential component of implementing ecosystem management.

Shortly after the 1990s explosion of interest in ecosystem management, most experts pointed to adaptive management as the preferred model for dealing with scientific uncertainties and making management changes over time. Even Congress has mandated use of adaptive management in high profile ecosystem restoration and management projects.[24] Despite its widespread acceptance, however, adaptive management is often misunderstood. Moreover, if used improperly in ecosystem management, adaptive management can lead to problems during the implementation process or even serve as a cover for inappropriate modifications of standards choices. On the other hand, effective application of adaptive management requires deliberately connecting this concept to choices about ecosystem management standards and science process.

Depending upon how it is used, adaptive management can either allow for modifications to management measures during the implementation process in order to more effectively meet the goals of an ecosystem management plan, or it can make implementation decisions a source of controversy and a means for making or reopening choices that should be made during the standard-setting process. Quite clearly, success in ecosystem management follows from promoting the former and taking steps to avoid the latter.

Adaptive management is often mistakenly seen as making adjustments to management based on lessons learned during implementation, or as a scheme that allows for 'flexibility' in which management measures to apply. When applied based on these ideas, decision makers have used adaptive management as a cover for rationalizing indefinite standards choices or inappropriate risk-taking during plan formulation. Rather than deal with the challenges and controversy that often inevitably accompany setting standards, decision makers all too commonly sidestep this uncomfortable but necessary process by adopting broad, discretionary standards (often in the form of undefined general benchmarks such as ecosystem 'health,' 'integrity,' or 'sustainability') and assuring stakeholders and the public that managers will resolve any problems that arise during implementation by making adaptive changes to management measures. But such a course of action is actually a blatant misuse of adaptive management. This approach conflates establishing standards and goals for an ecosystem management scheme with choosing the specific measures necessary to implement the plan's standards, and inserts into the latter the controversy that often accompanies standard-setting decisions. More insidiously, however, this improper use of adaptive management can allow decision makers to preclude a transparent dialog over standards decisions by inaccurately portraying choices about which management measures to implement—a step which now requires making the value judgments not made during standard-setting—as merely a technical exercise appropriately left to scientists' and managers' expertise.

Adaptive management also sometimes inappropriately serves as a means to obfuscate—rather than explicitly acknowledge—uncertainty as to how species or resources may respond to management. Agencies or other actors can be reluctant to admit uncertainties, particularly when there is a chance that resolving those uncertainties may push management in a direction more likely to curtail economic resource uses or other politically popular actions. Under such circumstances, vague assurances of future changes if indicated by 'adaptive management' become a means to de-emphasize uncertainty and rationalize risky or unproven management measures. However, true adaptive management involves identifying and explicitly acknowledging uncertainty from the beginning of formulating an ecosystem management strategy, identifying alternative management treatments as active experiments, then using science-based predictions and carefully planned monitoring of the results to reduce identified uncertainties and thereby adjust a plan's management measures to better meet identified standards.[25] This form of active adaptive management is a powerful tool for gaining information and making science-driven adjustments to management actions, but it requires embracing uncertainty, a willingness to experiment with resources targeted for protection or management, and the funds, expertise, and long-term commitment to implement rigorous science-based monitoring, evaluate results, and feed new information into a process for adjusting management actions. Unfortunately, these qualities are often the exception rather than the rule in ecosystem management regimes.

It is crucial for decision makers and managers to view adaptive management as a sophisticated and rigorous implementation strategy, not merely a way to change management practices on the fly. Best practices in adaptive management include the following:[26]

Tailor the adaptive strategy to the problem: An ecosystem management strategy should include clear standards decisions, and adaptive management should be a technical tool for addressing explicitly identified uncertainties, not for revisiting policy-based choices reflected in standards. The adaptive management strategy should attempt to identify the types of management changes that may be needed to respond to increased understanding; providing information about potential changes may enable federal agencies to better comply with NEPA obligations for environmental assessment and public involvement up-front, and thus avoid potentially disruptive requirements for additional analyses (or legal challenges seeking such analyses) if and when management adjustments become necessary.

Ensure accountability and enforceability: Decision makers often go through lengthy and contentious processes to choose an initial course of action in an ecosystem management strategy, and thus may be reluctant to make adjustments even in light of important new information. An adaptive management strategy should explicitly identify uncertainties or contingencies, specify applicable triggers for making management changes, and set forth a clear and transparent process for monitoring, evaluation, and making adjustments to management actions. A clear and explicit process will enhance the ability of interested parties to hold managers accountable for implementing adaptive management.

Promote directed learning: Decision makers and managers should explicitly identify uncertainties germane to achieving the plan's standards and goals, and formulate specific management 'experiments' to help reduce or resolve those uncertainties. To the extent possible, experts should develop explicit models of an ecosystem or target resource to make predictions and facilitate learning. Stakeholders and the public must be informed that learning may entail some level of risk to target resources, though managers should take all possible steps to minimize irreversible harms.

Ensure sufficient resources: True adaptive management often requires significant investment of funds, expertise, and time. It does not make sense (and is likely impossible) to attempt to resolve every uncertainty, so identifying key areas to be targeted for learning must be an important and explicit part of an adaptive management approach. Once identified, up-front commitment of the resources needed to implement an adaptive management strategy is essential to ensure that it is actually carried out.

Adaptive management is a key component of implementing an ecosystem management strategy, but it has close connections to the other two principal elements of integrating law and science in managing ecosystems, i.e. setting standards and developing a workable science process. Adaptive management is a way to use science to better achieve, not usurp, standards choices made through a transparent and inclusive initial decision-making process, and the investigative and analytical elements of adaptive management should employ a science process consistent with that set forth for making scientific determinations in carrying out a plan.

Beyond effective adaptive management, there are other essential ingredients for successful implementation of ecosystem management strategies. Third-party enforcement mechanisms allow the public to take an active role in encouraging landowners to consider ecosystem-based conservation schemes such as habitat conservation plans under the ESA to avoid potential legal liability; judicial review is also an important tool for holding federal agencies accountable to implement ecosystem management or restoration plans, as well as other legal protections for ecosystems and ecosystem services. Accordingly, moves to weaken the public's ability to enforce legal requirements—such as efforts in Congress to eliminate attorney fee awards to plaintiffs that successfully bring suit against federal agencies under environmental statutes—will likewise weaken implementation of ecosystem management. But ultimately, social will is a crucial driver of successfully implementing plans for ecosystem management and protecting ecosystems and their resources. If the public views these steps—and these resources—as important, governments and even private landowners will be much more likely to follow through with their commitments to ecosystem management. Educating the public about the value of ecosystems, their components, and their services to humans is therefore an important implementation tool.

IV. CONCLUSION

The term 'ecosystem management' conjures up a variety of different images. Many think of large-scale species protection strategies or ecosystem restoration efforts, such as the Northwest Forest Plan's protections for spotted owl habitat or the initiative to restore the Everglades. Others think of management across administrative boundaries, such as attempts to coordinate land and resource management in the greater Yellowstone area. Perhaps most see a concept that incorporates an environmental-protection spin, such as constraining human actions within ecologically defined limitations.

I suggest here that these views of ecosystem management, like Dr. McCoy's vision in that long-ago TV trek into space, are in fact largely a projection of what the viewer wishes to see. It is certainly true that the law requires protection of northern spotted owls, and Congress has deemed Everglades restoration a goal worthy of substantial federal investment. And one can make a compelling argument that it is in our best interests to recognize important ecological limitations to our actions. However, the concept of ecosystem management does not compel these outcomes—though it may well be able to facilitate them. We should instead see ecosystem management for what it really is, an integration of science—represented by the technical term 'ecosystem'—with law and policy, the tools for determining how to manage natural resources. Said another way, to manage ecosystems in a manner that achieves socially desired outcomes, we need to effectively integrate science with law and policy.

Unfortunately, effective interdisciplinary thinking and practice has rarely been a hallmark of natural resources management, nor of natural resources education. That must change.

The three-part framework explained above—standards, science process, and implementation—provides a systematic way to better understand the interactions between science, law, and policy. Applying this framework to designing and applying strategies for ecosystem management will substantially enhance our ability to establish and meet society's goals for protecting and using natural resources, as well as reduce—or at least properly focus—the debates and conflicts over striking a balance between competing visions for these resources.

Ultimately, I submit that ecosystem management is not itself the answer to vexing questions surrounding how to manage declining species, scarce natural resources, fragile rural economies, scenic vistas, or ecosystem services. We still must do that heavy lifting on our own, using ecosystem management as a tool for putting our answers to those questions into practice. But while focus on this concept has waned to some degree since its zenith in the 1990s, it is no less relevant or important today. When seen and practiced as an imperative to think, plan, and manage across disciplinary lines, ecosystem management is ready to boldly go into the future of natural resources management.

NOTES

1. *Star Trek: The Man Trap* (NBC television broadcast Sept. 8, 1966).

2. *See* 16 U.S.C.S. § 6703(1)(A) (2006) (subtitled "Southwest Forest Health and Wildfire Prevention"), which calls for application of "adaptive ecosystem management."

3. D.S. Wilcove & R.B. Blair, *The Ecosystem Management Bandwagon*, 10 Trends in Ecology & Evolution 345 (1995).

4. The U.S. Forest Service and Bureau of Land Management collaborated on the Interior Columbia Basin Ecosystem Management Project (ICBEMP) for a decade beginning in 1993. Though the agencies formulated a management strategy and Environmental Impact Statement covering management of tens of millions of acres of public land in portions of six Northwest states, the plan was never formally adopted. For more information on this project, including useful publications and extensive scientific data, *see* http://www.icbemp.gov/.

5. Oliver Houck, *Tales From a Troubles Marriage: Science and Law in Environmental Policy*, 302 Sci. 1926 (2003).

6. Endangered Species Act, Pub. L. No. 96-159, 93 Stat. 1225 (1979).

7. Tennessee Valley Authority v. Hill, 437 U.S. 153 (1978).

8. Robert T. Lackey, *Seven Pillars of Ecosystem Management*, 40 Landscape & Urb. Plan. 21 (1998).

9. U.S. Forest Service, Draft All-Lands Approach for the Proposed Forest Service Planning Rule, *available at* http://www.fs.usda.gov/Internet/FSE_DOCUMENTS/stelprdb5182029.pdf (last visited Dec. 14, 2011).

10. M. Lynne Corn, Cong. Research Serv., R93-655 ENR, Ecosystems, Biomes, and Watersheds: Definitions and Use (1993).

11. U.S. Forest Service, Record of Decision for Amendments to Forest Service and Bureau of Land Management Planning Documents within the Range of Northern Spotted Owls (1994), *available at* http://www.reo.gov/library/reports/newroda.pdf (last visited Dec. 14, 2011).

12. U.S. Dep't of Interior, U.S. Army Corps of Eng'rs, The Comprehensive Everglades Restoration Plan: The First Five Years (2005), *available at* http://www.evergladesplan.org/docs/fs_first_5_yrs_english.pdf (last visited Dec. 14, 2011).

13. R. Edward Grumbine, *What is Ecosystem Management?*, 8 Conserv. Biology 27, 30–31 (1994).

14. *Id.* at 31.

15. U.S. Forest Service and Bureau of Land Management, *supra* note 11.

16. U.S. Forest Service, Record of Decision for Amendments to Forest Service and Bureau of Land Management Planning Documents within the Range of Northern Spotted Owls (1994), *available at* http://www.reo.gov/library/reports/newroda.pdf (last visited Dec. 14, 2011).

17. Idaho Dep't of Fish and Game v. Nat'l Marine Fisheries Serv., 850 F. Supp. 886, 893, 900 (D. Oregon 1994).

18. Northwest Power Act, 16 U.S.C. § 839(h)(10)(D).

19. Comprehensive Everglades Restoration Plan (CERP), Implementation Plan: Central and Southern Florida Comprehensive Review Study Final Integrated Feasibility Report and Programmatic Environmental Impact Statement (April 1999), *available at* http://www .evergladesplan.org/pub/restudy_eis.aspx#mainreport (last visited Dec. 14, 2011).

20. *See* Interior Columbia Basin Ecosystem Management Project, *available at* http://www .icbemp.gov/html/icbhome.html (last visited Dec. 11, 2011).

21. U.S. Dep't of Interior, Dep't of Agric., Report to Congress on the Interior Columbia Basin Ecosystem Management Project 17 (2000), *available at* http://www .icbemp.gov/pdfs/sdeis/congressreport/congressrpt.pdf (last visited Dec. 14, 2011).

22. National Academies, Report in Brief, Progress Toward Restoring the Everglades: Second Biennial Review (2008), *available at* http://dels.nas.edu/resources/static-assets/materials -based-on-reports/reports-in-brief/everglades_brief_final.pdf (last visited Dec. 11, 2011).

23. *See* 16 U.S.C. § 1539(a)(2)(B)(iii) (2006).

24. *See, e.g.*, 16 U.S.C.S. § 6703(1)(A) (2006); Water Resources Development Act of 2000, Pub. L. No. 106-541, 114 State. 2572.

25. For a good discussion of adaptive management, *see* Byron K. Williams, Robert C. Szaro & Carl D. Shapiro, U.S. Dep't of Interior, Adaptive Management: U.S. Dep't of Interior Technical Guide (2009), *available at* http://www.doi.gov/initiatives/AdaptiveManagement/ TechGuide.pdf.

26. These practices are adapted from Center for Progressive Reform, Making Good Use of Adaptive Management (White Paper No. 1104, April 2011), *available at* http://www .progressivereform.org/articles/Adaptive_Management_1104.pdf (last visited Dec. 11, 2011).

4

Whatever Happened to Ecosystem Management and Federal Land Planning?

Martin Nie, The University of Montana

To be honest, I did not know what to make of the invitation to write a chapter on ecosystem management (EM) and federal lands. My cynical side questioned the relevance of doing so, as a lot of thought has been given to the topic over the years. Yet here we are, roughly two decades after the term became popularized and the same problems remain largely unresolved. So do I write the chapter in the past tense, as a sort of obituary? More than a few colleagues of mine rolled their eyes when asked "whatever happened to ecosystem management on federal lands." Now, the language du jour is adaptation, ecosystem services, resiliency, landscape-scale restoration, and other fashionable terms. And like EM, some of these terms are useful rhetorical devices that are malleable enough to become multiple things to multiple constituencies. Like 'sustainability,' some language becomes so politically appropriated that it loses its original meaning.

But this cynical narrative is too simple. Part of the problem in assessing EM lies in the difficulty of tracing and measuring political change and transformation. In this case, there is no single law, regulation, or policy statement about ecosystem management that has forced change in a neat and linear fashion from the top-down. The story is much messier but one with some hope. My argument, as explained in the following pages, is that the central components of EM have undoubtedly made their way onto the federal lands. This includes adaptive management, collaboration, and restoration. All were commonly associated with the EM paradigm, and now all figure more prominently in federal land politics and planning. All the talk about EM was not in vain. The problem, however, is that the same barriers to practicing a more ecosystem-based approach to planning are still in place. The legal and institutional challenges identified years ago as hindrances to EM now frustrate efforts in adaptive management, collaboration, landscape-scale restoration, and other related initiatives.

I. BACKGROUND

Where federal lands predominate, so too does a confusing jumble of different land ownerships. From the checkerboarded forests of the Pacific Northwest to the 'blue rash' of state trust lands scattered throughout the West, intermixed ownership presents multiple challenges to ecosystem management. Put simply, while federal, state, tribal, and private properties often come in squares, ecosystems do not. There is no lack of creative ways in which to describe the "cartographic chaos" found within the federal estate.[1] Some see the mess as a "crazy quilt" of land ownership,[2] while for others it signifies a "tragedy of fragmentation."[3] In previous writing, I suggested that alcohol may have been involved in the original design, as one is tempted to see the chaos as the result of a bad joke played on future generations by a group of drunken legislators.[4]

Numbers help place the amount of land ownership fragmentation in perspective. Consider, for example, that although 41 percent of the Rocky Mountain West consists of public lands, less than 25 percent of that land is more than 1.2 miles removed from private land.[5] And these private lands are quickly becoming developed. From 1982 to 1997, 3.2 million acres of rangeland were converted to developed land,[6] and some studies estimate that another 25 million acres of "strategic ranch lands" are at risk of residential development by 2020.[7] Or take forest lands: between 1982 and 1997, more than 10 million acres were converted to something else, with another 26 million acres projected to be developed by 2030.[8] This development, moreover, is taking place at the problematic interface of wild and urban lands. Estimates show that from 1990–2000, 60 percent of new housing units were built in the wildland-urban interface (WUI).[9] These and other numbers show that the fragmentation problem has, if anything, become only more acute since the emergence of EM.

Stories are another way to emphasize the importance of boundaries and the challenges they pose to planning. Take the story of grizzly bears in the Greater Yellowstone Ecosystem (GYE), a term coined by bear biologists Frank and John

Craighead. Suffice it to say that the bear has habitat needs going well beyond the boundaries of Yellowstone National Park. Thus, we saw the political emergence of the roughly twenty million acre GYE, a place dominated by borders and inter-jurisdictional complexity. The ecosystem includes three states (Montana, Wyoming, and Idaho) and nearly 2,500 miles of administrative boundaries among more than twenty-five federal, state, and local agencies.[10] Surrounding Yellowstone and Grand Teton National Parks are seven national forests, three national wildlife refuges, two Indian reservations, and various Bureau of Land Management (BLM), Bureau of Reclamation, and state and private holdings.

Now enter *Ursus arctos horribilis*. Though uncommon in so many ways, the grizzly bear is similar to other species and ecosystem processes that are trans-boundary in nature. Grizzlies—like fire, weeds, water, wildlife, and most conser-vation issues—require a boundary-spanning planning approach. This point is now obvious, at least to those reading their Leopold and EM literature. But the question of *how* to adequately plan at appropriate spatial scales is tougher than it sounds, as bears in Yellowstone demonstrate. Multiple planning processes are involved in the recovery of grizzly bears, such as the writing of an Endangered Species Act (ESA) recovery plan and conservation strategy by the U.S. Fish and Wildlife Service, three state bear management plans, and amending six national forest plans in the GYE, among others.[11] Planning at such a scale is no cheap trick. Complicating matters is how to analyze the cumulative effects on bear habitat in the area, as required by the National Environmental Policy Act (NEPA), and the general lack of planning (and resulting development) on adjacent private lands. Then there are questions of how to effectively coordinate, implement, adapt, and enforce these plans—assuming, that is, they survive the judiciary.[12]

Planning is ubiquitous in federal land management. Laws governing the national forests, rangelands, parks, and wildlife refuges include planning man-dates. All generally require the writing of management plans for particular administrative units, such as a plan for a single national forest or park. Basic resource allocation decisions are made at this level of administration. A national forest plan, for example, typically includes the designation of various manage-ment areas or zones, along with their permitted and prohibited uses, at the indi-vidual forest level. Some federal land planning laws include provisions requiring opportunities to coordinate with state and local governments in the development of land use plans.[13] And agencies can identify adjacent nonfederal parcels that might be acquired or exchanged in order to rationalize ownership patterns. But by-and-large, first-generation planning efforts were mostly inward-looking affairs.

This insularity was shaken by the spotted owl crisis in the Pacific Northwest. The ESA, and its enforcement by the courts, best explains the emergence of ecosystem management on federal lands. The purpose of the ESA—"to provide a means whereby the ecosystems upon which endangered species and threatened species depend may be conserved"—accounts for why agencies started talking and planning differently.[14] *Seattle Audubon Society v. Lyons* (1994) was a major catalyst in this regard, as Judge Dwyer invoked the ESA, NEPA, National Forest Management Act (NFMA), and Federal Land Policy and Management Act (FLPMA) in endorsing ecosystem management and the large-scale Northwest Forest Plan. Like the habitat needs of the grizzly bear, the spotted owl required the Forest Service (USFS) and BLM to look beyond their respective units and plan more on an ecosystem basis.

The ESA, and the enforcement of other environmental laws, continues to drive ecosystem-based approaches to federal land management and planning.[15] But sometimes the story is more complicated than an owl necessitating the writing of a massively-scaled Northwest Forest Plan. Federal land agencies like the National Park Service are often still reluctant to directly confront external threats and plan accordingly. Instead, National Park policy generally favors cooperative approaches where "[s]uperintendents will encourage compatible adjacent land uses and seek to avoid and mitigate potential adverse impacts on park resources and values" by participating in other planning processes.[16]

But a de facto ecosystem management plan can take shape notwithstanding agency reticence to more explicitly plan at the landscape level (or dare use a word such as 'buffer' in their planning documents).[17] Take, for example, the case of Glacier National Park and the slow but steady 'regionalism' that has emerged around it. Like the GYE, Glacier is now politically and strategically viewed as part of the much larger Crown of the Continent Ecosystem which is comprised of one state, two Canadian provinces, multiple national forests, tribal lands, state lands, private property and some boundary-challenged endangered species.

Despite this fragmentation, there has been some movement toward more regionally based management in the area. Serious threats in the region remain, but they are not as widespread and unrelenting as they once were. And the remaining threats now face a more unified and coordinated political backlash. That, according to law professors Joseph Sax and Robert Keiter, is not really due to some top-down policy statement about ecosystem management, but rather a more complicated constellation of changes forcing a larger ecosystem view.

One indispensable factor is "the law and its enforcement, which has played a pivotal role in promoting management across formal boundaries."[18] These laws have been aggressively used by environmental groups to protect species like the grizzly bear and bull trout and to stymie various road building, timber sales, and oil and gas projects that threaten the park and the Crown. Some form of regionalism is the result, even though it is not centrally-driven or memorialized in a single planning document.

One particularly noteworthy example of an ecosystem-based approach in the Crown region is the Blackfoot Challenge, a well-oiled and nationally-recognized grassroots collaborative group focused on the Norman Maclean-famous Blackfoot River in western Montana; "[e]ventually, all things merge into one, and a river runs through it."[19] Like Leopold before him, Maclean saw the big picture, and so does the Challenge. The U.S. Fish and Wildlife Service supplies important leadership along with other federal and state agencies, ranchers, environmental groups, and dozens of other partners. All of them are focused on a simple mission, "to coordinate efforts that conserve and enhance the natural resources and rural way of life throughout the watershed."[20]

Property boundaries dominate the 1.5 million acre watershed, and private property and corporate (Plum Creek) timber lands are some of the most ecologically significant in the valley. To protect them, the Challenge has used every conceivable policy tool and funding source, from Land and Water Conservation Act funds to the deep pockets of the Nature Conservancy and the Trust for Public Lands. The coordination among federal, state, and private partners in the watershed is remarkable; the organization grew into something that is greater than the sum of its parts. And most impressive of all is the quantifiable environmental achievements made by the Challenge. This is more than a feel-good collaboration story, but rather one with a bottom line: ranches have been protected through conservation easements, corporate timber lands at risk of development have been acquired with public and private funds, streams have been restored, water quality has been improved, wildlife conflicts reduced, and so on. Of course, not all is smiles and sunshine in the Blackfoot, but the Challenge demonstrates the type of conservation imagined decades before in the name of ecosystem management.

These examples offer a better way of explaining what happened to EM than to count the number of large-scale planning initiatives that have succeeded or crashed and burned under the banner of ecosystem management. Nor is it enough to focus solely on high profile planning endeavors such as the Northwest

Forest Plan, the Sierra Nevada Framework, the Everglades Restoration, or the Interior Columbia Basin Ecosystem Management Project. These are important, but so too are other initiatives advancing the ideas of EM without all the limelight.

Consider, for example, the first USFS designation of a wildlife corridor on the Bridger-Teton National Forest, on the southern end of the Greater Yellowstone Ecosystem in Wyoming.[21] In 2008, Bridger-Teton amended its forest plan in order to help protect the migration of pronghorn through multiple land ownerships, including a 47,000 acre swath of national forest lands. Given the threats to one of the longest remaining wildlife migrations in North America, one might write off such piecemeal efforts as inconsequential. It is certainly not enough; nor is the resolution by the Western Governors' Association calling for the identification and protection of wildlife corridors.[22] But examples like this demonstrate how the ideas of EM continue to influence federal land planning.

This example proves little in isolation. But research suggests that it might be part of a pattern showing that some haphazard progress towards EM has been made over the years. One study, for instance, shows that selected forest plans written before the emergence of EM are measurably different from those written afterwards.[23] Not only does the language differ but so do the types of decisions made, such as the amount of timber harvested by clear-cutting. Another study shows that USFS line officers believe that the agency has been relatively successful in attaining various EM objectives.[24] (But such optimism is not universal, as other work shows a general lack of progress and that stakeholders outside the agency are more critical).[25] Complicating matters is the fact that talking about EM is different from practicing it.[26] The phrase is no longer popular as it once was, but as shown below, the central principles of EM endure.

II. ECOSYSTEM MANAGEMENT AND ITS PROGENY

There are multiple definitions of EM, but most share some common principles. Some basic themes often identified in the literature include: (1) socially defined goals and objectives, (2) holistic, integrated science, (3) adaptable institutions, and (4) collaborative decision making.[27] One popular study emphasizes key traits, including a focus on biodiversity and ecological integrity, the importance of working across administrative-political boundaries in order to manage at appropriate ecological scales, the need for organizational change and interagency cooperation, ecological restoration, and the benefits of a more monitoring-intensive adaptive management.[28] More recent work on EM identifies similar core attributes, such as addressing problems at a landscape or regional

scale, collaborative planning, and a heavy reliance on flexible, adaptive implementation of planning goals.[29]

Of course, there are major differences in what parts of EM get emphasized.
So it is unsurprising to find some interests invoking EM as a way to better protect
biodiversity and ecological processes, while the former chief of the USFS sees it
as "enhanced multiple use planning."[30] For some managers, EM is primarily
about process, hence the focus on adaptive management and collaboration. Yet
for others it is about objectives such as ecological restoration.

The amorphous term also had its political virtues for some agencies, such
as the USFS, which used the concept to strategically reinvent itself following the
tumultuous timber wars. Viewed through this lens, "[e]cosystem management
was an ambiguous, undefined concept that the agency could shape in the context
of political events . . . In the volatile and politicized atmosphere of forest policy,
the Forest Service attempted to change its image by adopting a new name for its
practices, resorting to the common practice of meeting conflict and crisis with
vague, sensationalist political imagery and drama."[31] Like the Rorschach test of
multiple use, EM came to represent different things to different people. Or as
the Congressional Research Service put it, "[t]here is not enough agreement on
the meaning of the concept to hinder its popularity."[32]

These different definitions and applications of EM complicate things. But
under most definitions, the central pillars of EM pose a challenge to federal land
agencies and their disparate missions and planning requirements.[33] Several
studies analyze the institutional obstacles to implementing EM on federal lands.
Commonly identified impediments include such things as rigid budgetary
systems, insufficient funding, deficiencies in leadership, and an assortment of
organizational biases and legal challenges.[34] Rare is the study failing to emphasize the importance of law because it simultaneously promotes and hinders a
more ecosystem-based approach to management.[35] While instability and dis-
equilibrium characterize ecosystems, most property and natural resource laws
emphasize boundaries, stability, and the pursuit of certainty.

The take home point is that implementing EM proved even more difficult than
defining it. Little has changed in this regard. Some of the major barriers impeding
EM now frustrate its progeny, including adaptive management, collaboration, and
landscape-scale restoration. While the terminology differs, the challenges do not.
And while progress has been made on various fronts, the following discussion
shows how federal land agencies continue muddling through the same challenges
presented to them decades ago under the rubric of ecosystem management.

A. Adaptive Management

In theory, EM is fundamentally different from the traditional resource management paradigm. One difference is how each model approaches science, with EM tackling problems in a more integrated fashion and at larger spatial scales. Doing so underscored the importance of uncertainty and adaptive management was put forth as "one way to address the staggering information requirements of ecosystem management while allowing management to move forward in the face of uncertainty."[36] Adaptive management was thus a central principle in most conceptions of EM and it remains at the forefront.

Federal land planning is in the midst of a messy paradigm shift. The core challenge is how to practice adaptive management and planning in the modern regulatory state. The question is one of governance: how to plan adaptively while ensuring accountability, transparency, inclusiveness, and other democratic principles and processes? More adaptive planning models are being advanced as the problems and pathologies of rational comprehensive planning become more apparent.[37] This is the 'synoptic' ideal in which a decision maker collects all the information relevant to a decision, considers all reasonable alternatives and possible consequences of each, and then chooses the alternative with the highest probability of achieving the agreed-upon goals in the most efficient way possible.

The truth is that the theory of synoptic planning is trumped regularly by the practice of politics. The planning model is practiced in the messy world of countervailing political pressures, layered legislative mandates, muddled court decisions, and insecure agency budgets. And this happens against a backdrop rich in environmental, political, and stochastic uncertainty. As the American screenwriter Woody Allen famously said, "If you want to make God laugh, tell him about your plans." Things like large-scale fires, drought, international commodity markets, and changes in political leadership regularly frustrate long-range rational planning endeavors.

Adaptive management and planning has therefore been supported by federal land agencies as a way to cope with such rampant uncertainty. In the context of federal lands, a popular definition, as adapted from the National Research Council, is as follows:

> Adaptive management [is a decision process that] promotes flexible decision making that can be adjusted in the face of uncertainties as outcomes from management actions and other events become better understood. Careful monitoring of these outcomes both advances scientific understanding and helps adjust policies or operations as part of an iterative learning process. Adaptive management also recognizes the importance of natural variability

in contributing to ecological resilience and productivity. It is not a 'trial and error' process, but rather emphasizes learning while doing. Adaptive management does not represent an end in itself, but rather a means to more effective decisions and enhanced benefits. Its true measure is in how well it helps meet environmental, social, and economic goals, increases scientific knowledge, and reduces tensions among stakeholders.[38]

Most scientific and scholarly definitions include a similar set of components, all designed to deal with the inherent uncertainty of natural resources management. This question-driven approach is about learning by doing—approaching management as an experiment upon which to learn and reduce uncertainty. It is a systematic, iterative, incremental approach requiring the continuous monitoring, evaluation, and adjustment of management actions. Adaptive management can also be understood in the negative, as it is different from more typical front-ended approaches to planning whereby assumptions and predictions are made in the beginning of the process, but then not necessarily adjusted according to what actually happens as a result. A NEPA Task Force, for example, contrasts the status quo "predict-mitigate-implement" NEPA-based model with a "predict-mitigate-implement-monitor-adapt model."[39]

The innate administrative tendency to prioritize discretion helps explain how some agencies have implemented adaptive management and some of the backlash that has ensued. In some cases, agencies have interpreted adaptive management in a way that puts a premium on flexibility, discretion, and expedited decision making. They have embraced parts of the adaptive management model while eschewing others. Some interests are concerned that the perceived need for flexibility, discretion, and expedited decision making can be easily abused by agencies and make it harder to hold them accountable for their actions.

These fears are exacerbated by the lack of specificity given to adaptive management in law or regulation. Most administrative definitions are actually vaguer than those found in the academic literature. No statute defines the term and agency regulations doing so usually provide more platitudes than detail. As law professor J.B. Ruhl points out, "[o]ne has to be concerned when legal text becomes even more obscure than the theory on which it is based."[40] The problem, as Ruhl sees it, is that "[m]ushy definitions of adaptive management are likely to make for mushy standards of implementation."[41]

Lots of examples can be used to demonstrate this problem. To take one, consider how the USFS approached adaptive management in its 2005 and 2008 planning regulations.[42] The agency emphasized the problems and challenges of

NEPA-based rational comprehensive planning and proposed in its stead a "paradigm shift in land management planning."[43] The 2005/2008 regulations embraced the language and some of the core principles of adaptive management. But to be truly adaptive the agency believed it had to free itself from some NEPA obligations so to be able to respond to new science, information, and problems more quickly.[44] Forest plans would thus not be decision-making documents, but rather "strategic and aspirational" in nature, one tentative step in a more adaptive planning process.[45] Also gone from the regulations were some of the sharpest standards and legal hooks holding the agency accountable, such as the wildlife viability standard.[46] Put simply, the USFS believed that it needed more flexibility and discretion in order to practice adaptive management.

The USFS's discretion-based approach to adaptive planning did not sit well with environmental groups and their lawyers. Some critics believed that these regulations simply used the rhetoric of adaptive management as a means to remove standards, undermine NEPA and NFMA, and maximize agency discretion.[47] At the time of this writing, the USFS continues to grapple with how to practice adaptive management while lawfully implementing its other substantive and procedural obligations.[48]

The BLM is also trying to implement more adaptive approaches to planning and management.[49] One of the most closely watched experiments in this regard has been the agency's approach to oil and gas development in Wyoming, most notably the Pinedale Anticline oil and gas exploration and development project.[50] In this case, the BLM adopted an adaptive approach because of the possible impacts of energy development to wildlife in the area, including sage grouse, mule deer, and pronghorn antelope.

But the BLM met some of the same legal challenges to adaptive management as did the USFS—legal challenges that have long been identified as hindering EM. Two such impediments include the Federal Advisory Committee Act (FACA) and NEPA. As discussed below, FACA can present problems because of its lengthy procedural obligations that can have the unintended consequence of limiting public participation in an adaptive/collaborative management process. This was the case on the Pinedale Anticline, where a collaborative group was formed to help implement the process. The problem was that it took two years for the group to obtain its FACA charter once the process was initiated, and this was four years into the project.[51]

NEPA also presented challenges in the Pinedale case, ones that are compounded by the complexity and multistaged nature of oil and gas planning and development. One challenge, for example, is NEPA's requirement to prepare a supplemental environmental impact statement (EIS) upon the discovery of "significant new information."[52] The challenge here is that agencies implementing adaptive management may continuously trigger the need for more (supplemental) NEPA analysis. Unlike the traditionally front-ended NEPA process, adaptive management is about the continuous collection of information and the making of corresponding adjustments throughout the life of a plan or project.

The USFS and BLM examples are not anomalies. Much of the policy and legal scholarship on adaptive management (and governance) goes so far as to suggest that modern environmental problems require a fundamental reorientation of environmental law and planning.[53] The examples above show why NEPA is an important part of this puzzle, as a key question is whether NEPA facilitates or obstructs this sort of adaptive learning.[54] Some people believe that NEPA must be changed or clarified in order to make it more adaptive management-ready.[55] On the other side we see those believing that NEPA can already accommodate more adaptive-based planning approaches. This is because NEPA requires a forward-looking approach—a requirement to assess possible environmental impacts *before* undertaking a major new action.[56] And once that general course is set, adaptive management becomes a means to an end. Put differently, once a program, plan, or project is established using NEPA, adaptive management can be used as a way to ensure the goals are being met. After all, adaptive management is not about experimenting for the sake of experimenting. It needs a purpose and hopefully NEPA will be used as a way to define it.

B. Collaboration

Collaboration plays a significant role in ecosystem management. Planning at such scales requires greater coordination among landowners and "the acquiescence, if not active support, of a broad cross section of society."[57] So entwined was collaboration with EM that some started calling it "cooperative ecosystem management" or "grassroots ecosystem management."[58] Some inherent tensions between the two were also apparent from the get-go, for how does an agency reconcile the scientific expertise needed for what some began to call 'ego-system management' with increased demands for more widespread participation in planning?

Contradiction or not, the language and ideas of collaboration are now firmly established in federal land planning and management. Growth and interest in the 'movement' is extraordinary, with dozens of advocates, scholars, think-tanks, clearinghouses, and government officials promoting its beneficial use.[59] It would be an overstatement to say that EM catalyzed this development, partly because calls for more participatory planning were being made well before spotted owls changed the game. In fact, one of the dominant criticisms of the USFS before NFMA was enacted was the lack of meaningful public participation in agency decision making.[60] Nevertheless, EM certainly helped push things along.

Take the case of the BLM for example. Former Secretary of the Interior Bruce Babbitt strongly embraced EM for scientific and political purposes. Its blending of science with collaboration offered a new path for BLM politics, providing a way to broaden the scope of conflict, both spatially and in the number of new values and interests that would be brought into BLM planning. James Skillen explained in his history of the agency that,

> For Babbitt, ecosystem management was promising despite its inherent ambiguity and tension. First, it changed the nature and politics of land use planning by shifting the scale and boundaries of planning debates. By tackling land use planning on a broader, ecosystem scale, Babbitt could increase competition among conflicting interest groups and political subsystems. Forcing interest groups and public lands users into a different political arena could break iron triangles that had developed over particular resources or particular geographic areas.[61]

Collaboration was also advanced during Babbitt's tenure by creating multistakeholder resource advisory councils (RACs) that provide advice to the BLM regarding the "preparation, amendment and implementation of land use plans and the development of standards and guidelines."[62] In contrast to the old rancher-dominated advisory committees established under the Taylor Grazing Act, RACs were endorsed as a way to bring new voices into the fold and to 'institutionalize' more collaboration in rangeland planning.

The executive branch pendulum swing from Clinton to Bush best explains what happened to EM and collaboration on public rangelands. "Sustaining Working Landscapes" eclipsed EM and became the new mantra for the BLM under Interior Secretary Gale Norton.[63] Collaboration became even more politically attractive and it was often contrasted to more regulatory and adversarial approaches to conservation. Irony notwithstanding, Executive Order 13,352 (Aug. 26, 2005) aimed to facilitate the bottom-up use of "cooperative conservation" and a White House-sponsored conference on the matter convened in 2006.[64] Secretary Norton also espoused a "Four Cs" agenda: "consultation, cooperation, and communication, all in the name of conservation."[65]

But this emphasis on collaboration during the Bush administration did not translate into all realms of range management. In 2006, for example, the BLM tried to amend its grazing regulations,[66] though they were eventually set aside by the courts.[67] One of the more controversial amendments included removing the requirement that the agency consult, cooperate, and coordinate with the 'interested public' regarding various management decisions. The regulations also proposed to no longer require the involvement of interested members of the public when issuing or renewing individual grazing permits. Instead, the BLM would only "consult, cooperate, and coordinate" with "affected permittees and lessees, and the state."[68] This example demonstrates how collaboration, like EM before it, is subject to changing priorities in the White House.

Collaboration was also part of the EM package advanced on the national forests. As commonly done elsewhere, they were linked together by the Committee of Scientists in 1999, which recommended more ecosystem and collaborative-based approaches to forest planning.[69] The collaborative focus endures, with just about everyone asking for more or better collaboration with the USFS. Congress entered the fray in 1998 by requiring the USFS to use a "multiparty monitoring and evaluation process" when using stewardship contracts.[70] And in 2003, Congress required collaboratively-written community wildfire protection plans as part of the Healthy Forests Restoration Act.[71] This law provides incentives for the writing of such plans while providing more participation and community engagement earlier in the planning process. As discussed below, Congress again endorsed more collaborative approaches to forest restoration in passing the Collaborative Forest Landscape Restoration Act in 2009.[72]

Top Bush administration officials in the USFS also encouraged more collaborative approaches to planning, with some going so far as to suggest that collaboration was the future of conservation and national forest management.[73] Little wonder, then, that collaboration was emphasized by the agency in its rewriting of forest planning regulations. The Clinton, Bush, and Obama administrations have all taken on the Sisyphean task of rewriting the NFMA planning rule. A full accounting of this interminable process is (thankfully) impossible here. But a common thread running through each proposed rule is the importance of collaboration. Though dismissed by the courts, the 2005 and 2008 planning regulations called for a "collaborative and participatory approach to land management planning."[74] The Obama planning rule took collaboration even further, starting with an unprecedented outreach effort and proposing a rule in which collaboration and science were the two basic "anchor points."[75]

Collaboration proceeds in the national forests despite the travails of the NFMA planning rule. Throughout the country, divergent interests are collaborating about how they would like particular national forests to be managed. Some of these initiatives are seeking place-based legislation as a way to secure their agreements, outside of forest-planning processes, while others use an assortment of different approaches and memoranda of understanding with the USFS.[76] What is most remarkable about these collaborative is what they have in common. One of their most defining characteristics, for example, is the shared belief that the USFS should be planning at much larger spatial scales than currently practiced (as discussed below). There is also a widespread desire to be continuously engaged in forest management, not just during the limited time frames offered by rulemaking, NEPA, and the forest-planning process.

This quick review demonstrates a top-down and bottom-up embrace of collaboration by the USFS in the national forests. But embracing the idea of collaboration is different from practicing it. Some of the same institutional barriers to implementing EM are now commonly identified as impeding more collaborative planning in the national forests. First of all, there remain philosophical concerns about collaboration and more decentralized approaches to federal land management.[77] There is a persistent fear that collaboration can be politically exploited and used as cover to undermine the national interest and rule of law. Recall, for instance, how the Department of the Interior sang the praises of collaboration while simultaneously making it harder for the nonranching public to participate in range management and planning decisions. For some critics, then, 'impediments' to collaboration may in fact be necessary sideboards and safeguards.

There are also some familiar legal barriers to more collaborative planning. Some people for instance, complain that the Federal Advisory Committee Act (FACA) can have a chilling effect on collaborative planning.[78] FACA's purpose is to open agency decision making to the public and to check the improper use of experts, industry, and advisory committees in agency decision making. The law applies to groups providing "advice or recommendations" to the federal government that are either "established" or "utilized" by the government.[79] A few prominent initiatives in EM violated FACA's procedural requirements, including the Forest Ecosystem Management Assessment Team (FEMAT), the Sierra Nevada Ecosystem Project, and the Southern Everglades Restoration Alliance.[80] These and other cases led to a 'FACA-phobia' amongst some agencies who sometimes decided to simply forego collaboration instead of working through the complexities of this well-intentioned law.

NEPA is also often cited by collaborative groups as being an unnecessary hurdle to more collaborative planning by the USFS. This should not be the case, as one of NEPA's most enduring legacies is the public participation and transparency demanded by the statute.[81] Nevertheless, some practitioners believe that "many landscape-scale forest restoration efforts are hindered by agency and stakeholder assumptions that collaboration must be narrow and limited once project planning enters a formal NEPA process."[82] Part of the problem here is the predictive-based structure of NEPA, as discussed above in the context of adaptive management. Instead of being solely engaged pre-decision—from scoping to commenting on draft EISs—there are increasing calls by some planning participants to be included throughout the entire NEPA process, from scoping through project implementation and postproject monitoring. Other groups are asking for a more pro-active role in helping shape projects from the most initial design stage of the NEPA process. In doing so, these groups hope that common zones of agreement can be found early on, saving the agency and various interests the costs and troubles of a postdecision legal challenge.[83]

C. Landscape-Scale Restoration

The word restoration is usually found in most definitions of ecosystem management. A 1994 BLM definition is typical; "[t]he primary goal of ecosystem management is to conserve, restore, and maintain the ecological integrity, productivity, and biological diversity of public lands."[84] While the term 'ecosystem management' is not as prevalent as it once was, the restoration focus is now more widespread, with federal land agencies and others working on several 'landscape-scale' initiatives. 'Landscape-scale' or an 'all lands' approach to conservation is now the popular vernacular, replacing EM in various agency rules and policy statements.[85] But all are basically premised on the same thing when it comes to restoration: planning and management should occur at scales commensurate with natural disturbances, such as large wildfires, invasions of cheat grass, and so forth.

As discussed below, in some cases the restoration agenda has been pushed by Congress or the executive branch and in other cases pressure comes from the bottom up. But regardless of where it comes from, efforts in landscape-scale restoration face an array of familiar complications. For the purposes here I focus mostly on the national forests because the USFS is once again trying to fit a new and exciting management approach into an old and unexciting statutory, planning, and budgetary framework.

Like EM, restoration is subject to multiple and sometimes competing interpretations; that is part of its political allure. But this lack of clarity and a common definition can also be problematic, from a political and managerial standpoint.[86] Engaging in such a dialogue quickly reveals how differently people define the restoration 'problem' and what, if anything, should be done about it. There is also no avoiding the political choices that must be made in restoration decisions, from determining a historic baseline and desired future conditions, to whether it is best to do something or nothing in various places.

Congress has supported restoration in different federal land laws since the emergence of EM.[87] In 1998, for example, Congress authorized the USFS and BLM to use stewardship contracting to achieve various land management goals such as restoring forest and rangeland health and water quality, improving fish and wildlife habitat, and reducing hazardous fuels.[88] To achieve these goals, stewardship contracting allows the exchange of goods for services.[89] In other words, the timber commodities produced through a contract are exchanged for requested restoration services, like decommissioning roads or replacing culverts. Stewardship contracting allows the USFS and BLM to retain the receipts generated by selling timber for use in future stewardship projects.[90]

Congress then passed the Healthy Forests Restoration Act (HFRA) in 2003, which mostly aims to expedite hazardous fuel reduction projects on USFS and BLM lands.[91] Some common EM themes are found in the law, including collaborative planning, multiparty monitoring, and some safeguards pertaining to biodiversity and old-growth trees. But the statute basically emphasizes hazardous fuels reduction and tries to ease the procedural burdens of getting it done quickly. To do so, HFRA created new administrative and judicial review procedures and modified NEPA compliance requirements. Under HFRA, agencies consider fewer EIS alternatives and their decisions are subject to a 'predecisional administrative review process' rather than traditional administrative appeals. The law also exempts community wildfire protection plans from FACA. Of course, HFRA hardly counts as restoration in the eyes of its critics, who believe that "HFRA's more ecologically sensitive and prescriptive restoration provisions . . . are counterbalanced by a renewed congressional commitment to timber cutting and to minimizing the law's role in this process."[92]

Congress endorsed a more experimental approach to restoration in 2009, creating the Collaborative Forest Landscape Restoration Program.[93] The program selects and funds landscape-level forest restoration projects that are screened by a committee. To be eligible, restoration projects must be at least 50,000 acres and be done in scales to improve wildfire management, reduce management costs, restore ecosystem functions, and to facilitate the use of biomass and small-diameter trees. Such projects must comply with existing environmental laws and be

developed and implemented through a collaborative process.[94] Up to ten proposals can be funded per year (with only two proposals in any one region of the
national forest system), and each project is evaluated based on several criteria.[95]
The program authorizes $40,000,000 per year (FY 2009–2010) to be used to pay
for up to 50 percent of selected restoration projects.[96]

There are also increasing demands for landscape-scale restoration coming
from the bottom up. A common complaint about the USFS is that the agency
manages and implements restoration projects at too small a scale. This is probably due in part to the agency's fear of administrative appeals and litigation and
perceptions of risk. These legal challenges are believed to be easier as the projects
get larger in scope and scale, which explains why the agency has sometimes
moved away from large, multifaceted projects to smaller and more isolated ones.
Whatever the reasons, there is now widespread frustration with small-bore and
disjointed approaches to restoration.[97] Many groups want restoration to be
planned at much larger spatial scales, for both ecological and economic reasons.

Arizona's Four Forests Restoration Initiative (4FRI) provides a case-in-point.
This collaboration between industry and environmental groups seeks to restore
ponderosa pine forests in four national forests in Arizona. The partnership marks
the 467,000 acre Rodeo-Chediski fire in 2002 as an important turning point, with
several interests recognizing that fires and other events of such magnitude necessitate a larger scale approach to planning. The 4FRI partnership believes the USFS
should be planning at scales twenty to thirty times larger than they currently do.
In order to do landscape-scale restoration across roughly 2.4 million acres of
ponderosa pine forests, the 4FRI anticipates that "the first large-scale planning
area will cover ~750,000 acres, which will identify roughly ~300,000 acres for
thinning over 10 years at a rate of up to 30,000 acres of treatment per year."[98]

Increasing calls for restoration have not gone unheard by the USFS. The
agency tried to once again reframe the forest management debate during the
Bush administration, with Chief Bosworth and others claiming the 'forest wars'
over once and for all. "Community-based Stewardship" and restoration were
put forth as the new direction for the agency.[99] As proof, agency leaders pointed
to facts like "roughly 75–80% of the timber from national forest land now comes
from projects for other purposes, such as fuels reduction, habitat improvement,
and ecological restoration."[100] The restoration focus is also easily squared with
the open-ended statutory mandate given to the USFS, including the 1897 Organic
Act, which specifies that the USFS "improve and protect the forest" and secure
"favorable conditions of water flows."[101]

Restoration became even more paramount within the USFS under President Obama. The Secretary of Agriculture Tom Vilsack made waves at the beginning of his tenure by saying that restoration is the agency's vision of the future and that "[r]estoration means managing forest lands first and foremost to protect our water resources, while making our forests more resilient to climate change."[102] Forest-planning regulations introduced by the Obama administration also emphasized other central principles of EM (including adaptability, collaboration, and an 'all-lands' approach to forest management). The 2012 planning rule aimed to facilitate the writing of forest plans that would "protect, reconnect, and restore" national forests and grasslands.[103]

All of this goes to show that the flow of forest politics is heading to restoration. But this current faces some all-too-familiar blockages to planning and implementing landscape-scale restoration. To begin, planning is difficult absent an agreed-upon purpose and there remain significant disagreements about what most needs restoring and conserving in the national forests. There are concerns, for example, that the USFS approaches restoration with an organizational timber bias. Some of these worries stem from the timber-centric approach to restoration as found in HFRA and President Bush's Healthy Forests Initiative. And some concern is due to the scale of the perceived problem; a vision made clear when Obama's Undersecretary of Agriculture stated that 110 million acres (out of 193 million acres in the National Forest System) are in need of restoration.[104] Some people also believe that the USFS conflates watershed restoration with forest restoration, and this means that initiatives by the agency tend to prioritize things like biomass while not doing enough road decommissioning and the like.[105]

How to fund restoration projects at large scales is another recurring question. Several interests believe that stewardship-contracting authority is a promising yet underutilized approach by the USFS and BLM. One problem is that while the law authorizes agencies to retain receipts from stewardship contracts, administrative policy prohibits the use of this money for planning and monitoring purposes.[106] There are also concerns that the timber-goods for restoration-services structure of stewardship contracting will inevitably lead to increased pressure to cut more trees in order to restore more things. If so, this funding mechanism leads us back to some familiar political terrain.

Another recognizable challenge to restoration is the programmatically structured nature of the USFS budget. Generally speaking, money for programs in the national forests is based on a limited set of resource-specific line-items that get 'stovepiped' from national headquarters to the individual national forests. The problem is that this approach does not align well with the integrated or ecosystem-based nature of forest management, because some prioritized activities, such as restoration, do not have their own line items. Budget lines that currently fund restoration in the national forests "are not coordinated with one another, have individual targets that drive work plans, and are allocated in ways that constrain agency flexibility, efficiency and adaptability."[107] What often happens is that one aspect of a restoration project gets funded, such as hazardous fuels reduction, while companion efforts like road decommissioning do not. As the Western Governor's Association sees it, "[w]hen activities are 'stovepiped' into separate programs with their own funding, targets and accomplishment reporting, the large-scale treatment objectives are not achieved."[108]

Landscape-scale restoration projects have also encountered some NEPA-related challenges. At one end of the spectrum are those simply unhappy with the time and resources it takes to comply with the law. For some, restoration requires expedited action and NEPA unnecessarily bogs things down—the "analysis paralysis" and "process predicament" complaint.[109] This is a standard NEPA critique and restoration provides yet another opportunity to make this case. But more moderate and even green interests have struggled to find the right application of NEPA for large-landscape restoration. One of the challenges is how to make proper assessments at such massive scales, such as the proposal to restore some 750,000 acres of ponderosa pine forests in Arizona. Would one broad-based programmatic EIS suffice? Or must the traditional approach of tiering project-level EISs to programmatic plans still apply? What is the most effective way to analyze cumulative effects? And how might NEPA work be more strategically timed to the awarding of stewardship contracts? Some of these questions, such as how to effectively tier plans, are standard fare in federal land management, but they can become more complicated when applied at the large-landscape level.

III. CONCLUSION

The question of whatever happened to ecosystem management is kind of like asking whatever happened to someone with a multiple personality disorder. There are multiple conceptions of EM, from the theoretically sound to the politically advantageous, so it is necessary to first be sure of whom we are speaking. Once certain, it is then possible to trace what happened to various ideas important to EM, such as how they were used politically, to how they manifested themselves in routine agency actions. The term EM is no longer used with the regular-ity it once was, but its basic principles and view of management and planning persist. The language has changed, but the core ideas and challenges have not. Adaptive management, collaboration, and restoration: all were basic ingredients of EM and all are even more relevant today than they were twenty years ago.

In some cases, progress has been made on these fronts. Much of it is due to the enforcement of laws such as the ESA and the NFMA's wildlife viability standard. More than anything else, legal standards such as these have brought change to the federal lands. EM would be little more than a vacuous slogan without such laws backing it up. Congress has also facilitated change by providing new tools, such as stewardship contracting authority and the Collaborative Forest Landscape and Restoration Program, among others. There is also a lot of innovation happening at the grassroots level, with several collaborative groups advancing more adaptive, participatory, and landscape-level approaches to federal land management. A collaborative group making some headway in resolving conflicts and getting agencies to think about the big picture and all of its interconnections is becoming a common story.

Yet even more common are stories about initiatives in adaptive management, collaboration, and landscape-scale restoration meeting some familiar impediments. These include disparate agency missions and planning processes, shifting political priorities, problematic budgets, and an assortment of other challenges. Laws such as NEPA and FACA can also present dilemmas, though they also help ensure that environmental and democratic values are not subverted in the name of EM.

It is beyond the scope of this chapter to analyze what is next for EM and its descendants. Doing so would entail an analysis of federal lands governance writ large. Nonetheless, there are some general ideas worth consideration. First is the status quo option of muddling through; making incremental progress towards EM and its core principles baby step-by-step. Painful as it may be to watch, this evolutionary approach is relatively safe and politically feasible. It is possible to imagine, for example, that Congress might one day bring the USFS budgetary structure into the twenty-first century,[110] or that an Executive Order might mandate an "interagency coordination statement" in all planning documents,[111] or that the Council on Environmental Quality might soon provide guidance or new regulations pertaining to NEPA's application to adaptive management, collaboration, and restoration.[112]

Instead of shooting for the stars, perhaps the best way to proceed is to focus on some relatively feasible proposals. The danger, though, is the fatigue and cynicism that comes from repeatedly hitting the same walls. The proverbial definition of insanity comes to mind: doing the same things over and over again, expecting different results. How many times, for example, will the USFS try to rewrite NFMA planning regulations? Or how many times will legitimate efforts in collaborative restoration be stymied by underfunded agencies and inflexible budgets?

A related approach to consider is the possibility of more legislative or regulatory 'nudges' by Congress and the executive branch. These laws and regulations would not systematically overhaul the system, but rather continue prodding agencies to plan in a more adaptive, collaborative, and landscape-scale fashion. Post spotted-owl laws such as the Community Forest Landscape Restoration Act and stewardship-contracting show that Congress is capable of passing new federal land laws. Controversy arises, however, when Congress passes a new law that changes the application of older ones, such as NEPA. The Healthy Forests Restoration Act is worth thinking about in this context because it is a rare case in which Congress confronted legal barriers and actually modified the preexisting statutory framework, from exempting FACA to streamlining NEPA. The substance of the law is controversial, and it is certainly no paragon of EM or restoration, but the law's design demonstrates how Congress might act again in the future—this time in order to facilitate other EM principles.

A harder nudge would be for Congress to pass a new ecosystem management or restoration law that would supplement existing federal land statutes. Professor Keiter observed long ago that "until Congress speaks, ecosystem management can only claim a tenuous legitimacy, which also leaves the concept undefined for legal purposes."[113] In subsequent work, Keiter recommends a law that would establish clear priorities among multiple uses, acknowledge the need for coordinated landscape-level planning, and include at least two statutory standards: a nonimpairment standard establishing a threshold for evaluating management proposals, and a biodiversity conservation standard imposing "an affirmative obligation on the agencies to protect and restore species diversity."[114] "Framed as management standards rather than hard-and-fast rules," says Keiter, "the proposal seeks to protect ecological components and processes without placing land managers in a straitjacket, rendering them unable to respond to unique local conditions or exceptional circumstances."[115] As Keiter sees it, "[t]he statutory proposal does not envision a radical restructuring of agencies or boundaries; the proposed legal standards are not new, nor do the procedural or enforcement mechanisms depart from existing law. By linking the nonimpairment standard with an ecosystem restoration obligation, the proposal should help promote truly sustainable resource management policies, thus enhancing community stability and perhaps restoring some peace on the public domain."[116]

The third broad option to consider is a more comprehensive review of federal land planning and management. Reconvening another Public Lands Law Review Commission, or something like it, could provide an opportunity to bring federal land planning into the twenty-first century.[117] With appropriate sideboards and a clearly defined charter, a comprehensive review has the potential of providing more enduring solutions to problems reviewed in this chapter. Bear in mind that more than thirty years have passed since NFMA and FLPMA were enacted. Gone are the halcyon days of rational comprehensive planning—or illusions thereof. The world is a different place; perhaps the crux of the matter is that these planning statutes are simply not designed to deal with today's problems. Say what you will about the NFMA, for example, but the statute is clearly timber-focused. It was born from the clear-cutting and 'get out the cut' controversies of the 1960s and 70s—and the law reads as such, with most of its provisions related to logging and its constraints. But today's planning challenges go well beyond timber. Restoration, ecosystem services, fire management, motorized recreation, biomass and renewable energy, climate change, and wildlife viability; these and similar issues are the problems confronting planners and advocates alike. Given the enduring nature of EM and its offspring, perhaps it is time to finally codify EM principles in a more coherent and contemporary set of laws.

NOTES

1. Robert L. Glicksman & George Cameron Coggins, Modern Public Land Law in a Nutshell 25 (3d ed. 2006).

2. U.S. Dep't of Agric., Committee of Scientists, Sustaining the People's Lands: Recommendations for Stewardship of the National Forests and Grasslands into the Next Century 4 (March 15, 1999).

3. Eric T. Freyfogle, The Land We Share (2003).

4. Martin A. Nie, The Governance of Western Public Lands 39 (2008).

5. William R. Travis, David M. Theobald & Daniel B. Fagre, *Transforming the Rockies: Human Forces, Settlement Patterns, and Ecosystem Effects, in* Rocky Mountain Futures (Jill S. Baron ed., 2003).

6. NRCS, Summary Report: 1997 National Resources Inventory (2000).

7. American Farmland Trust, Strategic Ranchland in the Rocky Mountain West, *available at* http://www.farmland.org/resources/rockymtn/default.asp (last visited Dec. 8, 2011).

8. USDA FS, Four Threats to the Health of the Nation's Forests and Grasslands, *available at* http://www.fs.fed.us/projects/four-threats/ (last visited Dec. 8, 2011).

9. U.S. Gov't Accountability Office, GAO-07-427T, Wildland Fire Management: Lack of a Cohesive Strategy Hinders Agencies' Cost Containment Efforts (2007).

10. Dennis A. Glick & Tim W. Clark, *Overcoming Boundaries: The Greater Yellowstone Ecosystem, in* Stewardship Across Boundaries 237 (Richard L. Knight & Peter B. Landres eds., 1998).

11. For an overview of these planning processes *see* 72 Fed. Reg. 14, 866 (March 29, 2007).

12. *See, e.g.,* Greater Yellowstone Coalition v. Servheen, 672 F. Supp. 2d 1105 (D. Mont. 2009) (vacating the delisting of the Greater Yellowstone Grizzly Bear Distinct Population Segment (DPS) from the Endangered Species Act due in part to inadequate plans by the FWS and USFS).

13. NFMA, for example, requires forest plans be "coordinated with the land and resource management planning processes of State and local governments and other Federal agencies," 16 U.S.C. §1604(a) (2006), and FLPMA requires the BLM to coordinate with state and local governments in the development of land use plans "to the extent consistent with the laws governing the administration of the public lands," and to consider input from states and other nonfederal entities in developing plans. 43 U.S.C. §1712(c) (2006).

14. 16 U.S.C. §1531 (2006)

15. *See, e.g.,* Steven L. Yaffee et al., Ecosystem Management in the United States (1996); Lara D. Guercio & Timothy P. Duane, *Grizzly Bears, Gray Wolves, and Federalism, Oh My! The Role of Endangered Species Act in De Facto Ecosystem-Based Management in the Greater Glacier Region of Northwest Montana,* 24 J. Envtl. L. & Litig. 285, 285–366 (2009); Oliver A. Houck, *On the Law of Biodiversity and Ecosystem Management,* 81 Minn. L. Rev 869–978 (1997); Craig W. Thomas, Bureaucratic Landscapes (2003).

16. National Park Service, Management Policies §1.6, §3.4 (2006), *available at* http://www.nps.gov/policy/mp2006.pdf (last visited Dec. 8, 2011).

17. *See generally* Craig L. Shafer, *U.S. National Park Buffer Zones: Historic, Scientific, Social, and Legal Aspects,* 23 Envtl. Mgmt. 49 (1999)

18. Joseph L. Sax & Robert B. Keiter, *The Realities of Regional Resource Management: Glacier National Park and Its Neighbors Revisited,* 33 Ecology L.Q. 307 (2006). *See also* Guercio & Duane, *supra* note 15.

19. Norman Maclean, A River Runs Through It and Other Stories (1976).

20. For more on the Blackfoot Challenge *see* http://blackfootchallenge.org (last visited May 30, 2011).

21. U.S. Forest Service, Bridger-Teton National Forest, Decision Notice and Finding of No Significant Impact: Pronghorn Migration Corridor Forest Plan Amendment (May 31, 2008).

22. Western Governor's Association, Policy Resolution 07-01, Protecting Wildlife Migration Corridors and Crucial Wildlife Habitat in the West (2007).

23. Claudia Goetz Phillips & John Randolph, *Has Ecosystem Management Really Changed Practices on the National Forests?,* 96 J. Forestry 40 (1998).

24. Kelly F. Butler and Tomas M. Koontz, *Theory into Practice: Implementing Ecosystem Management Objectives in the USDA Forest Service,* 35 Envtl. Mgmt. 138 (2005).

25. Catherine M. Rigg, *Orchestrating Ecosystem Management: Challenges and Lessons from Sequoia National Forest,* 15 Conserv. Biology 78 (2001).

26. S. Andrew Predmore, Carolyn A. Copenheaver & Michael J. Mortimer, *Ecosystem Management in the U.S. Forest Service: A Persistent Process but Dying Discourse*, 106 J. Forestry 339 (2008).

27. Hanna J. Cortner & Margaret A. Moote, The Politics of Ecosystem Management 40 (1999).

28. R. Edward Grumbine, *What Is Ecosystem Management?*, 8 Conserv. Biology 27, 30–31 (1994).

29. Judith A. Layzer, Natural Experiments 22, 22–23 (2008).

30. Jack Ward Thomas, *Forest Service Perspective on Ecosystem Management*, 6 Ecological App. 703 (1996).

31. Richard Freeman, *The EcoFactory: The United States Forest Service and the Political Construction of Ecosystem Management*, 7 Envtl. Hist. 633 (2002).

32. M. Lynne Corn, Cong. Research Serv., R93-655 ENR, Ecosystems, Biomes, and Watersheds: Definitions and Use 2 (1993).

33. U.S. Gov't Accountability Office, GAO/T-RECED-94-308, Ecosystem Management: Additional Actions Needed to Adequately Test a Promising Approach (1994).

34. *See, e.g.,* Cortner & Moote, *supra* note 27; Rigg, *supra* note 25, at 60–69.

35. Robert B. Keiter, *Ecosystems and the Law: Toward an Integrated Approach*, 8 Ecological App. 332–41 (1998). *See also* John Copeland Nagle & J.B. Ruhl, The Law of Biodiversity and Ecosystem Management (2002).

36. Cortner & Moote, *supra* note 27, at 44.

37. Paul R. LaChapelle, Stephen F. McCool & Michael E. Patterson, *Barriers to Effective Natural Resource Planning in a 'Messy' World*, 16 Soc'y & Nat. Resources 473 (2003).

38. U.S. Dep't of the Interior, Adaptive Management: The U.S. Department of the Interior Technical Guide, at v (2009). *See also* National Research Council, Adaptive Management for Water Resources Planning (2004).

39. The NEPA Task Force, Report to the Council on Environmental Quality: Modernizing NEPA Implementation, ch. 4 (2003).

40. J.B. Ruhl, *Adaptive Management for Natural Resources—Inevitable, Impossible, or Both?*, 54 Rocky Mountain Mineral L. Inst. Proc. 11-1 (2008).

41. *Id.*

42. The 2008 regulations are basically the same as the 2005 regulations, though the 2008 iteration went through the NEPA process, as ordered by a District Court who found the 2005 planning regulations in violation of the APA, NEPA, and ESA. *See* Citizens for Better Forestry v. Dep't of Agric., 481 F. Supp. 2d 1089 (N.D. Cal. 2007). *Compare* 73 Fed. Reg. 21, 468 (Apr. 21, 2008) *and* 70 Fed. Reg. 1023 (Jan. 5, 2005).

43. 70 Fed. Reg. 1023, 1024 (Jan. 5, 2005).

44. 70 Fed. Reg. 1023 (stating that the "intended effects of the final rule are to streamline and improve the planning process by making plans more adaptable to changes in social, economic, and environmental conditions..."). *See also* Deann Zwight, *Smokey and The EMS*, 21 Envtl. F. 28 (2004) (discussing the need for a more adaptive forest planning process).

45. Emphasized throughout the rule, and in subsequent forest plans using it, is that the rule and plans "will not contain final decisions that approve projects or activities except under extraordinary circumstances." 70 Fed. Reg. 1023, 1024 (Jan. 5, 2005).

46. In its stead the USFS put forth a much less prescriptive 'ecosystem approach' to diversity. 70 Fed. Reg. 1023, 1028 (Jan. 5, 2005).

47. *See, e.g.*, Earthjustice and Defenders of Wildlife, Complaint for Declaratory and Injunctive Relief, Defenders of Wildlife et al, v. Schafer, Case No. C08-02326 (D. N. Cal. 2008); Alyson Flournoy, Robert L. Glicksman & Margaret Clune, *Regulations in Name Only: How the Bush Administration's National Forest Planning Rule Frees the Forest Service from Mandatory Standards and Public Accountability* (Center for Progressive Reform White Paper, June 2005); Nathaniel S.W. Lawrence, *A Forest of Objections: The Effort to Drop NEPA Review for National Forest Management Act Plans*, 39 Envtl. L. Rep. 10651 (2009); Society for Conservation Biology, Comments on Proposed Changes to the National Forest System Land and Resources Management Planning Rule (no date provided); and WildLaw, Review of the New NFMA Planning Regulations (2005), *available at* http://www.sierraforestlegacy.org/Resources/TakeAction/ActionAlertArchives/WildLaw_NFMA_Regs_White_Paper.pdf.

48. *See* 74 Fed. Reg. 67, 165 (Dec. 18, 2009) (notice of intent asking how the USFS's new planning rule can be more adaptive and address uncertainty).

49. Melinda Harm Benson, *Adaptive Management Approaches by Resource Management Agencies in the United States: Implications for Energy Development in the Interior West*, 28 J. Energy & Nat. Resources L., 87–118 (2010).

50. Melinda Harm Benson, *Integrating Adaptive Management and Oil and Gas Development: Existing Obstacles and Opportunities for Reform*, 39 Envtl. L. Rep. 10962–10978 (2009).

51. *Id*, at 10970.

52. 40 C.F.R. §1502.9(c)(1)(ii).

53. *See, e.g.*, Ronald D. Brunner, et al., Adaptive Governance (2005); Bradley C. Karkkainen, *Collaborative Ecosystem Governance: Scale, Complexity, and Dynamism*, 21 VA. Envtl. L.J. 189 (2002); Jamison E. Colburn, *The Indignity of Federal Wildlife Habitat Law*, 57 Ala. L. Rev. 417 (2005). For a thorough review and critique of the "new governance" literature *see* Annecoos Wiersema, *A Train Without Tracks: Rethinking the Place of Law and Goals in Environmental and Natural Resources Law*, 38 Envtl. L. 1239 (2008).

54. Martin Nie & Courtney Schultz, Decision Making Triggers in Adaptive Management: Report to USDA Pacific Northwest Research Station, NEPA for the 21st Century (2011) (analyzing how decision-making triggers might be used to bridge the science of adaptive management with the need for political accountability)

55. R.G. Dreher, Georgetown Envt'l Law and Pol'y Inst., *NEPA Under Siege: The Political Assault on the National Environmental Policy Act* (2005).

56. *Id.*

57. Cortner & Moote, *supra* note 27, at 44–45.

58. Steven L. Yaffee, et al., Ecosystem Management in the United States (1996); Edward P. Weber, Bringing Society Back In (2003).

59. For a review *see* Martin Nie, *The Underappreciated Role of Regulatory Enforcement in Natural Resource Conservation*, 41 Pol'y Sci. 139 (2008).

60. *See, e.g.,* Arnold W. Bolle, *Public Participation and Environmental Quality,* 11 Nat. Res. J. 497–505 (1971).

61. James R. Skillen, The Nation's Largest Landlord 143 (2009).

62. 60 Fed. Reg. 9894, 9900 (Feb. 22, 1995).

63. Skillen, *supra* note 60, at 176.

64. For conference background and documents *see* http://cooperativeconservation.gov/conference805home.html (accessed May 2, 2011).

65. Lynn P. Scarlett, *A New Approach to Conservation: The Case of the Four Cs,* 17 Nat. Res. Env't. 73–113 (2002).

66. 71 Fed. Reg. 39, 402 (July 12, 2006).

67. Western Watersheds Project v. Kraayenbrink, 620 F.3d 1187 (9th Cir. 2010).

68. *Id.* at 1194.

69. Committee of Scientists, *supra* note 2.

70. Pub L. No. 105-277, § 347, 112 Stat. 2681-298 (1998). This authority was initially implemented on a pilot basis, but Congress extended and expanded the authority in 2003. Pub. L. No. 108-7, div. F, tit. III, § 323, 117 Stat. 11, 275 (2003).

71. 16 U.S.C. §6511(3) (2006).

72. Pub. L. No. 111-11, tit .IV, 123 Stat. 991 (2009).

73. *See, e.g.,* Mark Rey, *A New Chapter in the History of American Conservation, in* Challenges Facing the U.S. Forest Service 22–25 (Daniel Kemmis ed. 2008) (saying that a fortunate trend in collaborative conservation is evolving into a fourth chapter in the history of American conservation); Dale Bosworth & Hutch Brown, *After the Timber Wars: Community-Based Stewardship,* 105 J. Forestry 271 (2007) ("The future of national forest management lies in Community-based stewardship").

74. *See* 73 Fed. Reg. 21,468, 21,508 (Apr. 21, 2008); 70 Fed. Reg. 1023 (Jan. 5, 2005); and Citizens for Better Forestry v. Dep't of Agric., 481 F. Supp. 2d 1089 (N.D. Cal. 2007).

75. 76 Fed. Reg. 8480 (Feb. 14, 2011). *See also* comments of Tony Tooke, USFS Director of Ecosystem Management Coordination, U.S. Forest Service Science Forum, Washington, D.C. (March 29, 2011) (on file with author).

76. Martin Nie, *Place-Based National Forest Legislation and Agreements: Common Characteristics and Policy Recommendations,* 41 Envtl. L. Rep. 10229–246 (2011).

77. *See, e.g.,* David J. Sousa & Christopher McGrory Klyza, *New Directions in Environmental Policy Making: An Emerging Collaborative Regime or Reinventing Interest Group Liberalism?,* 47 Nat. Resources J. 377 (2007); George C. Coggins, *Regulating Federal Natural Resources: A Summary Case against Devolved Collaboration,* 25 Ecology L.Q. 602 (1998).

78. *See, e.g.,* Thomas C. Beierle & Rebecca J. Long, *Chilling Collaboration: The Federal Advisory Committee Act and Stakeholder Involvement in Environmental Decisionmaking,* 29 Envtl. L. Rep. 10399 (1999).

79. 5 U.S.C. app. §3(2) (2006).

80. Sarah Bates Van de Wetering, Public Pol'y Research Inst., The Legal Framework for Cooperative Conservation 15 (2006).

81. *See, e.g.*, Council on Environmental Quality, Executive Office of the President, The National Environmental Policy Act: A Study of Its Effectiveness After Twenty-Five Years (1997); Claudia Goetz Phillips & John Randolph, *The Relationship of Ecosystem Management to NEPA and Its Goals*, 26 Envtl. Mgmt. 1 (2000).

82. Western Governors' Association Forest Health Advisory Committee, Forest Health Landscape-Scale Restoration Recommendations 3 (2010).

83. Martin Nie, *supra* note 75.

84. Cortner and Moote, *supra* note 27, at 41 (reviewing multiple definitions of EM).

85. Matthew McKinney, Lynn Scarlett & Daniel Kemmis, Lincoln Inst. of Land Policy, Large Landscape Conservation: A Strategic Framework for Policy and Action (2010); 76 Fed. Reg. 8480 (Feb. 14, 2011) (proposing an "all lands" approach to forest management).

86. Evan Hjerpe et al., *Socioeconomic Barriers and the Role of Biomass Utilization in Southwestern Ponderosa Pine Restoration*, 27 Ecological Restoration 169 (2009).

87. The National Wildlife Refuge System Improvement Act of 1997, for example, sets forth a mission "to administer a national network of lands and waters for the conservation, management, and where appropriate, restoration of the fish, wildlife, and plant resources and their habitats within the United States for the benefit of present and future generations of Americans." Pub. L. No. 105-57, §4, 111 Stat 1252 (1997).

88. Pub L. No. 105-277, § 347, 112 Stat. 2681-298 (1998). This authority was initially implemented on a pilot basis, but Congress extended and expanded the authority in 2003. Pub. L. No. 108-7, div. F, tit. III, § 323, 117 Stat. 11, 275 (2003). For more background *see* Stewardship End Result Contracting, 68 Fed. Reg. 38, 285, 38, 286 (June 27, 2003).

89. 16 U.S.C. § 2104(d) (2006).

90. *Id.*

91. 16 U.S.C. §§ 6501-6591.

92. Robert B. Keiter, *The Law of Fire: Reshaping Public Land Policy In an Era of Ecology and Litigation*, 36 Envtl. L .348 (2006).

93. Pub. L. No. 111-11, Title IV, §4001.

94. *Id.* §4003(b) [74].

95. *Id.* [74].

96. *Id.* 4003§ (f) [74].

97. Martin Nie, *supra* note 75.

98. The 4 Forest Restoration Initiative: Promoting Ecological Restoration, Wildfire Risk Reduction, and Sustainable Wood Products Industries: A Proposal for Funding Under the Collaborative Forest Landscape Restoration Program, at 3, *available at* http://www.fs.fed.us/restoration/CFLR/documents/2010Proposals/Region3/R3_4FRI/R3_4FRI_CFLRP_Proposal_05142010.pdf. For more on the proposed action *see* 76 Fed. Reg. 4279 (Jan. 25, 2011).

99. *See* Bosworth & Brown, *supra* note 72; Dale Bosworth & Hutch Brown, *Investing in the Future: Ecological Restoration and the USDA Forest Service*, 105 J. Forestry 208 (June 2007).

100. Bosworth & Brown, *Investing in the Future*, *supra* note 97, at 210.

101. 16 U.S.C. §475 (2006).

102. U.S. Dep't of Agric., Secretary of Agriculture, Tom Vilsack, Seattle, WA, August 14, 2009 (transcript on file with author).

103. 76 Fed. Reg. 8480, 8482 (Feb. 14, 2011).

104. Comments made by Undersecretary of Agriculture Harris Sherman, at the U.S. Forest Service Science Forum, Washington, D.C., Mar. 29, 2010.

105. *See* Bethanie Walder, The Forest Service's Fatal Flaw, *available at* http://ncfp.wordpress .com/2010/07/01/the-forest-services-fatal-flaw (last visited May 20, 2011).

106. Western Governors' Association Forest Health Advisory Committee, *supra* note 81, at 7.

107. *Id.* at 10.

108. *Id.*

109. U.S. Forest Service, Dep't of Agric., The Process Predicament: How Statutory, Regulatory, and Administrative Factors Affect National Forest Management(2002).

110. This possibility emerged in 2011 with the USFS proposing a new integrated resource restoration program and budget. *See* Forest Service Budget: Senate Hearing Before the Comm. On Energy and Natural Resources, 111th Cong. 4-5 (statement of Tim Tidwell, Chief of USFS), *See also* U.S. Forest Service, Fiscal Year 2011 President's Budget in Brief (2010).

111. *See* Robert B. Keiter, Testimony Before House Committee on Natural Resources, The Role of National Parks in Combating Climate Change (Apr. 7, 2009) (on file with author).

112. For analysis of these matters *see* NEPA Task Force, Modernizing NEPA Implementation: The NEPA Task Force Report to the Council on Environmental Quality (2003).

113. Robert Keiter, *Toward Legitimizing Ecosystem Management on the Public Domain*, 6 Ecological App. 727 (1996).

114. Robert B. Keiter, Keeping Faith With Nature 309 (2003).

115. *Id.* at 309.

116. *Id.* at 310.

117. For a discussion *see* Nie, *supra* note 4; and Jim Burchfield & Martin Nie, University of Montana, College of Forestry and Conservation, National Forests Policy Assessment: Report to Senator Jon Tester (2008).

II
Letting Theory Inform Practice

5

Ecosystem Services and Ecosystem Management—How Good a Fit?

J.B. Ruhl, Vanderbilt University

By the mid-1990s, the evolving concept of ecosystem management had become the subject of intense debate in natural resources policy.[1] In a landmark 1994 article in *Conservation Biology*, R. Edward Grumbine captured the state of play of that debate and synthesized what he drew from the literature on ecosystem management to define its central tenets.[2] At the core of ecosystem management, he concluded, was the overarching theme of sustaining ecological integrity in order to reduce and eventually eliminate the staggering losses of biodiversity ecologists had begun documenting since the middle of the century.[3] According to Grumbine, this goal sharply contrasted with the goal of traditional resource management to maximize the production of traditional goods (such as timber and minerals) and services (such as hunting and camping) from natural resources.[4] Grumbine thus warned that "if ecosystem management is to take hold and flourish, the relationship between the new goal of protecting ecological integrity and the old standard of providing goods and services for humans must be reconciled."[5]

A few years after Grumbine published his study of ecosystem management theory, another evolving concept, ecosystem services, hit the scene with a splash in a 1997 *Nature* magazine article[6] and also a more comprehensive book.[7] Ecosystem services are the economic benefits humans derive from the ecosystem structure and processes that form what might be thought of as natural capital.[8] Ecosystem services flow to human communities in four streams: (1) *provisioning services* are commodities such as food, wood, fiber, and water; (2) *regulating services* moderate or control environmental conditions, such as flood control by wetlands, water purification by aquifers, and carbon sequestration by forests; (3) *cultural services* include recreation, education, and aesthetics; and (4) *supporting services*, such as nutrient cycling, soil formation, and primary production, make the previous three service streams possible.[9] In the *Nature* article, lead author Robert Costanza and his research team estimated the global economic value of ecosystem services at over $30 trillion annually,[10] and the book, edited by Gretchen Daily, surveyed an array of ecological systems such as wetlands and forests as sources of numerous ecosystem service streams.[11]

Fast forward to today and the concepts of ecosystem management and ecosystem services are both firmly implanted in natural resources policy dialogue and as focal points of scientific research, though each has had its detractors and difficulties gaining traction in concrete regulatory programs.[12] Yet little attention has been paid to how they relate, particularly in a way relevant to Grumbine's call for reconciliation between ecological integrity and the use of ecosystems for human prosperity. More specifically, does the concept of ecosystem services light the way for harmonizing the goal of sustaining ecological integrity and the goal of providing goods and services to humans? The concept of ecosystem services is, after all, unequivocally about delivering economic value from ecosystems to humans, and in that sense is inherently anthropocentric. That does not sound promising for reconciliation. Is there something about the ecosystem services approach that changes the calculus from that used in traditional resource management in such a way that makes it more likely that ecosystem services concepts will promote sustainable ecological integrity? Or is the concept of ecosystem services just a wolf in sheep's clothing, appealing in name, but no less threatening to ecological integrity in practice?

This chapter explores those questions from two perspectives. The first section of the chapter examines the fit between ecosystem management and ecosystem services in theory. Is the concept of ecosystem services compatible with, even supportive of, what Grumbine laid out as the themes and goals of ecosystem management, or will it put ecosystem management even farther behind? The second section explores the topic in an applied context through a case study of ecosystem services and federal public land management. Is there anything to be gained for ecosystem management on federal public land regimes through the use of ecosystem services concepts? The analysis from both perspectives is that the concept of ecosystem services offers significant potential to support the application of ecosystem management principles generally, but that care must be taken to ensure it does not distort or undermine ecosystem management goals in local contexts.

I. THEORY: MATCHING ECOSYSTEM MANAGEMENT WITH ECOSYSTEM SERVICES

How compatible are the concepts of ecosystem management and ecosystem services? One way to explore that question is to put the principles of ecosystem services theory up against Grumbine's themes and goals of ecosystem management to see how well they match up. Grumbine outlined ten such themes and five overarching goals for ecosystem management. The following discussion compares each with the themes and goals of ecosystem services theory.

A. Ecosystem Services and the Themes of Ecosystem Management

Grumbine's synthesis of the ecosystem management literature as it stood in 1994 led him to the following definition of the term: "[e]cosystem management integrates scientific knowledge of ecological relationships within a complex sociopolitical and values framework toward the general goal of protecting native ecosystem integrity over the long term."[13] He then unpacked this definition into ten key components. Each is summarized below, followed by discussion of how ecosystem services theory is complementary or in conflict with the ecosystem management theme.[14]

1. Hierarchical Context. To be effective, ecosystem management must operate at multiple levels of the biodiversity hierarchy (genes, species, populations, ecosystems, landscapes), with managers seeking the connections between all levels through what Grumbine called a "systems" perspective.[15] Ecosystem services theory has at its core a geographical perspective that wholly embraces such scalar and systems models.[16] Formulating ecosystem services policy requires a thorough understanding of the complex delivery chain from natural capital sources to human population beneficiaries. These delivery chains operate at multiple scales. For example, a forest may provide groundwater recharge services to one local population, surface water purification services to numerous distant downstream populations, and carbon sequestration services to the global population. The connections between these services—how managing one might affect another—are complex, reflecting the underlying complexity of the ecological processes responsible for producing the services.[17]

2. Ecological Boundaries. Grumbine emphasized that ecosystem management must operate across political and other administrative boundaries, such as counties and national parks, to define appropriate ecological management boundaries.[18] Given the vast distances over which many ecosystem services delivery chains span, it is inevitable that managing for ecosystem services will demand crossing political and administrative boundaries—indeed, in some cases even far more so than Grumbine envisioned for ecosystem management.[19]

3. Ecological Integrity. At its heart, ecosystem management focuses on protecting the ecological patterns and processes that maintain total native biodiversity.[20] Managing total native biodiversity, or any level of native biodiversity, provides a set of natural capital resources which in turn provide a flow of associated ecosystem services. Looking at it from the other end, managing for a particular suite of ecosystem services and levels will drive managers toward conservation of a particular set of natural capital resources. The question, taken up in more detail below, is whether it matters which end managers start from—managing for levels of biodiversity or managing for a suite of ecosystem services. Given the complexity of ecosystems and the connections between ecological processes, natural capital resources, and flows of different services, one does not have to work hard to imagine how tradeoffs might arise between a goal of, say, total native biodiversity and a goal of, say, maximizing carbon sequestration services or groundwater recharge services.[21] These potential tradeoffs could have profound management policy implications.[22]

4. Data Collection. Grumbine recognized that to achieve ecosystem management as so defined, more research and data collection would be required.[23] Ecosystem services theory also points to the need for far more robust data collection in order to support the ecological, economic, and geographic analyses needed for decision making.[24]

5. Monitoring. Part of the data collection demand of ecosystem management will be data from tracking the results of management decisions to create a feedback loop of useful information.[25] Ecosystem services management will also require an effective monitoring feedback process.[26]

6. Adaptive Management. Grumbine called for managers to use the data feedback loop of their monitoring to flexibly incorporate the results of prior actions into a learning process for decision making.[27] Ecosystem services theory also focuses on identifying drivers of change and developing models for making adaptive management decisions.[28]

7. Interagency Cooperation. Grumbine observed that the ecological boundary approach of ecosystem management requires that managers working within political and administrative units, as well as private resource owners, cooperate and coordinate.[29] For the same reasons, this will be true of ecosystem services management as well.[30]

8. Organizational Change. As envisioned, Grumbine acknowledged that ecosystem management is very different from conventional resource management and could require potentially significant changes in the structure of land management agencies.[31] While for the same reasons this will be true of ecosystem services management as well,[32] it is by no means clear that the organizational changes for ecosystem management will be compatible with those for ecosystem services management. Ecosystem management focuses exclusively on management of the natural capital resource, whereas ecosystem services management extends the focus to include the delivery chains and end user beneficiaries. The broader scope of ecosystem services management could lead to different organizational demands. Market institutions, for example, may be more suited for ecosystem services management than for the inward perspective of ecosystem management of, say, a national forest or wilderness area.

9. Humans as a Part of Nature. As much as ecological integrity sits at the core of ecosystem management, Grumbine recognized that ecosystem management ultimately is a human undertaking—that humans fundamentally influence and are influenced by ecosystems.[33] Ecosystem services theory is all about how humans derive benefits from ecosystems, and thus does focus on the relationship of humans with nature. Grumbine's central premise of ecosystem management, however, is that it is fundamentally a biocentric perspective—it is about nature first and only then asks where and how humans fit in. Ecosystem services theory, by contrast, is fundamentally anthropocentric—it measures ecosystem services as end points of ecological processes, and measures them for their economic value to humans. While it is entirely possible that these two different starting points may lead to the same management decisions, it is also possible that they would lead to conflicting management decisions.

10. Values. Grumbine recognized the normative dimension of ecosystem management, necessitating that managers work out the relationship between scientific knowledge and human values.[34] Of course this is true for ecosystem services management as well, with numerous policy-relevant, value-laden decisions to be made before managers can begin to manage.[35] The contrasting perspectives of ecosystem management theory (biocentric) and ecosystem services theory (anthropocentric) suggest, however, that different value systems may arise. The potential for conflicting management perspectives and decisions thus cannot be discounted.

Table 1. Summary of Themes of Ecosystem Management and Ecosystem Services Theory

Ecosystem Management Theme	Fit with Ecosystem Services	Explanation
Hierarchical context	Strongly complementary	Ecosystem services flow at many scales and thus require sound understanding of cross-scalar and systems effects of management decisions
Ecological boundaries	Strongly complementary	Ecosystem services flow from natural capital to human beneficiaries across political and administrative boundaries
Ecological integrity	Potential conflicts	Managing for particular ecosystem services could undermine ecological integrity, depending on how ecological integrity is defined
Data collection	Strongly complementary	Ecosystem services management will demand intensive data collection to inventory natural capital and trace flows to human beneficiaries
Monitoring	Strongly complementary	Ecosystem services monitoring will demand robust monitoring of natural capital capacity and flows of services
Interagency cooperation	Strongly complementary	Ecosystem services could flow across land units managed by many different agencies and landowners
Organizational change	Potential conflicts	Organizational changes required for ecosystem management may or may not be compatible with management for ecosystem services
Humans as a part of nature	Potential conflicts	Ecosystem services theory is inherently anthropocentric, but recognizes the importance of natural capital to achieving ecosystem service values
Values	Potential conflicts	Values of ecosystem management regime may or may not be compatible with management for ecosystem services

B. Ecosystem Services and the Goals of Ecosystem Management

As the values theme of ecosystem management implies, there is a strong normative dimension to be defined before managers can know how and what to manage. Indeed, by inserting ecological integrity into the ten themes and defining it as "total native biodiversity," Grumbine put the cart before the horse. None of the first six themes contains this kind of goal constraint—each is a methodological perspective with the normative details left to be worked out. Even the other four themes are goal neutral as Grumbine described them. But "total native biodiversity" is a goal, not a methodology. To say that the ecological integrity necessitates managing for "total native biodiversity" goes beyond methodology, ruling out the possibility that ecosystem management could be used to manage grazing leases, farms, parks, and other working landscapes. If we remove that goal from the ecological integrity theme, to say ecosystem management is about managing for ecological integrity begs the question: integrity for what purpose, for what kind of ecosystem end point? The ultimate question for ecosystem management, therefore, is toward what end are managers to manage?

Grumbine offered his vision of those management goals, which, given his theme of "total native biodiversity," look strikingly as if they are designed to produce that end point:

1. Maintain viable populations of all native species *in situ*.
2. Represent, within protected areas, all native ecosystem types across their natural range of variation.
3. Maintain evolutionary and ecological processes.
4. Manage over periods of time long enough to maintain the evolutionary potential of species and ecosystems.
5. Accommodate human use and occupancy within these constraints.

As well he had to, Grumbine recognized that "these fundamental goals provide a striking contrast to the goals of traditional resource management," which in his description are "based on maximizing production of goods and services, whether these involve number of board feet (commodities) or wilderness recreation visitor days (amenities)."[36] Putting aside whether Grumbine forced that contrast by embedding "total native biodiversity" in the methodology themes of ecosystem management, it is clear that most advocates of ecosystem management theory endorse Grumbine's theme and goals, and thus see the same contrast he saw with traditional resource management.[37]

Herein lies the conundrum for fitting ecosystem management with ecosystem services. Put into the lexicon of ecosystem services theory, Grumbine's critique of traditional resource management is that it has focused on maximizing *provisioning* and *cultural* services. It has all been about board feet and fishing licenses. To check that trend, he and other ecosystem management theorists would reset management policy to use total native biodiversity as the management goal, allowing humans to benefit from provisioning and cultural services only within that constraint.

But where does this leave *regulating* and *supporting* services? Most ecosystem service theorists agree that the real challenge is in how to get humans to pay attention to regulating and supporting services.[38] There are well-established markets for provisioning and cultural services, which generally are consumed or enjoyed directly—people are used to buying corn and paying to enter parks. Regulating and supporting services, by contrast, behave more like public goods, in that they are consumed or enjoyed indirectly and leak off of the lands where their natural capital sources are found.[39] How could a forest owner charge for groundwater recharge benefits a local population receives, or for pollination, or flood control, and so on? Modern American property rights, regulatory regimes, and social norms have for the most part failed to account for this feature of regulating and supporting services.[40] So, compatible with ecosystem management themes, ecosystem services theory focuses on how to enhance regulating and supporting services through a variety of institutional mechanics such as regulation, property rights, planning, and so on.[41] But even if that program of reforms is successful, this does not necessarily mean the two theories will merge in terms of the reconciliation between ecological integrity and providing benefits to humans.

There are two scenarios to consider. In the first, enhancing total native biodiversity becomes the management driver, and the result is that humans have less access to provisioning and cultural services but will benefit from the boost in regulating and supporting services that flow inherently, albeit incidentally, from increased total native biodiversity. In the other scenario, enhancing regulating and supporting services becomes the management driver, and the result is that humans have less access to provisioning and cultural services but total native biodiversity improves inherently, albeit incidentally, from the ecosystem services management goal.

The two scenarios are not simply two sides of the same coin, and the differences between them all come down to values—what we are trying to achieve and why. In the first scenario—the biocentric values scenario—total native biodiversity dictates what types and levels of ecosystem services benefit humans. Provisioning and cultural services will likely be constrained through regulation on extraction, use, and access. Regulating and supporting services will not be constrained directly, but will be limited by the total native biodiversity goal. In other words, we will get of them whatever total native biodiversity delivers.

In the second scenario—the anthropocentric scenario—the relationship is flipped. Enhancing the delivery of regulating and supporting services to human populations dictates the types and level of total native biodiversity in the management unit. As in the first scenario, provisioning and cultural services will likely be constrained through regulation on extraction, use, and access activities that are incompatible with sustained flows of regulating and supporting services. Total native biodiversity will not be constrained directly, but will be influenced by how managers manage for regulating and supporting services.

One might ask whether there truly is much difference between the two scenarios. After all, if total native biodiversity supports the provision of regulating and supporting services, and if managing for regulating and supporting services means restoring the ecological integrity needed to produce them, then so long as we want more of either we get more of both, right? Not necessarily. It all depends on which regulating and supporting services managers choose to manage for in the second scenario. In the first scenario, managers have decided to let total native biodiversity dictate the suite of ecosystem services that flow from the management unit and their respective levels. Essentially, managing for total native biodiversity sets the natural capital available to produce ecosystem services. In the second scenario, managers choose the services to manage for and their desired levels, which in turn drives decisions about what natural capital to favor. This could lead to management decisions for natural capital that stray from achieving total native biodiversity. Choosing carbon sequestration as the regulating service management goal, for example, does not necessarily support achieving total native biodiversity, and potential tradeoffs such as this need to be explored and understood.[42]

In theory, therefore, the methodological fit between ecosystem management and ecosystem services is potentially very strong, but the normative fit has the potential for tension. In other words, what managers would do to implement ecosystem management concepts and to implement ecosystem services concepts looks very similar—work across scales, collect data, monitor, and so on—but *why* they are doing it could differ in significant ways. To give more concrete grounding to this potential conflict (or happy unison), the next section puts the theories into an applied setting.

II. APPLICATION: A CASE STUDY OF FEDERAL LAND MANAGEMENT POLICY

One of the gaping holes in ecosystem management literature when Grumbine conducted his 1994 review was the treatment of how to implement ecosystem management themes and goals through *law*. Of the articles and authors he cited and reviewed, only one author, Robert Keiter of the University of Utah Law School, was producing work on how ecosystem management could fit into legal regimes. To be sure, since then legal academics and practitioners have generated a robust dialogue on the topic,[43] and yet there remains little actual coherent law of ecosystem management.[44] One question, therefore, is whether the concept of ecosystem services could help propel ecosystem management from theory into applied law and policy. This section of the chapter uses a case study of federal public lands to probe that inquiry.

It makes sense that the federal government, as the largest landowner in the nation,[45] would begin to consider as a policy matter how it might manage the flow of ecosystem services on and off of its landholdings; however, it has only recently begun to do so in a coherent policy framework.[46] Ecologists and economists have been forging the theory and application of the ecosystem services concept since the mid-1990s,[47] but only in the past few years has the concept begun to register in federal public lands policy in any meaningful way.[48]

This case study[49] examines the emerging policy intersection between ecosystem services and ecosystem management on federal public lands and proposes a set of key policy questions, research needs, and options for building on the policy work completed to date. The first section outlines the basic context for thinking about the role federal public lands might play in the management of ecosystem services, and why using the ecosystem services concept in public land management is worth considering. The discussion then turns to proposing several key research paths that must be addressed before federal lands can be managed effectively for ecosystem service flows. The final section bears down on the different roles federal lands might play in promoting or participating in markets for ecosystem services. The goal is not to propose any particular policy for federal land management and ecosystem services, but rather to suggest how federal public land management agencies should go about formulating and implementing such policies.

A. Key Threshold Considerations

Three disciplines merge at the core of the concept of ecosystem services: ecology, to understand the ecological structures and processes that produce and deliver ecosystem services; economics, to understand how those delivered ecosystem services provide value to human beneficiaries; and geography, to understand where the 'natural capital' providing services is located, where the beneficiaries of ecosystem services are located, and how the services flow from the former to the latter.[50] The federal land management agencies already deploy expertise in each of these fields to carry out their statutory duties, such as deciding where to allow recreation, timber harvesting, and mineral exploration in national forests. The agencies also already include providing values to the public as part of their respective missions. Indeed, without calling them such, the agencies have been providing ecosystem services to the public for many decades in the form of provisioning services (e.g., timber from national forests and water from reclamation projects), regulating services (e.g., watershed protection from national forests), cultural services (e.g., recreation and education in national parks), and supporting services (e.g., nutrient cycling in wetlands on federal lands).

So what difference will it make to think explicitly about the concept of ecosystem services when formulating federal land management policy? Good question. To answer it—to appreciate how the concept of ecosystem services can reorient and clarify federal land management policy—we need to step back and consider how the central properties of ecosystem services connect with the context of federal public land policy.

1. DEFINING MANAGEMENT MISSIONS

A fundamental starting point for designing ecosystem services policy is that the concept of ecosystem services is anthropogenic in focus—it is about delivering economic value to humans. As noted, federal land policy already does so in many ways through the management responsibilities and goals assigned to the major land management agencies.

> Each of the four major federal land management agencies manages its lands and the resources they contain on the basis of its mission and responsibilities. The Forest Service and the Bureau of Land Management manage lands for a variety of uses, including recreation, timber harvesting, livestock grazing, oil and gas production, mining, and wilderness protection. The Fish and Wildlife Service manages lands primarily to conserve and protect fish and wildlife and their habitat, although other uses, such as hunting and fishing, are allowed when they are compatible with the primary purposes for which the lands are managed. The National Park Service manages lands to conserve, preserve, protect, and interpret the nation's natural, cultural, and historic resources.[51]

If the land management agencies were to orient their missions around eco-system services, however, this description of what the agencies do would look quite different. For example, the description might employ the four typology categories to arrange the different management goals. Also, in each case it would be necessary to identify the intended human beneficiaries. For example, the Bureau of Land Management would be described as focused on delivering pro-visioning services to the public in general through access to commodity produc-ers, whereas the Fish and Wildlife Service would provide on-site cultural services, such as education about endangered species and hunting. Using ecosystem ser-vices to inform the agencies' work also would supply a new metric for the agen-cies' missions and performance evaluations, as well as making the economic value to society of the federal public lands more explicit to the public. Clearly, therefore, employing the ecosystem services concept in federal land policy would lead to a different way of describing what the land management agencies do and how well they do it.

2. PUBLIC GOALS AND PUBLIC LANDS

The consequence of using the ecosystem services approach as described above, however, is that there must be economic value delivered to humans for there to be ecosystem services. Yet providing economic value to humans may not be the only goal we have for federal public lands. Or, to put it more in focus, *maximizing* economic value to humans is likely not the overarching goal we have for all of our federal public lands. For example, setting aside land for wilderness or managing land to protect an endangered species might provide some ecosys-tem services, such as benefits to local human populations from the watershed functions of the conserved lands, but only as an incidental effect of implement-ing the conservation goal.

In this sense, what *public* land policy does with the concept of ecosystem services will be fundamentally different from how the concept can be employed to improve the use of private lands. In the private lands context, the concept of ecosystem services improves the market information available to landowners to decide what constitutes the most efficient use of the land and its associated resources.[52] Of course, to take advantage of that information, *private* landown-ers need some way of capturing the value of the services in markets, which is difficult for services like pollination from wild pollinators and groundwater recharge of aquifers from wetlands.[53] Thus the challenge in the private lands context is how to integrate ecosystem service values into market contexts.[54]

The point of having public lands, by contrast, is that we don't have to manage them like private lands—that is, the public can decide, using nonmarket decision mechanisms if we so choose, to suspend the goal of achieving the most efficient economic outcome.[55] When considering how to incorporate the concept of ecosystem services into federal land management policy, it will be important for Congress and the agencies to define precisely how far to take the concept and in what contexts the concept is not a desirable medium for expressing agency mission and assessing agency performance. It might be useful, for example, for agencies to describe the incidental ecosystem service benefits of preserving habitat for endangered species, but if the goal is preserving habitat for endangered species, the concept of ecosystem services has little if any useful direct role to play, and could even be counterproductive if used to define the policy means and outcome.[56]

3. OFF-SITE DELIVERY OF REGULATING AND SUPPORTING SERVICES

If we were to describe what benefits federal lands provide under current policy through the lens of ecosystem services, two forms of services would be well represented in the inventory. One is on-site delivery of cultural services, such as recreation in parks and hunting in wildlife refuges. The other is delivery of provisioning services for off-site use, such as timber and water supply. Along with these, of course, are the regulating and supporting services provided *within* the federal public lands to facilitate and complement policies focused on delivering the cultural and provisioning services. The federal public lands have been managed for decades to include delivering these cultural and provisioning services, and it is not clear how calling hunting a cultural service and timber a provisioning service will fundamentally alter how the land management agencies go about their work.[57]

But what about delivery of regulating and supporting services to *off-site* human populations? This is fertile ground for using the concept of ecosystem services to reorient and clarify federal land policy. This is the context in which ecosystem services offer the greatest opportunity to define agency mission, communicate the value of the federal lands to the public, and measure agency performance. Presumably, it would not be news to most people that federal public lands can benefit surrounding and even distant human populations, including in ways consistent with ecosystem services theory. But the existing and potential flow of services is vast and has not been coherently managed and communicated as such. This context, it strikes me, is where the greatest payoffs and challenges lie for incorporating ecosystem services into federal public land management policy. The next section explores how to manage that start-up process for regulating and supporting services.

B. Policy Questions and Research Needs

Before we can formulate policy for ecosystem services on federal land, the agencies will need to define which resources will provide or potentially provide which services, who benefits or potentially benefits from the services, and the extent of policy discretion allowed in the relevant legal framework applicable to the lands under study.

1. ESTABLISH BASELINES

One can download countless maps of federal public lands showing all sorts of different attributes—elevation, land cover, species distribution, and so on—yet there is no map of the regulating and supporting services they provide to off-site human communities. Before the land management agencies can begin to think clearly about developing and implementing ecosystem service based policies, such a baseline representation is needed. The baseline must accomplish the following:

- Inventory on-site and off-site natural capital that can be supported.
- Identify off-site flows of current and potential regulating and supporting services.
- Identify off-site human populations receiving current and potential service values.
- Inventory service values to those populations with appropriate valuation methods.

Research of this scope is only beginning to gain funding and attention in the federal agencies,[58] and is even less developed with respect to federal public lands. There is a strong consensus that "the science of ecosystem services needs to advance rapidly" through these and other initiatives.[59]

2. IDENTIFY TRADEOFFS

As the baseline is pulled together, federal land management agencies can begin to answer three foundational questions: what services can the land unit provide, what populations can be benefitted, and when can they be benefitted? The answers, however, are in many instances likely to reveal a complex multi-scalar mosaic of ecosystem service potentials. For example, table 2 shows how a hypothetical national forest might compile the following inventory of ecosystem service possibilities:[60]

Table 2. Forest Ecosystem Services

Forest Ecosystem Service	Population Benefitted and Timing
Carbon sequestration	Global; lagged
Surface water quality	Region A; immediate
Groundwater recharge	Region B; fluctuating
Microclimate	Locality C; fluctuating
Pollination	Farm D; immediate

It is immediately apparent from this example that policy decisions about how to manage the production of ecosystem services from this land unit face a suite of five potential tradeoffs:[61]

1. Service tradeoffs: It may not be possible to manage land units for all the potential services, as the underlying ecological processes that are the source of the services may put one service in competition with another. For example, maximizing the hypothetical land unit for carbon sequestration may involve vegetative management practices that diminish the supply of habitat for pollinators, and vice versa. Which service should be the primary management goal?

2. Spatial policy tradeoffs: As the example suggests, moreover, different services operate at different spatial policy scales. Carbon sequestration serves national climate mitigation policy goals, whereas pollination operates primarily on local or parcel scales. If services flowing at different spatial scales experience tradeoffs, then so too do the corresponding policies. Which policy scale should the agency target?

3. Temporal policy tradeoffs: Similarly, different services have different delivery mechanisms and timing. It may be, for example, that habitat manipulation on the land unit can boost pollination services rather quickly, whereas doing so to promote carbon sequestration produces results only relatively far into the future. Should the agency seek immediate payoffs or pursue the long-term strategy?

4. Goal tradeoffs: As suggested above, managing public lands for ecosystem services may not always be compatible with other goals for public lands. It is unlikely, for example, that the most effective way to manage lands for the benefit of endangered species will align well with the most effective way to maximize ecosystem service flows from those lands to human populations. Which goal should the agency pursue?

5. Population tradeoffs: Inherent in all of these tradeoffs is the possibility that very different populations may benefit depending on the policy decisions about which service, spatial scale, and temporal scale to favor. Carbon sequestration on a national forest benefits a global population; pollination services from the forest might directly benefit just a few area farms and indirectly benefit the consumers of their crops; and endangered species conservation benefits, in addition to the species, people interested in endangered species conservation. Which population should the agency favor if the services and scales are not compatible? This could become a particularly acute problem if the wealth transfer—i.e., the enhancement of one service benefitting a particular population at the expense of another service benefitting another population—favors the wealthy over the poor.

Federal public lands management already faces numerous policy and scale tradeoffs—it is difficult to hike in an active timber cut or conserve endangered species habitat in an active grazing lease. Whole courses in law schools are devoted to studying how Congress and the land management agencies wrestle with these difficult choices.[62] Ironically, integrating ecosystem services into public land management policy and building robust baseline inventories will only reveal more tradeoffs, and likely on larger scales given the focus on off-site benefits.[63]

3. IDENTIFY LEGAL AUTHORITIES AND CONSTRAINTS

Some of the tradeoff challenges of ecosystem services policy formulation will be mooted or amplified depending on the legal constraints associated with different land units in the federal public land system. Broadly speaking, federal public lands can be lumped into three categories based on the range of uses allowed under applicable statutes.[64] *Single use* lands such as wilderness areas have a narrowly defined purpose that cannot be violated.[65] A wilderness area, therefore, is not a candidate for managing for services such as pollination or carbon sequestration, though those or other services may flow incidentally from management as a wilderness area. *Primary use* lands such as national wildlife refuges have a defined priority use or uses, but others are allowed if compatible with the primary use or uses.[66] There may be many such opportunities on a refuge to enhance off-site service flows while not impeding purposes such as on-site waterfowl habitat conservation. *Multiple use* lands such as national forests require the managing agency to fulfill a range of uses, some of which may be conflicting and impossible to accomplish in the same area of the land unit.[67] As summarized in table 3, multiple use lands thus hold the greatest potential for management focusing on off-site delivery of regulating and supporting services, but as a consequence they also present the greatest chance of facing tradeoffs between competing services, policies, and populations. With discretion comes the heat from making choices that favor one interest group over another.

Table 3. Categories of Land Management Mandates

Land Management Mandate	Legal Constraints	Ecosystem Service Policy Options
Single Use (e.g., Wilderness)	All management actions must satisfy single use purpose; restricted agency discretion	Manage for wilderness and inventory the baseline services and beneficiaries
Primary Use (e.g., National Wildlife Refuges)	Other uses must not be inconsistent with the primary use; limited agency discretion	Identify services that can be enhanced within the primary purpose constraint; manage for them
Multiple Use (e.g., National Forests)	The multiple uses must be balanced; extensive agency discretion	Integrate service valuation more explicitly in multiple use decision making; manage for them

C. Federal Lands and Ecosystem Service Markets

One sense in which ecosystem services approaches could lead to significant reorientation of public land management is if the market theme of ecosystem services gains traction. Indeed, it already seems to be. Section 2709 of the Food, Conservation, and Energy Act of 2008 requires the Department of Agriculture to "establish technical guidelines that outline science-based methods to measure the environmental services benefits from conservation and land management activities in order to facilitate the participation of farmers, ranchers, and forest landowners in emerging environmental services markets" and to establish guidelines to develop a procedure to measure environmental services benefits, a protocol to report environmental services benefits, and a registry to collect, record, and maintain the benefits measured.[68] To implement section 2709 of the Farm Bill, the Forest Service established an Office of Ecosystem Services and Markets, now known as the Office of Environmental Markets. A multiagency Conservation and Land Management Environmental Services Board was established in December 2008 to assist the Secretary of Agriculture in adopting the technical guidelines to assess ecosystem services provided by conservation and land management activities.[69]

The board's guidelines are intended to focus on scientifically rigorous and economically sound methods for quantifying carbon, air and water quality, wetlands, and endangered species benefits in an effort to facilitate the participation of farmers, ranchers, and forest landowners in emerging ecosystem markets.[70]

While the Farm Bill focused on how the Department of Agriculture can promote ecosystem service markets for farmers, ranchers, and forest landowners, it tantalizingly opened the door to thinking about the broader role of federal public lands as an integral part of ecosystem services markets. Take, for example, a national forest unit that could deliver groundwater recharge services to a regional population or carbon sequestration services to a national population. Assume that private lands near the national forest also can supply those services. Assume also that there is an emerging or even robust demand in the region or nation for those services—enough to potentially give rise to a market for them. What policy options are available to the national forest management team? The various speakers at the conference converged around the following set of options, arranged in ascending order of active market facilitation and participation:[71]

1. Do nothing: One option, of course, is to do nothing with respect to potential or active markets for ecosystem services, but rather to manage public land units for ecosystem services according to applicable policy goals. To the extent that the policy decision is to provide off-site regulating and supporting services at a particular level, they will be provided essentially for free to the beneficiaries and thus will decrease the market prices that private suppliers can demand.

2. Provide research subsidies: To the extent that the federal government wishes to promote the emergence of active private markets in ecosystem services, one policy approach could be to use public lands and agency budgets to conduct the research necessary to define crucial market parameters, such as how to promote groundwater recharge through vegetative management or how much wetland surface area is needed to control particular flood levels. Such research may be expensive to conduct and thus may operate as a barrier to emergence of what might, with the knowledge the research could reveal, become efficiently operating private markets.

3. Conduct demonstration projects: Going a step further, an agency could decide to use a public land unit to experiment with different methods of delivering off-site ecosystem services and measuring the economic benefits, in essence acting as a surrogate for a first mover in the potential market.

4. Provide market stability through standards and risk assurance: Support for private markets could also come in the form of more direct involvement, such as by promulgating or endorsing practices and standards, and even backing private market obligations as a market insurance mechanism. For example, if market development is hindered by risks associated with private supplier failure due to drought (leading to failure to supply recharge) or fire (leading to leakage of carbon sequestration), the federal government could use appropriately located public lands to offer supply assurance.

5. *Provide a third-party market platform*: Once active markets for ecosystem services exist, federal public lands could also be leased or licensed to third parties to produce and market off-site ecosystem services. In practice, this is how federal lands produce provisioning services such as timber and cattle— private producers obtain permits and leases to occupy the public lands and produce commodities for sale in private markets. Federal lands also are venues for private recreation concessions such as ski centers and outfitters. Third party markets for regulating and supporting services could be developed on federal public lands in much the same way.

6. *Act as a full market participant*: The most aggressive form of market support for federal public lands would be for the federal agencies to assume a proprietary role and enter the market as full participants using the federal land units as the production capital.

Which of these options best meets and balances public land policy goals and national goals for ecosystem services and for ecological conservation will depend on the services involved, the public land unit, and the policy tradeoffs discussed above. One concern in this respect, which could be fueled by the drive toward market approaches, is the mounting tension between carbon sequestration—currently the major player in ecosystem service market policy development—and the other services and values that could be delivered from public lands, whether through markets or not.[72] On the one hand, the carbon markets and the role of federal public lands in them could swamp policy regarding other service flows.[73] On the other hand, carbon sequestration is the only ecosystem services market game in town, so to speak, and helping it emerge and prosper could ignite similar markets in other ecosystem services.[74] At the dawn of federal public lands policy for ecosystem services, navigating between these two perspectives will be a profound policy challenge.

III. CONCLUSION

As Associate Chief of the Forest Service, Sally Collins (later head of the USDA's Office of Environmental Markets) cautioned that the agency must "resist the impulse to jump on the ecosystem services 'bandwagon' without some thinking."[75] This chapter has outlined at least a piece of what that "some thinking" should address—how will turning to an ecosystem services approach influence how ecosystem management is designed and implemented? Ecosystem management and ecosystem services management share many methodological traits. Both take a multiscalar, cross-boundary perspective. Both will demand robust data, performance monitoring, and adaptive decision-making capacity. Both depend on managing ecosystem integrity to sustain their goals.

It is their respective goals, however, that could cause ecosystem management and ecosystem services management to depart. Ecosystem management is biocentric in perspective, positioning total native biodiversity as the driver and accommodating human benefits only to the extent compatible with that goal. Ecosystem services management is anthropocentric in perspective, with ecosystem service benefits as the desired end product and management of natural capital to meet that goal as the conservation strategy.

To be sure, ecosystem management, even if fully oriented toward securing total native biodiversity, can use ecosystem services theory to enhance public awareness of the economic benefits that flow from achieving that goal. In this sense, more knowledge about ecosystem services can be a useful implementation strategy for ecosystem management, making its hard line on nature first more appealing to a broader public. Indeed, it may be difficult for ecosystem management to flourish if it cannot convey to the public the ecosystem service benefits they will enjoy. But more knowledge about ecosystem services also means more knowledge about the tradeoffs that may be inherent in managing for total native biodiversity and the ecosystem service benefits that may be sacrificed by staying true to that central tenet of ecosystem management.

These are value choices that must be worked out, and can only be worked out in concrete settings where the tradeoffs are known. This chapter has not suggested what the right choice is, only that one will have to be made many times across the complex social-ecological landscape given the different starting points of ecosystem management and ecosystem services management. To extend Sally Collins' maxim, it may be wise, therefore, to resist the impulse to jump on either the ecosystem services or the ecosystem management bandwagon without some thinking about how they fit.

NOTES

1. The history and content of ecosystem management are covered in this volume in the opening chapter. Further elucidation of ecosystem management principles is supplied in this chapter only to advance the discussion of how ecosystem services concepts fit with ecosystem management in theory and in application.

2. *See* R. Edward Grumbine, *What is Ecosystem Management?*, 8 Conserv. Biology 27 (1994).

3. *See id.* at 27–29.

4. *See id.* at 31.

5. *Id.*

6. Robert Costanza et al., *The Value of the World's Ecosystem Services and Natural Capital*, 387 Nature 253 (1997).

7. Nature's Services (Gretchen C. Daily ed., 1997) [hereinafter Nature's Services].

8. Ecosystem services are economically valuable benefits humans derive from ecological resources directly, such as storm surge mitigation provided by coastal dunes and marshes, and indirectly, such as nutrient cycling that supports crop production. Natural capital consists of the ecological resources that produce these service values, such as forests, riparian habitat, and wetlands. For descriptions of natural capital and ecosystem services, *see* Millennium Ecosystem Assessment, Ecosystems and Human Well-Being: Synthesis (2005), *available at* http://www.millenniumassessment.org/documents/document.356.aspx.pdf.

9. This typology of ecosystem services is developed in the Millennium Ecosystem Assessment, *supra* note 8, at vi.

10. *See* Costanza et al., *supra* note 6.

11. *See* Nature's Services, *supra* note 7.

12. For a survey of ecosystem management policy in the context of different ecological settings, *see* John Copeland Nagle & J.B. Ruhl, The Law of Biodiversity and Ecosystem Management 313–1035 (2d ed. 2006). For coverage of the emergence of the ecosystem services concept in law and policy, *see* J.B. Ruhl et al., The Law and Policy of Ecosystem Services (2007); J.B. Ruhl & James Salzman, *The Law and Policy Beginnings of Ecosystem Services*, 22 J. Land Use & Envtl. L. 157 (2007); James Salzman, *A Field of Green? The Past and Future of Ecosystem Services*, 21 J. Land Use & Envtl. L. 133 (2006).

13. *See* Grumbine, *supra* note 2, at 31. For collections of additional definitions, all similar in thrust to Grumbine's, *see* Hanna J. Cortner & Margaret A. Moote, The Politics of Ecosystem Management 41 (1999); Gary K. Meffe et al., Ecosystem Management 71 (2002).

14. The discussion of ecosystem services theory in this section is derived from Ruhl et al., *supra* note 12, which provides extensive references for support and further reading.

15. *See* Grumbine, *supra* note 2, at 29; *see also* Cortner & Moote, *supra* note 13, at 42–43; Meffe et al., *supra* note 13, at 115–39.

16. *See* Ruhl et al., *supra* note 12, at 36–56.

17. *See* Ruhl et al., *supra* note 12, at 15–33.

18. *See* Grumbine, *supra* note 2, at 29–30.

19. *See* Ruhl et al., *supra* note 12, at 42–47.

20. *See* Grumbine, *supra* note 2, at 30–31.

21. *See* Ruhl et al., *supra* note 12, at 34–35.

22. *See* Ruhl et al., *supra* note 12 at 258–64.

23. *See* Grumbine, *supra* note 2, at 31.

24. *See* Ruhl et al., *supra* note 12, at 273–75.

25. *See* Grumbine, *supra* note 2, at 31.

26. *See* Ruhl et al., *supra* note 12, at 275–77.

27. *Id. See also* Cortner & Moote, *supra* note 13, at 43–44; Meffe et al., *supra* note 13, at 60–66, 95–111.

28. *See* Ruhl et al., *supra* note 12, at 251–57.

29. *See* Grumbine, *supra* note 2, at 31. *See also* Cortner & Moote, *supra* note 13, at 44–45.

30. *See* Ruhl et al., *supra* note 12, at 281–83.

31. *See* Grumbine, *supra* note 2, at 31. *See also* Meffe et al., *supra* note 13, at 74–76.

32. *See* Ruhl et al., *supra* note 12, at 281–92.

33. *See* Grumbine, *supra* note 2, at 31.

34. *Id.*

35. *See* Ruhl et al., *supra* note 12, at 255–57.

36. Ruhl et al., *supra* note 12, at 31.

37. *See, e.g.*, Cortner & Moote, *supra* note 13, at 38 (providing detailed comparison); Meffe et al., *supra* note 13, at 59–60 (same).

38. *See* Ruhl et al., *supra* note 12, at 23–30.

39. *See* Ruhl et al., *supra* note 12, at 57–83.

40. *See* Ruhl et al., *supra* note 12, at 87–168.

41. *See* Ruhl et al., *supra* note 12, at 265–81.

42. *See* Bhaskar Vira & William M. Adams, Ecosystem Services and Conservation Strategy: Beware the Silver Bullet, 2 Conserv. Letters 158 (2009).

43. *See, e.g.*, Law & Ecology (Richard O. Brooks & Ross A. Virginia eds., 2002).

44. *See* Nagle & Ruhl, *supra* note 12.

45. "The federal government owns about 30 percent of the nation's total surface area (about 650 million acres). Four major federal land management agencies—the Forest Service, Bureau of Land Management, Fish and Wildlife Service, and National Park Service—are responsible for managing about 95 percent of these lands. The Department of Defense manages most of the remainder." U.S. Gov't Accountability Office, RCED-96-40, Land Ownership: Information on the Acreage, Management, and Use of Federal and Other Lands 2 (1996) [hereinafter Land Ownership].

46. Important ecosystem services also flow within and from the marine environment over which the federal government has dominion. *See* Charles H. Paterson & Jane Lubchenco, Marine Ecosystem Services, in Nature's Services, *supra* note 7, at 215. This chapter focuses on the federal government's inland holdings and their associated resources. Coastal regions, which fit this focus as the boundary of inland and marine environments, also are tremendous sources of ecosystem services. *See* Elise F. Granek et al., *Ecosystem Services as a Common Language for Coastal Ecosystem-Based Management*, 24 Conserv. Biology 207 (2010).

47. *See* Ruhl & Salzman, *supra* note 12, at 158–61; Harold A. Mooney & Paul R. Erlich, *Ecosystem Services: A Fragmentary History*, in Nature's Services, *supra* note 7, at 11.

48. *See* Ruhl et al., *supra* note 12, at 127–57; Ruhl & Salzman, *supra* note 12, at 163–64.

49. This section of the chapter is based on a presentation the author made at Duke Law School's 2009 symposium, *Next Generation Conservation: The Government's Role in Emerging Ecosystem Service Markets*. *See* J.B. Ruhl, Address at the Duke Law & Policy Forum Symposium, Next Generation Conservation: The Government's Role in Emerging Ecosystem Service Markets (Oct. 23, 2009), *available at* http://www.law.duke.edu/webcast. The discussion here is adapted with permission from a previously published version of those remarks. *See* J.B. Ruhl, *Ecosystem Services and Federal Public Lands: Start-Up Policy Questions and Research Needs*, 20 Duke Envtl. L. & Pol'y F. 275 (2010).

50. *See* Ruhl et al., *supra* note 12, at 15–83.

51. Land Ownership, *supra* note 47, at 2.

52. *See* Christopher L. Lant, J.B. Ruhl & Steven E. Kraft, *The Tragedy of Ecosystem Services*, 58 BioScience 969, 970–71 (2008).

53. The extensive literature on the economics of ecosystem services given their status as public goods is surveyed in Ruhl et al., *supra* note 12, at 57–83.

54. *See* James Salzman, *Creating Markets for Ecosystem Services: Notes from the Field*, 80 N.Y.U. L. Rev. 870, 883 (2005).

55. Of course, private land managers are often free to do the same, as is the case with land trusts and other private conservation uses of land.

56. Going even further, some commentators express deep concern over the effect the concept of ecosystem services may have on public perceptions of ecological function and conservation of ecological integrity as a sufficient policy goal, in the sense that commodifying the ecosystem function into the metrics of economic service value may decouple the public's perception of the service from the underlying ecological processes. *See* Marcus J. Peterson et al., *Obscuring Ecosystem Function with Application of the Ecosystem Services Concept*, 24 Conserv. Biology 113, 114 (2009); Kent H. Redford & William M. Adams, Payment for Ecosystem Services and the Challenge of Saving Nature, 23 Conserv. Biology 785, 785 (2009). Others counter that the concept of ecosystem services has been employed to rebut economic justifications for activities antithetical to conservation and to open up new conservation opportunities that would not likely be accomplished by relying on purely intrinsic and scientific justifications for conservation. *See* Matt Skroch & Laura López-Hoffman, *Saving Nature Under the Big Tent of Ecosystem Services: A Response to Adams and Redford*, 24 Conserv. Biology 325, 325 (2009).

57. Indeed, in general there is little to be gained in domestic public or private land management contexts by describing commodities such as corn or timber as provisioning services and activities such as hunting and fishing as cultural services. Markets obviously already exist for these services in the private lands context, and public policy has for decades hashed out how they are delivered on public lands. *See* George Cameron Coggins et al., Federal Public Land and Resources Law *passim* (6th ed. 2007).

58. For example, in 2007 the EPA's Office of Research and Development began planning such studies on wetlands as a major component of its Ecosystem Services Research Program (ESRP). *See* U.S. Envtl. Protection Agency, Research to Value Ecosystem Services Identifying, Quantifying, and Assessing Nature's Benefits (2007) (discussing the importance of ecosystem services in researching wetlands), *available at* http://epa.gov/ord/esrp/pdfs/ESRP-overview -fact-sheet-final.pdf. This research provides a foundation to enable the assessment of an

array of core ecosystem services provided by freshwater and coastal wetlands. *See id.* (stating that this new wetland research will determine how the position of wetlands on the landscape alters the provision of ecosystem services). In addition, ESRP research is developing methods to quantitatively assess other regulating and supporting services from wetlands, including flood control and storm surge protection, maintenance of water quality, nutrient cycling, and carbon storage and sequestration. *See* U.S. Envtl. Protection Agency, Ecosystem Services Research Focuses on Wetlands (2007), *available at* http://www.mawwg.psu.edu/resources/ESRP.pdf (discussing the range of benefits gained from wetland ecosystems that contribute to human well-being). This line of research is expected to prove very useful in private lands regulatory contexts such as wetlands conservation, *see* J.B. Ruhl, James Salzman & Iris Goodman, *Implementing the New Ecosystem Services Mandate of Section 404 of the Compensatory Mitigation Program—A Catalyst for Advancing Science and Policy*, 38 Stetson L. Rev. 251, 269–70 (2009), and there is reason to believe the same for public lands.

59. Gretchen C. Daily, *Ecosystem Services in Decision Making: Time to Deliver*, 7 Frontiers in Ecology & Environment 21, 21 (2009).

60. For background on forest ecosystem services, *see* Norman Myers, *The World's Forests and Their Ecosystem Services, in* Nature's Services, *supra* note 7, at 215.

61. For a general discussion of these tradeoffs, *see* Ruhl et al., *supra* note 12, at 32–33; Millennium Ecosystem Assessment, *supra* note 8, at 6–20; Erik Nelson et al., *Modeling Multiple Ecosystem Services, Biodiversity Conservation, Commodity Production, and Tradeoffs at Landscape Scales*, 7 Frontiers in Ecology & Environment 4, 4–10 (2009). The five tradeoff categories summarized in the text are derived from this set of sources.

62. *See, e.g.,* Coggins et al., *supra* note 59.

63. Environmental markets such as wetlands mitigation banking have been shown, for example, to shift ecosystem services across the landscape if the service values are not accounted for in the market 'currency' system, which in the case of wetlands mitigation banking focuses on acres and ecological function. *See* J.B. Ruhl & James Salzman, *The Effect of Wetlands Mitigation Banking on People*, 28 Nat'l Wetlands Newsl. 1, 1 (2006).

64. This typology of federal public land management mandates is discussed in more detail in Nagle & Ruhl, *supra* note 12, at 393–95.

65. Nagle & Ruhl, *supra* note 12, at 393–95.

66. *Id.*

67. *Id.*

68. Food, Conservation & Energy Act of 2008, H.R. 2419, 110th Cong. § 2709 (2008).

69. U.S. Dep't of Agric. Office of Environmental Markets, Ecosystems and the Farm Bill, *available at* http://www.fs.fed.us/ecosystemservices/Farm_Bill/index.shtml (last visited Dec. 11, 2012).

70. *See* Conservation and Land Management Environmental Services Board Charter 1-2, *available at* http://www.fs.fed.us/ecosystemservices/pdf/farmbill/ESB_Charter.pdf (last visited Dec. 11, 2012).

71. This list is a synthesis of notes from the Duke Law School conference discussed in note 51 *supra* and includes observations made by many of the speakers, none of whom the author purports to identify with any particular aspect of the list. The full panel of speakers can be

viewed at 2009 DELPF Symposium, part 1 (Oct. 23, 2009), *available at* http://www.law.duke
.edu/webcast.

72. *Id.*

73. *Id.*

74. *Id.*

75. Sally Collins & Elizabeth Larry, U.S. Forest Service, Caring for Our Natural Assets: An
Ecosystem Services Perspective 8 (2007), *available at* http://www.fs.fed.us/ecosystemservices/
pdf/collins_larry.pdf.

6 Ecosystem Management
A Policy-Oriented Jurisprudence Perspective

Susan G. Clark, Yale University &
David N. Cherney, University of Colorado

The environment is the "fragile envelope of our planet in which we all live," and is presently "under unimaginable stress as industrial and science-based civilizations use the resources of the planet ever more intensively."[1] To address the unintended consequences of our intensive natural resource use (e.g., biodiversity loss, depletion of ocean fisheries, pollution, climate change), humans have divided the world into categories to aid our ability to solve these problems. Ecosystems are one such construct that allow us to address natural resource problems at large scales. The ecosystem concept gives us a tractable way to understand our relationship with nature and allows us to alter that relationship, if we so choose.[2]

Currently, no comprehensive policy or law for ecosystem management exists; perhaps one is not needed. As such, there is no single formally codified goal for ecosystem management across management and political contexts. There are, however, many policies and laws that speak to environmental management directly and indirectly. Through these policies and laws, we can logically infer that the goal of ecosystem management is sustainability of the environment in relation to human uses of natural resources. While an admirable goal overall, it is clear that our policy and management responses are lagging well behind the loss and degradation of ecosystems throughout the world.[3] Nevertheless, ecosystem management is a widely popular construct and the dominant approach to nature and resource management. Ecosystem management often rests on the principles of positivism and "scientific management," yet many people are calling for a more effective foundation to management (e.g., "adaptive governance").[4] This chapter looks at ecosystems and their management as an evolving concept and set of practices, examines the process of ecosystem management using cases from the Greater Yellowstone Ecosystem and elsewhere, and offers recommendations for more effective policy and law.

I. METHODS AND STANDPOINT

Our method and standpoint rests in a policy-oriented jurisprudence that focuses on social choices in real world contexts.[5] It is a genuinely interdisciplinary approach to public and civic order and decision making. The policy-oriented approach that we use was established as a means to help advance human dignity (and sustainability) for all. In everyday understanding (and in many jurisprudence theories), the word 'decision' often refers to a judge applying rules to a specific dispute in a highly organized situation (i.e., the courtroom). Policy-oriented jurisprudence goes well beyond convention, as described below.

The *locus classicus* of this interdisciplinary jurisprudential school is H.D. Lasswell and M.S. McDougal's 1992 two-volume treatise, *Jurisprudence for a Free Society: Studies in Law, Science, and Policy*.[6] This outlook was developed in Yale University's Law School and often referred to by other names, including "The School of Law, Science, and Policy," "Policy Sciences," and the "New Haven School of Jurisprudence." This jurisprudence is concerned with understanding both decision making and real-world choices. The activities of choice-making involve many elements beyond the legislature and courtroom.[7] A more complete conception of decision includes a series of interrelated, dynamic functions.[8] In short, it is not possible to design or amend existing law or create new policy and law without attending to these extra-legal functions of decision making.

As described by Reisman et al., this jurisprudence "adapts analytic methods of the social sciences to the prescriptive purposes of law," deploys "multiple methods," and seeks to "develop tools to bring about changes in public and civic order that will make them more closely approximate the goals of human dignity."[9] This philosophy stands in marked contrast to the jurisprudence of positivism. Positivism's "common focus is on existing rules emanating solely from entities deemed equally 'sovereign,'" and "does not adequately reflect the reality of how law is made, applied and changed."[10] Positivism has limits as noted above and detailed below. Postpositivism (with its deconstructionism) rejects absolutes, views all social and political discourses as saturated with power or dominance, and celebrates "difference."[11] As Burbules and Rice noted, postmodern analysis does "not attempt to judge or prioritize the explanatory significance" of differences.[12] Postpositivism makes it clear what we should not stand for.[13] Neither positivism nor postmodern deconstructionism is sufficient to understand ecosystem management or decision making in the common interest.

We go beyond positivism and postmodernism's deconstructism, to a 'reconstructive' epistemology, theory, and method (the policy-oriented jurisprudence, an interdisciplinarity). Our jurisprudence is a position of integration and reconnection. Our 'reconstructive postmodernism (a postideological view)' addresses the possibility that not every idea, generalization, or proposed group universal is necessarily a form of power domination. Instead, it is a form of making meaning directed at the goal of human dignity for all. This view, we maintain, is essential for sustaining a healthy world. It is not *ipso facto*, absolutistic or ideological in the conventional sense. It is a developmental part of the "long conversation," a multigenerational one about the "meaning of life" and the future of the human enterprise.[14]

II. ECOSYSTEMS—AN EVOLVING POLICY AND LEGAL CONCEPT

Ecosystems and their management are the objects of attention for diverse communities—government, business, and private entities.[15] These objects figure strongly in our social choice making about the environment, sustainability, and our notion of progress. Consequently, these objects have scientific, management, and policy significance. However, the concept of ecosystems is not a static entity in modern political discourse. Ecosystems mean different things to different political actors. Improving the practice of ecosystem management must take these differences into account, as a unified vision for ecosystem management is unlikely to occur. In this section, we explore trends related to the evolution of the term 'ecosystem' in the United States by various political actors.

A. Ecosystems and Epistemology—'Fact' or 'Construction?'

What are ecosystems? Depending on one's philosophy of knowledge (epistemology),[16] ecosystems are either real things, like rocks (the positivists view), or entities that we humans have invented for our own understanding and meaning (the postpositivists view).[17] Different versions of science, law, and policy employ these epistemologies and notions of ecosystems as 'objects' for management in quite different ways.

A positivistic view, the dominant perspective today, sees ecosystems as 'objective' entities. For them, ecosystems are definable concrete objects that exist independent of human perceptions. Positivists view the existence of ecosystems as the way "things really are."[18] The role of ecological (positivistic) science "is not to judge, but rather illuminate the ecological consequences of different potential choices that might be made."[19] The positivistic view has led to diverse ways of thinking about ecosystems in recent years, including as complex, emergent systems,[20] units in sustainability science,[21] in terms of ecosystem services,[22] and other ways, all of which have direct implications for the welfare of the human enterprise.

This perspective argues that ecosystems can be manipulated through management, at least potentially, like manipulating levers on a machine to meet society's goals. So ecologists ask questions, such as: "What factors control the assembly of ecosystems and determine their response to various stressors?"; "How does fragmentation of the landscape affect the persistence of species on the landscape?"; and "How does biological diversity influence ecosystem process?"[23] From this perspective, science serves as the primary guidance for management. Because positivism dominates today, most discussion about ecosystem law, science, and policy takes place within the positivistic discourse.[24] However, many environmental problems today are much more complex, with unidentifiable causal factors, than positivistic science can address.

In contrast, under various postpositivists views, ecosystems are entities that can only be understood from layered contextual meaning (the interpretistic view), and related to as a product of how humans construct reality through social and natural processes (the constructistic view).[25] A 'constructionist' views ecosystems as socially constructed by communities of people held together by shared philosophic principles, identities, norms of behavior, or shared terms of discourse.[26] Constructionists argue that ecosystems or nature do not set the 'rules,' but that 'rules' [behavior of ecosystems] are determined by the people. Depending on the theory of knowledge used, the work of postpositivist science is to predict as positivists do,[27] as well as to understand[28] and explain.[29] In turn, each epistemological camp employs different methods—experimental, statistical, comparative, ethnographic, and triangulation methods in their ecosystem work. For example, positivists tend to use experimental methods to engage their view of ecosystems.[30] Interpretistic people tend to use ethnographic methods, supported by comparative and triangulation methods, in their work.[31] And constructionists tend to use comparative and 'deconstruction' methods in their work.[32]

B. Ecosystems as Objects of Management

What is to be managed? Since the 1990s, paradigms for ecosystem management have proliferated.[33] Despite all falling under the umbrella of a single term, Cortner and Moote provide seven different definitions of ecosystem management that are currently utilized in practice.[34] Each account provides drastically different descriptions of what it means to manage an ecosystem. Cortner and Moote note that definitions and management targets are typically positivistic in content. However, Cortner and Moote conclude that all definitions of ecosystem management focus on: "(1) socially defined goals and objectives, (2) holistic, integrated science, (3) adaptable institutions, and (4) collaborative decision making."[35]

The term 'ecosystem' was coined in 1935 by Tansley.[36] However, the use of the term ecosystem management proliferated when President Clinton announced in the early 1990s that all U.S. natural resource management would be based on recognition of ecosystems and scientific principles. Meffe et al. note that the initial mainstream approach to ecosystem management failed or met significant "public resistance and resentment."[37] The efforts Meffe et al. describe are top-down, expert-driven, and government-mandated. The authors' notion of ecosystem management is that it is "shared *decision making*, [emphasis added] cooperation rather than confrontation, and grass-roots, community-based..."[38]

Other frameworks are in use to understand ecosystems and manage them. For example, Chapin et al., who wrote *Principles of Ecosystem Stewardship: Resilience-Based Natural Resource Management in a Changing World*, contrasted three management paradigms: "steady-state resource management," "ecosystem management," and "resilience-based ecosystem stewardship." Among the many differences, people use ecosystems, are part of the social-ecological system, or have responsibility to sustain future options (respectively by paradigm).[39] Chapin et al. are adherents of the "resilience-based ecosystem stewardship" approach.[40]

Another example is in Ash et al.'s *Ecosystems and Human Well-Being: A Manual for Assessment Practitioners*.[41] The authors take a loose problem-oriented approach to getting "intelligence" about ecosystems.[42] They draw on notions about "ecosystem services," a popular way to address ecosystem management and list indicators and possible proxies for the maintenance of ecosystem services.[43] They divide services into provisioning, regulating, cultural, and supporting categories. A provisioning example for food crops and livestock production is yield of crop production and off take of animals or their products. For regulating, an example is net CO_2 flux out of atmosphere. And for cultural, an example is the recreational opportunities provided. Finally, for supporting, an indicator is net primary productivity. Overall, they list twenty-two services and scores of potential indicators and possible proxies. Typically, the approach to the work is positivistic modeling and scenario-building with the idea of valuing/monetizing services.[44] Still other approaches are in currency, some from social sciences, politics, or other perspectives.[45] Overall, a diversity of notions about what ecosystem management is; targets, indicators, surrogates, and approaches abound.

C. What Do Managers See Themselves as Managing?

The previous two sections demonstrate that there are multiple ways to understand ecosystems and their management. The varying views of the concept bring up practical questions, such as: "What is it that ecosystem managers are supposed to manage, how, and when?" We are not asking the normative question of what managers should manage. In contrast, we are asking the empirical question of what ecosystem managers imagine themselves to be currently managing.

The shifting category of 'management' over time justifies the importance of this question.[46] For example, a few decades ago, ecosystems were not managed as they did not exist (at least in our vocabulary). Clearly, natural resource management has changed over the years as new spheres of work and life emerged. Most styles of management have their origins in early manufacturing sectors, focusing on developing new categories of things.[47] Once targets are identified, then an appropriate assemblage of managers can do their work efficiently. Today, because ecosystems are complex, diverse specialists are needed to manage their many components and processes.

Overall, managers see themselves as managing natural resources, the 'resource' out there in nature, which we see as a positivistic construction of ecosystems. However, managers focus on significantly different aspects of the resources such as structure (soils, water, plants, animals), processes (fire, predation, disease, migration), restoration (e.g., Endangered Species Act, fisheries, clean water), corridors/connectivity (for dispersal, migrations), ecosystem services (resource outflows and markets), sustainability (long-term productivity), and more (e.g., watersheds, landscape coordination across boundaries, stewardship, such as resilience (resistance to perturbations)). In short, there is no comprehensive definition and set of inclusive practices that constitute ecosystem management from the perspective of managers.

Different conceptions of management and activities demonstrate that there is little unified conceptual framework or identity among ecosystem managers. While not inherently problematic, the broad definition of 'managing ecosystems' allows virtually all forms of large-scale conservation to be subsumed by the symbol of ecosystem management. This means that ecosystem management in practice focuses on not just different methodologies, but also a wide variety of substantive objects of conservation.

D. Dynamic Decision-Making Contexts

Regardless of whether one is a positivist or postpositivist, a deconstructionist or reconstructionist—or how one sees ecosystems, the targets of management, and the manager's role—in the real world ecosystem managers must make choices that impact both the environment and people. These choices occur though a wide range of contexts that can be generally summarized in seven functions: (1) intelligence gathering, (2) debate and promotion about the nature and status of the problems, (3) deciding on the plan to solve the problems (in other words, setting new rules), (4) invoking the new rules in specific cases, (5) applying the rules through administrative activities or in courts, (6) appraising progress or lack of it, and finally (7) terminating the rules when they no longer apply.[48]

From this standpoint, ecosystem management refers to the decision-making process about a set of problem-solving tasks (i.e., goal clarification, problem identification, and creation of options). In general terms, choice or decision making (specifically for ecosystem management), also called the policy process, is a human social dynamic that determines how the 'good and bad things' in life are partitioned out, and who gets what, how, and why.

Unfortunately, many people misunderstand the policy process because it is "often treated as an abstraction, associated with the dry prose and dusty volumes of government documents" for "law and legal matters."[49] Such a view is highly misleading. Policy is not the same as legislation, government action, or law. Instead, policy is what government and private bodies do for or to citizens and the environment. What professionals and other people in the field do matters far more in the long run than what is said in formal government documents. Real policy is made in the field through the collective actions of many people. To compound matters, different people have different conceptions of just what the choice-making process actually is, whether it is working well or not, and what to do about it, if anything. Some processes work better than others.

III. ECOSYSTEM MANAGEMENT: PRESCRIPTIONS AND FORMULAS

The Greater Yellowstone Ecosystem provides a grounded case to understand the evolution of the ecosystem management construct in practice. We first provide a brief history of ecosystem management in the Greater Yellowstone Ecosystem and then highlight a case of pronghorn antelope migration as an example of two different ecosystem management paradigms (scientific management and adaptive governance). We conclude through this case that the central problem of ecosystem management is a tendency by managers to overlook or misconstrue important aspects of the management context. This oversight perhaps reflects a limited methodology.

A. *The Greater Yellowstone Ecosystem Experience*

Greater Yellowstone was one of the first regions to attempt ecosystem management in the world. However, the application of ecosystem management in this region has been far from constant or ideal. Over the last forty years, varying conceptions of ecosystem management have been attempted by a variety of different management agencies in the region.[50] What is clear from the Yellowstone experience, however, is the need to manage landscapes at landscape scales.

The history of modern ecosystem management in greater Yellowstone can be traced back to the management of grizzly bears within Yellowstone's boundaries. In the 1960s, researchers Frank and John Craighead began promoting the idea that to maintain a viable population of grizzly bears within Yellowstone National Park, it was required that grizzlies were appropriately managed outside of the park's boundaries, too.[51] In other words, grizzly bear management could not effectively occur within the context of a single protected area—an entire ecosystem needed to be protected. This is the dominant paradigm for grizzly management today.

Whereas this appeared to be a novel idea in the 1960s, the idea of managing Yellowstone at larger scales to protect wildlife is far from new. General Philip Sheridan advocated for—and ultimately failed in—doubling the size of Yellowstone to protect migrating wildlife such as bison and elk.[52] While his proposal and many others were unsuccessful, in 1891 President Benjamin Harrison created the Yellowstone Timberland Reserve (1.2 million acres) adjacent to the park, to help protect the region. The Yellowstone Timberland Reserve is now encompassed by the region's national forests. As early as 1919, the assemblage of Yellowstone and the surrounding public lands were thought of as a related unit, with the term "Greater Yellowstone" first occurring in print.[53]

In 1964, the National Park Service and U.S. Forest Service established the Greater Yellowstone Coordinating Committee (GYCC), a collaborative body that seeks "to pursue opportunities of mutual cooperation and coordination in the management of core federal lands in the Greater Yellowstone."[54] The committee consists of the superintendents of two national parks (Yellowstone and Grand

Teton), the forest supervisors of six national forests (Beaverhead-Deerlodge, Bridger-Teton, Caribou-Targhee, Custer, Gallatin, and Shoshone), and the managers of two national wildlife refuges (Red Rock Lakes and the National Elk Refuge). This committee's establishment signifies a major management shift to view the region as a single ecosystem.[55]

B. Management Paradigms (For Ecosystem Management)

One modern case of ecosystem management in the Greater Yellowstone Ecosystem is the management of a pronghorn antelope migration between Grand Teton National Park and the Upper Green River Basin in southwest Wyoming.[56] This 170-mile migration is the longest land mammal migration in North and South America, excluding the caribou in Alaska and the Yukon.[57] Evidence exists suggesting that portions of this migration have been used by migrating pronghorn for over six thousand years.[58]

Management of this migration is highly complex. Three major federal conservation agencies control land along the migration route (Park Service, Forest Service, and Bureau of Land Management). Additionally, three counties and numerous private land owners are able to make decisions that affect the future of this migration. Given the complex nature of the management of ecosystem processes (e.g., migration), it is often unclear which governmental agencies are responsible for maintaining ecosystem function. This has been a chronic problem in the management of this migration, as governmental and private actors have tended to shed responsibility for the goal of maintaining this migration in perpetuity.[59]

In 1999, conservationists became concerned about private housing and natural gas development occurring along the migration route and demanded action to protect the future of the migration. The transboundary nature of migrations forced many conservationists and managers to think of the issue as a form of ecosystem management—to protect the migration the administrative units needed to be thought of as a holistic unit. The two major policy responses were implemented over the following ten years and can be loosely classified as scientific management and adaptive governance.[60]

Scientific management is a management philosophy that relies on the linear model of science as a way to guide management action.[61] The linear model argues that good science (or getting the science right) is the foundation for effective public policy.[62] After the science is settled, managers can make an effective plan (in this case the protection of a migration) and the plan will be implemented by experienced professionals. In other words, scientific management views policy-making as a technical exercise best left to experts. As such, expert opinion is viewed as having a higher value than public opinion. Public managers are usually deemed the appropriate authoritative body to make controlling decisions.

In contrast, adaptive governance is a management philosophy where the exact model of policy change is much less defined.[63] Rather than rely exclusively on expert option and a narrow set of authoritative managers, adaptive governance calls for authority and control to be fragmented between different political actors. A single comprehensive plan is not a prerequisite for action. Rather, policy action occurs through the agreement of interested parties. This allows for greater flexibility in policy solutions by forcing interested parties to find ways to accommodate alternate interests. In other words, there is no overarching goal (through a plan). Rather, interested parties can all work towards their preferred outcomes.

In the pronghorn migration management, the first plan advocated for was steeped heavily in scientific management. Environmental groups began actively studying the migration, with the Wildlife Conservation Society taking the lead in producing ground-breaking and important scientific research on the migration. Using this knowledge as their authoritative basis, a coalition of environmental groups formulated an explicit plan to protect the migration in its entirety. The general goal was to create the first national migration corridor throughout the length of the migration. After devising a general plan, environmental groups began lobbying the major federal agencies to implement the environmental groups' concept.[64]

The promotion of this plan caused major public outcry by Wyoming citizens. While the plan was an extremely effective technical means to protect the migration, it failed to account for the political context of the region. Virtually all political perspectives (environmental and otherwise) appeared interested in protecting the pronghorn migration in perpetuity; no one was against maintaining a viable population of migrating pronghorn. In contrast, many local citizens were against the idea of creating a new protected area as a means to solve the problem. Arguments against the proposed corridor stemmed from issues of property rights and the need for government intervention. By 2008, it was clear that a comprehensive plan to protect the migration in its entirety was doomed to fail.

However, a second policy response emerged in 2008 that proved to be much more successful in protecting the future of this migration.[65] The second response more closely resembles adaptive governance. Rather than develop a single plan to protect the future of the migration, both public and private landowners along the migration route took it upon themselves to develop solutions to protect the migration in ways that met their individual and organizational interests. There was no powerful authoritative body mandating that these individuals and agencies adopt such plans, nor was there any future threat that future legislation would emerge.

In the north part of the migration route, Grand Teton National Park and the Bridger-Teton National Forest signed a memorandum committing to protect the migration route on the land they controlled. While no significant threats occurred within the national park's boundaries, the Forest Service committed that there would be no human activity along the route during the timing of the migration. While this agreement closely resembles the environmental group's original plan, two caveats must be acknowledged. First, the memorandum only protects the northern half of the migration. Second, virtually all threats to the migration's future occur in the land further south.

At the same time the Park Service and Forest Service protected the northern half of the migration, virtually all private landowners along the southern portion of the migration protected the future of the migration by modifying fencing on their land. Fences are a major barrier for pronghorn, as they have trouble jumping over fences due to their bone structure. The Upper Green River Valley Land Trust implemented the Corridor Conservation Campaign in 2008 to help landowners voluntarily eliminate this threat. By 2009, virtually all fencing along the pronghorn migratory corridor was converted to be conducive to maintaining the migration. In short, landowners were able to effectively address many of the threats to the migration in the best way they saw fit and not through a comprehensive expert-driven plan. Today, the protection of this migration is heralded by many conservationists as a preeminent example of both migration and ecosystem management.

C. Problem Definition

The Yellowstone experience demonstrated two differing ways to view ecosystem management, one as scientific management and the other as adaptive governance. The question we must ask is: "Why did scientific management fail and adaptive governance succeed?" We previously argued that ecosystem management is an evolving concept subject to multiple epistemologies, objects of management, views on management focus, and decision-making contexts. While scientific management and adaptive governance are only two formulae used by ecosystem management proponents, there is sufficient evidence to conclude that a broader conception of ecosystem management (adaptive governance) is more likely to achieve successful outcomes than a narrow conception (scientific management). Yet, scientific management dominates government agency management today.[66]

In the Greater Yellowstone Ecosystem pronghorn migration case, successful ecosystem management occurred through the integration of multiple viewpoints and a solution that met the need of all participants in the policy process. The initial policy formation of an expert-driven protected area failed to adequately capture the political realities of the region. In other words, the expert-driven model restricted the scope of choices available to the community in solving the problem of maintaining the migration in perpetuity. Rigid plans often fail to take into account evolving political circumstances; this finding appears robust among a number of different cases. For example, in a comprehensive study of organizational leadership, Heifetz and Linsky found "the single most common source of leadership failure we've been able to identify—in politics, community life, business, or the nonprofit sector—is that people, especially those in positions of authority treat adaptive challenges like technical problems."[67] By treating complex policy problems as a linear technical issue, it is easy to overlook or misconstrue important details (e.g., local politics) that are necessary to help resolve the issue.

In essence, our diagnosis of the fundamental problem with most formulae for ecosystem management is that the perspectives of ecosystem managers and advocates are too narrow. By narrowly focusing ecosystem plans and management techniques, the scope of choices available to managers is reduced. Given the complex and interactive nature of most ecosystem issues, we conclude that maximizing the scope of choices available to managers is a worthy goal. Doing so will require managers to consider the widest range of causal factors—including biophysical, social, and decision making—to effectively solve management issues. In the following section, we provide three recommendations geared toward expanding the perspectives of managers to a wider variety of relevant issues that are often overlooked.

IV. ECOSYSTEM MANAGEMENT IN POLICY
AND LAW—RECOMMENDATIONS

There are two types of challenges in ecosystem management: analytic and policy, especially when viewed from the vantage of a policy-oriented jurisprudential perspective. In fact, not surprisingly, we make several recommendations to address these challenges. First, use a policy-oriented jurisprudential standpoint to understand and influence choice-making about ecosystem management. Second, understand that the 'law' of ecosystem management is really a process of communication and one of participating appropriately and responsibly. And, third, learn to recognize the choice-points in the legal and policy system and use them to bring about effective ecosystem management in the common interest.

A. Use a Policy-Oriented Jurisprudence

Using a policy-oriented jurisprudence can help with the sorts of problems that professionals, the public, and policy makers encounter in ecosystem management—problems that otherwise may go unseen, misdiagnosed, and unattended. Experience shows that the way that a person characterizes a problem, the intellectual tools that they use to research, and the information that they think relevant for addressing problems will all determine their conception of problem solving and 'law.' Often people have conceptual blind spots that hinder them from successfully resolving the problems faced. Policy-oriented jurisprudence and its "comprehensive framework of inquiry" that we recommend can aid ecosystem management by helping shed light into areas that other styles of problem solving may have overlooked or misconstrued.[68] This form of problem solving was developed as a functional, yet practical way to understand and analyze the decision-making process and influence it for the common good.[69]

Policy-oriented jurisprudential theory and method are about making social (and technical) choices. The major "jurisprudential and intellectual tasks are the prescription and application of policy in ways that maintain community order and, simultaneously, achieve the best possible approximation of the community's goals."[70] The considerations essential to carrying out these tasks include: "(1) the way one looks at oneself; (2) the way one looks at social process one is trying to understand and influence; and (3) the way one tries to influence it."[71] This jurisprudence is committed to several principles: (1) It is a tool for critical analysis and improving matters; (2) It is committed to the study of process, by which authoritative and controlling decisions are made; (3) It is committed to normative values (human dignity); (4) It is committed to connecting law and policy.[72] Its guiding themes include interdisciplinarity, studying 'law' in both substance (content) and process (procedural) terms, and it is dedicated to public service and practice. This form of jurisprudence moves us away from the scientific management paradigm and towards adaptive governance by focusing on the resolution of problems through multimethod and contextual inquiry.

B. Carry Out Effective Communication in Making Law and Policy

The problem of ecosystem management has never been more urgent or more complicated than it is now. Problem solving in the service of ecosystem management is also a problem about communication. Fortunately, adequate theory and practice exists to improve communication in the service of ecosystem management. In its simplest terms, lawmaking or the prescribing of ecosystem management policy, for example, as authoritative for a community is a process of communication.

All communication involves the mediation of subjective messages, whether about ecosystem management or anything else. Communicators must convey policy content, and give an authority signal, their intention to control the situation, and all this falls on a target audience. Communications and message senders and receivers can be coordinated to good effect in the common interest with adequate attention to details and with sound leadership.[73]

Sound communications can overcome, perhaps, previous generations of positivistic myth that science and policy are two different spheres, as oppositional things, and also positivistic ideas about law. We now know these artificial distinctions lack realism and are counterproductive to real world problem solving. To overcome the many practical and professional problems brought about by these and other artificial dichotomies requires their resolution through an appropriate theory and method about law, decision making, and problem solving in both formal (legalistic) and informal (civic) settings. This is not an academic matter. The urgency of having and using an appropriate theory is underscored by the fact that ecosystem management is being prescribed continuously and that it is used in virtually all of the institutions of natural resource policy and management. It has paramount significance to the sustainability of the human enterprise. If we fail to learn how to prescribe (and implement), the individual and collective consequences may be dire indeed.

C. Choice-Points in the Legal and Policy System

The functional understanding of policy that we describe above breaks down the operation of decision making in ecosystem management into seven interrelated functions. They are: gathering intelligence, the promotion/lobbying of preferences, the prescribing of authoritative policy or lawmaking, the making of provisional characterization of deviations for prescribing or invoking, applying prescriptions, appraising the aggregate performance of a community's decision process in terms of community goals, and terminating prescriptions and moving on. We need to acknowledge that these seven functions all represent opportunities for improving decision making by providing us with multiple 'choice-points' beyond the legal system.

Lawyers and judges play important roles in decision making about ecosystem management, as well as legislators and other formal decision-making bodies. However, there are many other 'choice-points' besides those of policy application and prescription in nature conservation and resource management. Often these two functions—application and prescription—are viewed as the most important decision-making functions by those outside of policy-oriented jurisprudence. The policy-oriented jurisprudence, in contrast, find it useful to judge what functions are most important on a contextual basis, as these functions are typically highly interactive and in play at any one time in any actual case.

The *intelligence* is about the gathering, processing, and dissemination of information relevant to making social choices (e.g., policy on ecosystem management). Choices about what information is to be considered (or available) matters in determining the outcome of any decision-making process. *Promotion* is the process by which people and society come to see a discrepancy between what they want (desirable state of things) and what is about to take place. This function typically leads to some type of intervention and regulation. Who is allowed to participate and in what manner will affect the outcome. *Prescription* or law-making is about people with the proper degree of authority selecting and installing specific preferences regarding policy as community law. They may or may not be the legislature. Customary (informal) law is prescriptive too and often ignored. *Invocation* is the initial or provisional characterization of certain actions as inconsistent with a prescription or law. Invocation is typically associated with a community's demand that something be done. *Application* is about dispute resolution in which facts are organized around a dispute and norms are clarified. When this function takes place in a courtroom, it is called a judgment. It also occurs in less organized, informal situations, too. *Appraisal* is about evaluating whether the overall performance of all decision functions met community requirements. It is also about assigning responsibility and accountability for performance. Finally, *termination* is about ending prescriptive norms and the social and organizational arrangement based on them. It is about putting in place transitional mechanisms and responding to people who have acted in good faith (e.g., compensation programs) that the old regime would continue. Each of these functions of decision offers unique opportunities for making choices. They each hold jurisprudential questions about how those choices should be made. We need to move beyond the designation and enforcement of formal law.

D. Other Recommendations

These recommendations can be applied more conventionally through practice-based approaches,[74] prototyping (trial interventions for learning), and pilot projects (small scale improvements based on practice-based and prototyping experience).[75] These are all subsumed and detailed in the term 'adaptive governance.' The competition for an adequate formula for ecosystem management in science and policy can best be found using our recommendations. Ecosystem management needs to be thought of in broader terms—in terms of decision making and communication.[76] What is needed is an understanding of the aggregate of processes in a community by which political perspective at varying levels of consciousness are shaped and shared, in this case for sustainable environmental management. These subjectivities are often the substratum and the vanguard of 'lobbying and lawmaking.' People interested in ecosystem management can ill afford to ignore these matters.

V. CONCLUSIONS

We live in an age of rapid change, a world in which governance, freedom, and terror are globalizing. We face two main challenges in building a sustainable world of human dignity in healthy environments and ecosystems. One challenge is *analytic:* to understand these globalizing dynamics and to shape them. The second challenge is *policy:* content and process. We want this process to unfold in a way in which law, science, and policy help organize and serve as a creative medium for our efforts to build a humane and sustainable world, in the deepest sense. We certainly need a more enlightened national and global community. In addressing problems, the intellectual tools one uses to research them and the information one thinks is relevant for answering them will all be determined by one's conception of policy, decision making, and law. These conceptions influence the role one assumes, the method one uses, the ethic one adopts, and the outcome.

People trained in—and sometimes locked into—one perspective can scarcely believe that there may be others that are equally authentic and that, for some or all tasks, they may be even more useful than the one with which they were indoctrinated and, as a result, with which they are comfortable. Each perspective is the basis for a legal jurisprudence. Take two examples: positivism and policy-oriented jurisprudence. First, positivism views ecosystem management (and law) from the perspective that science and law are bodies of commands. This perspective assumes the independent moral value of obedience. The essential technical problem is properly identifying the content and meaning of the command and the circumstances and procedures for obedience to it. Policy-oriented jurisprudence, in contrast, takes the view of the person charged with making decisions. From the perspective of the decision maker, the technical and moral problems that are confronted are not framed in terms of obedience but rather in terms of making choices that are appropriate for the relevant community and situation. In short and in sum, the body of rules that serves to provide the positivist with strict commands requiring obedience does not disappear, but from the perspective of the decision maker, those rules are more complex communications, conveying authoritative information about community policies of varying weight that must be assessed, in each case, and then shaped into a decision. The technical and moral problems associated with obedience recede. Therefore, selecting a sound jurisprudence (using interdisciplinary methods) is the task at hand.

In sum, all decision making—including ecosystem management—is a process about "How will resources (natural and cultural) be used?" and "Who gets to decide?"[77] It is the process through which people interact to answer these two questions. Ideally, any and all decision-making processes should clarify and secure the common interest. This is a legitimate purpose and a requirement of good governance.[78] Our recommendations support this goal. Most simply understood, the common interest is the set of interests shared by members of a community as a whole.[79] In contrast, exclusive special interests are incompatible with the common interest and benefit (or are promoted by) only some members of a community at the expense of the whole community. Ecosystem management is just one of a long string of diverse developmental modernizations (of decision-making processes) to address or adjust our relationship with the environment and resources.

ACKNOWLEDGMENTS

We want to thank a great many people, too numerous to mention all by name, for the ideas in this chapter and their description. We are members of a growing worldwide movement to upgrade natural resource conservation, its science, management, and policy. We have benefitted from our employing organizations and many supporters. We would like to thank Denise Casey for providing critical review.

NOTES

1. W. Michael Reisman, *The Vision and Mission of* The Yale Journal of International Law, 25 Yale J. Int'l L. 263–69 (2000).

2. Yale School of Forestry & Environmental Studies, Large Scale Conservation: Integrating Science, Management, and Policy in the Common Interest (Susan G. Clark, Aaron Hohl, Catherine Picard & Darcy Newsome eds., 2010).

3. Millennium Ecosystem Assessment, Ecosystems and Human Well-being: Biodiversity Synthesis (2005).

4. Ronald D. Brunner et al., Finding Common Ground (2002) [hereinafter Finding Common Ground]; Ronald D. Brunner et al., Adaptive Governance (2005) [hereinafter Adaptive Governance].

5. See, e.g., Ronald D. Brunner & Tim W. Clark, A Practice-Based Approach to Ecosystem Management, 11 Conserv. Biology 1, 48–58 (1997) (practice-based ecosystem management).

6. See W. Michael Reisman, Siegfried Weissner & Andrew R. Willard, The New Haven School: A Brief Introduction, 32 Yale J. of Int'l L. 575, at 574 n.2 (2007); Ronald D. Brunner, A Milestone in the Policy Sciences, 29 Pol'y Sci. (1996); Harold D. Lasswell & Myres S. McDougal, Jurisprudence for a Free Society (1992).

7. Harold D. Lasswell, A Pre-View of Policy Sciences (1971).

8. W. Michael Reisman, The View from the New Haven School of International Law, 86 Am. Soc'y of Int'l L. Proc. 118 (1992).

9. Reisman et al., supra note 6, at 567.

10. Reisman et al., supra note 6, at 579 n.13.

11. Nicholas C. Burbules & Suzanne Rice, Dialogue Across Differences: Continuing the Conversation, 61 Harv. Educ. Rev. 4, 393–416 (1991) [hereinafter Dialogue Across Differences]; Nicholas C. Burbules & Suzanne Rice, Can We Be Heard? A Reply to Leach, 62 Harv. Educ. Rev. 2, 264–71 (1992); Mary S. Leach, Can We Talk? A Response to Burbules and Rice, 62 Harv. Educ. Rev. 2, 257–63 (1992); Ian Hacking, The Social Construction of What? (1999).

12. Dialogue Across Differences, supra note 11, at 393–416.

13. Robert Kegan, In Over Our Heads (1994).

14. Id.

15. E.g., J. Lubchenco et al., Ecological Society of America, The Sustainable Biosphere Initiative: An Ecological Research Agenda, 72 Ecology 2, 371–412 (1991); Millennium Ecosystem Assessment, supra note 3.

16. Sanjeev Khagram et al., Thinking About Knowledge: Conceptual Foundations for Interdisciplinary Environmental Research, 37 Envtl. Conserv. 4, 388–97 (2010).

17. Michael E. Soulé & Gary Lease, Reinventing Nature (1995); Bruno Latour, Politics of Nature (2004).

18. Gregory J. Feist, The Psychology of Science and The Origins of the Scientific Mind (2007).

19. Oswald J. Schmitz, Ecology and Ecosystem Conservation 5 (2007).

20. National Science Foundation Advisory Committee for Environmental Research and Education, Complex Environmental Systems: Synthesis for Earth, Life, and Society in the 21st Century: A 10-year Outlook for the National Science Foundation, at 1–68 (2003).

21. Hiroshi Komiyama & Kazuhiko Takeuchi, *Sustainability Science: Building a New Discipline*, 1 Sustainability Sci. 1, 1–6 (2006).

22. Neville Ash et al., Ecosystems and Human Well-Being (2010).

23. Schmitz, *supra* note 19, at 4–5.

24. Soulé & Lease, *supra* note 17.

25. Khagram et al., *supra* note 16, at 388–97.

26. Martha Finnemore, National Interests in International Society (1996).

27. Milton Friedman, Essays in Positive Economics (1953).

28. Hacking, *supra* note 11.

29. Barney Glaser & Anselm Strauss, The Discovery of Grounded Theory (1967).

30. Schmitz, *supra* note 19.

31. James Ferguson, The Anti-Politics Machine (1994).

32. Hacking, *supra* note 11; Latour, *supra* note 17.

33. R. Edward Grumbine, *What is Ecosystem Management?*, 8 Conserv. Biology 1, 27–38 (1994).

34. Hanna J. Cortner & Margaret A. Moote, The Politics of Ecosystem Management (1999).

35. *Id.* at 40.

36. Frank Benjamin Golley, A History of the Ecosystem Concept in Ecology (1996).

37. Gary K. Meffe et al., Ecosystem Management 4 (2002).

38. *Id.*

39. Principles of Ecosystem Stewardship: Resilience-Based Natural Resource Management in a Changing World 4 (F. Stuart Chapin, III, Gary P. Kofinas & Carl Folke eds., 2009).

40. *Id.*

41. Neville Ash et al., *supra* note 22.

42. *See* Susan G. Clark, The Policy Process (2011).

43. Neville Ash et al., *supra* note 22, at 128–29.

44. Erik Nelson et al., *Modeling Multiple Ecosystem Services, Biodiversity Conservation, Commodity Production, and Tradeoffs at Landscape Scales*, 7 Frontiers in Ecology and Env't 4 (2009), *available at* http://www.frontiersinecology.org.

45. *E.g.*, David Johns, A New Conservation Politics (2009); Nina Chambers, Shawn Johnson, Matt McKinney & Gary Tabor, Remarkable Beyond Boundaries: People and Landscapes in the Crown of the Continent (Sarah Bates ed., 2010); Esther Blanco, *A Social-Ecological Approach to Voluntary Environmental Initiatives: The Case of Nature-Based Tourism*, 44 Pol'y Sci. 1, 35–52 (2011).

46. William M. Sullivan, Work and Integrity (1995).

47. Finding Common Ground, *supra* note 4.; Adaptive Governance, *supra* note 4.

48. Lasswell, *supra* note 7; Clark, *supra* note 42.

49. Paul J. Culhane, Public Lands Politics 30 (1981).

50. The Greater Yellowstone Ecosystem (Mark S. Boyce & Robert B. Keiter eds., 1991).

51. John J. Craighead, *Yellowstone in Transition, in* Keiter, *supra* note 50.

52. Craighead, *supra* note 51.

53. Tim W. Clark & Dusty Zaunbrecher, *The Greater Yellowstone Ecosystem: The Ecosystem Concept in Natural Resource Policy and Management*, 5 Renewable Resources J., Summer 1987, at 8–15; Paul Schullery, *Greater Yellowstone Science: Past, Present, and Future*, 18 Yellowstone Sci. 2, 7–13 (2010).

54. The Greater Yellowstone Coordinating Committee, *available at* http://www.fedgycc.org/ (last visited Dec. 7, 2011).

55. Susan G. Clark, Ensuring Greater Yellowstone's Future (2008).

56. David N. Cherney & Susan G. Clark, *The American West's Longest Land Mammal Migration: Clarifying and Securing the Common Interest*, 42 Pol'y Sci. 2, 95–111 (2009).

57. Joel Berger, *The Last Mile: How to Sustain Long-Distance Migration in Mammals*, 18 Conserv. Biology 2, 320–31 (2004).

58. M.E. Miller & P.H. Saunders, *The Trapper's Point Site (48SU1006): Early Archaic Adaptations and Pronghorn Procurement in the Upper Green River Basin*, 45 Wyoming Plain's Anthropologist, 39–52 (2000).

59. Cherney & Clark, *supra* note 56, at 95–111.

60. David N. Cherney, *Securing the Free Movement of Wildlife: Revisiting the Lower 48's Longest Land Mammal Migration*, 41 Envtl. L. 2, 599–618.

61. Adaptive Governance, *supra* note 4, at 1–46.

62. Roger A. Pielke, Jr., The Honest Broker (2007).

63. Brunner & Steelman, *supra* note 61.

64. Cherney & Clark, *supra* note 56, at 95–111.

65. Cherney, *supra* note 60, at 599–618.

66. Clark, *supra* note 55.

67. Ronald A. Heifetz & Marty Linsky, Leadership on the Line 14 (2002).

68. Richard A. Falk, *Casting the Spell: The New Haven School of International Law*, 104 Yale L.J. 7, 1991–2008 (1995).

69. Harold D. Lasswell, *Clarifying Value Judgment: Principles of Content Ad Procedure*, 1 Inquiry, 87–99 (1958).

70. Reisman, *supra* note 8, at 120.

71. Reisman, *supra* note 8, at 120.

72. Ronald D. Brunner, *A Milestone in the Policy Sciences*, 29 Pol'y Sci., 45–68 (1996); Reisman et al., *supra* note 9, at 575–82; Hongju Koh, *Is There a "New" New Haven School of International Law?*, 32 Yale J. Int'l Law, 558–73 (2009).

73. W. Michael Reisman, *A Jurisprudence from the Perspective of the "Political Superior"*, 23 N. Ky. L. Rev. 23, 605–27 (1995).

74. Brunner & Clark, *supra* note 5, at 48–58.

75. Clark, *supra* note 42.

76. Clark, *supra* note 55.

77. Foundations of Natural Resource Policy and Management (Timothy W. Clark, Andrew R. Willard & Christina M. Cromley eds., 2000).

78. Daniel Kemmis, Community and the Politics of Place (1990); Robert A. Dahl, On Democracy (1998).

79. Lasswell & McDougal, *supra* note 6.

III

Making Better Use
of Existing Federal Law

7 Addition by Subtraction
NEPA Routines as Means to More Systemic Ends
Jamison E. Colburn

This volume on ecosystem management and law seems a good place for some thoughts on the National Environmental Policy Act (NEPA), "our basic national charter for protection of the environment."[1] NEPA is famously regarded in court as a 'procedural' statute, a statute that aims not for particular environmental outcomes but rather at the deliberative processes whereby federal agencies make their decisions which then impact the environment.[2] NEPA requires that "all agencies of the Federal Government . . . include in every recommendation or report on proposals for legislation and other major Federal actions significantly affecting the quality of the human environment a detailed statement by the responsible official,"[3] which predicts the environmental impact of the proposed action, any unavoidable adverse environmental impacts should the action be taken, alternatives to the proposed action, and other information useful in deciding whether the reasons for the proposal outweigh or otherwise defeat the reasons against it.[4]

Since 1978, when NEPA's Council on Environmental Quality (CEQ) finalized its 'regulations' implementing Section 102(2),[5] covered federal agencies have had to divide their NEPA practices into one of several discrete action-types that are oriented around this 'detailed statement' requirement.[6] Indeed, these NEPA categories have, in the intervening decades, hardened to become the principal content of NEPA law—the set of obligatory routines into which all covered agency 'actions' must fit and with which they must comply. It is these compliance-oriented routines that we must better structure if we are to have any hope of shaping NEPA into a selection of tools that we so desperately need to better manage the natural systems we depend upon. For reasons that will become clear, I argue that CEQ seems to be moving NEPA in the right direction, but that it is not doing so quickly or decisively enough.

In this chapter, I parse our established NEPA categories, some of the many precedents construing CEQ's rules, some of the hard choices administrators face in NEPA compliance, and the practical differences therein, all in an effort to suggest some modest reforms that could make NEPA more of an aid and less of an obstacle to the ecosystemic management of natural systems. Part I introduces the NEPA action types and the precedents construing their boundaries. Part II sketches the major issues that permeate NEPA practice today, regardless of the action-type which the agency has chosen (or is ordered) to take. Part III proposes a more causation- and less compliance-oriented interpretation of NEPA, an interpretation of NEPA itself as a system with multiple interacting layers and processes that can be managed, but only by the shrewdest and ablest of managers.

A serviceable model of NEPA as a system, and one particularly tailored to reworking NEPA as our emerging ecological consciousness would have it, is that of a 'joint cognitive system.' The focus of cognitive systems engineering is the ways in which humans can cope with and master the complexity of their environments.[9] Joint cognitive systems are those that cope with the self-reinforcing complexity of those environments by combining humans and machines into unified, adaptive systems.[10] They aim to achieve 'macrocognition' by combining the best of humanity and the machines they invent into the building blocks of extended, complementary, and effective networks of functionality.[9] Intentionally engineering such systems is massively complicated and prone to error, because the design challenges involved can lead to compounding the environmental complexity instead of reducing it. The first step is to recognize the elements of this system and their relationships. That is the task of Parts I and II.

I. NEPA'S PROCEDURAL ROUTINES: A NETWORK NOT YET IN BEING

The detailed statement requirement in NEPA § 102 has become a notorious hammer in environmental politics and law. Full-blown environmental impact statements (EISs) are arduous, costly, and time-consuming undertakings that tend to drive the agency decision-making process and all of the affected actors in turn.[11] As many have noted, the more the substance and sufficiency of an EIS is brought under question, the greater the incentive future NEPA actors have to 'bulletproof' their document and the processes behind it, perversely leading to an ever-more arduous ordeal in the creation of EISs.[11] Thus, for a plaintiff seeking to scuttle some particular program, project, plan, or other covered 'agency action'—including those where the agency's involvement is merely to give permission—the ability to force the EIS upon a reluctant agency has become a significant, often decisive victory.[12] An EIS is, in this connection, something of a 'penalty default.'[12] Moreover, "because a full scale EIS may take months or years to complete, the final work product typically arrives on the decisionmaker's desk at the end of a protracted, multistage project development and clearance process."[13] The EIS may not even be a very good means to one of NEPA's major ends: better-informed agency decisions.[14] EISs that are rare, gargantuan, strategized, and/or late cannot, in any practical sense, be 'integrated' with agency planning and decision making more generally.[15]

An even deeper tension than this practical reality exists in NEPA compliance today. Agencies, in making both their operational and programmatic decisions, must necessarily select the scales at which to assess the environmental risks they encounter. I shall call this the problem of bites and, in particular, how big a bite should be taken in any given NEPA routine. An agency like the Department of Interior, charged with developing offshore energy sources at acceptable costs, must decide whether and when to analyze the risk of oil spills, both as an entire plan of action and as to particular places with known and specific geologies, surface conditions, people in command, etc. In the real world, either the agency does so in-depth at one stage of the process or at another; its resources are normally constrained enough that its risk analyses will not often be repetitive in any real sense.[16] Either the agency takes a big bite of the risk analysis it should conduct or it takes smaller bites, perhaps even a series thereof. But the agency must choose—consciously or otherwise—how big a bite it should take in its chosen NEPA routines.[17] And although the 'scoping' of the NEPA process was intended to integrate NEPA with agency planning more generally,[18] the success of scoping in solving that particular bureaucratic problem has been mixed at best.[19] From its earliest suggestion that agencies 'tier' their NEPA analyses to better manage the scale and scope of the risks they were confronting, CEQ seems to have conceived of this problem in spatial terms,[20] even though it is just as much a matter of temporal and/or causal connection(s), whereby agencies must manage environmental risk.[21]

A subtype of EISs, known as the *programmatic* EIS (PEIS), aims to analyze environmental risks that are identifiable at the broadest scales of agency 'action' and cognition.[22] Agency *action* in this connection is the formative, animating concept; without the agency action itself as a guide, there is virtually no prin-cipled way to select an appropriately-sized bite of the agency's risk management problem(s).[23] As any administrative lawyer well knows, agency *action* is a noto-riously slippery thing. Agencies are diverse and institutionally corrigible, which means that discerning their 'actions' from their 'inaction' or other states of being has been and will remain a mighty struggle.[24] Typically, PEISs will involve time horizons like those used in strategic planning, a great deal of operational diver-sity, or both.[25] And because the reasons for preparing programmatic statements instead of other, more limited substitutes (or vice versa) are as diverse as the decision makers acting on them, the quest for the types of rationality to which NEPA aspires—a rationality born of concern for the 'whole environment' and the human impact thereon[26]—inevitably comes down to some more or less arbi-trary selection of scale(s) for the best risk assessment, management, and com-munication. There is no formula to resolve which environmental risks are better examined or managed in small bites, as opposed to those that are better handled as an aggregate. And there never will be. Risk is an inherently subjective condi-tion, both because the experience of harm/loss is so subjective and because the measures available for reducing risk are so actor-specific.[27]

The recognition of risks is closely tied to an agent's perceptions and habits and if one's perceptions and/or habits routinely occlude particular risks they will likely go unnoticed.[28] "Whatever the decision context, the goal of risk assess-ment is to describe the probability that adverse health or ecosystem effects of specific types will occur under specified conditions of exposure to an activity or an agent (chemical, biologic, radiologic, or physical), to describe the uncertainty in the probability estimate, and to describe how risk varies among popula-tions."[29] But for administrative agencies that do not normally carry out NEPA compliance routines—much like agencies wherein the traditional route of NEPA compliance is some form of general *exclusion*—risks can easily go completely unnoticed and therefore unmanaged.

CEQ's rules provide that an agency may categorically exclude actions "which do not individually or cumulatively have a significant effect on the human envi-ronment and which have been found to have no such effect in procedures adopted by a Federal agency in implementation of these regulations . . . and for which, therefore, [no other NEPA process] is required."[30] For agencies without much experience in risk assessment, and even for those where it is so routine that a resultant overconfidence in their own judgments about risks ensues, these exclusions are problematic.[31] They are essentially invisible, for one thing. No more systematic testing of such judgments, whether by CEQ or anyone else, has been forthcoming. Originally viewed as the means of excluding agency actions for which the prediction and analysis of environmental risk would be unwarranted,[32] these 'categorical exclusions' (CATEX) have taken on a whole new significance in a world of 'low probability/high-impact' risks. In this world, low probability/high-impact risks are often (and sometimes very quietly) handled via CATEX, much to our collective detriment.[33] The agency that takes a certain class of actions routinely, such as the permitting of environmentally risky extractive industries, may well seem blind to the fact that its PEISs, evaluating whole courses of action, normally do not consider specific actions in place with more determinate variables and concrete threats like particularized (human) agents.[34]

For risks to receive much deliberate consideration at all, they must be *recognized* as risks, whether under NEPA or some other decisional norm. And because agency actions will not always either *obviously* cause or not cause a significant impact on the quality of the human environment, since 1978 the CEQ rules have provided for a preliminary stage at which an agency can first decide whether it ought to prepare an EIS. At this preliminary stage the agency prepares what CEQ calls an 'environmental assessment' in order to determine at least roughly the significance of its action's impact on the quality of the human environment.[35] Unless the agency makes an affirmative "finding of no significant impact" (FONSI) once it commences this environmental assessment, it will likely have to complete an EIS.[36] But these findings of no significant impact have lately become one of the most heavily litigated and strategized NEPA compliance routines, especially in the Ninth Circuit. The factual uncertainties surrounding causation and causal attributions take on a particularly high significance in many of these cases. Thus, in this one regional circuit alone the go/no-go decisions in these preliminary NEPA stages have become enmeshed in a complicated web of conflicting precedents.

In *Blue Mountains Biodiversity Project v. Blackwood,*[37] the court confronted a record that raised 'substantial questions' as to the significance of the action's effects on the environment.[38] According to the panel, where the plaintiffs raise any substantial question about the risks underlying a decision to forego an EIS, that question itself gives the agency reason to prepare an EIS.[39] The action in question, a Forest Service decision to permit so-called salvage logging in the wake of an intense forest fire, garnered an EA that did not locate the specific sales, did not mention several other, similar actions in the vicinity, and that drew significant external critique while it was in draft form.[40] Of course, the plaintiffs' core objection to the Forest Service's analysis was the agency's judgment that the risks of logging after an intense forest fire were outweighed by its benefits.[41] And on such questions, the applied sciences of forestry were (and remain) far from definite or general conclusions.[42] Lying below the surface, in other words, was the bite-size of the Forest Service's salvage-logging decisions and the timing of those bites. And from the record it was hard to conclude that the service was intentionally engineering its NEPA compliance routines to enhance its own understanding of pest, fire, or other threats in the National Forest System.

In *National Parks & Conservation Ass'n v. Babbitt*,[43] the court heard a challenge to an EA that purported to find no significant impact from the agency's chosen course of action on the grounds that the actions in question—the permitting of more cruise ships into Glacier Bay—were of *unknown* significance and that the agency would manage the impacts adaptively as events unfolded.[44] According to the panel, an agency's FONSI could not be predicated on some a lack of "currently available information" because that is precisely the time to prepare a full EIS.[45] "Uncertainty," the court insisted, is no condition to leave unremedied by foregoing an EIS.[46] The problem with this reasoning is the tenuous relationship between NEPA processes—of whatever scope, duration, or depth—and the reduction of uncertainty. Nothing about the preparation of an EIS magically creates information or otherwise reduces uncertainty.[47] The better gauge of an agency's capacity to reduce uncertainty about the environment is the absolute *availability* of information thereon, and as the panel in *Babbitt* almost certainly knew, the Park Service had little choice but to learn about the effects of its 'vessel management plan' by implementing that very plan.[48] Thus, if the plan and its consequent actions were held up pending the completion of the necessary EIS, nothing about vessel traffic in Glacier Bay would necessarily grow *less* uncertain. The Ninth Circuit's ruling, in short, did nothing necessarily to match bite-size with available information.

Finally, in *Ocean Advocates v. Army Corps of Engineers*,[49] the court heard a challenge to the agency's decision to forego an EIS in connection with a new dock at an oil refinery in the Puget Sound.[50] The dock was proposed as an 'expansion,' but also as a technological enhancement of an existing facility—a facility located in a particularly sensitive environment.[51] Without knowing the likely future volumes of traffic through the facility, however, it was impossible to quantify precisely how oil spill risks would shift under the scenarios being considered.[52] Again, the Ninth Circuit panel hearing the case decided that an EIS was required because the agency's reasons for foregoing the EIS (and instead preparing an EA/FONSI) included its own uncertainty about the probable effects of the proposed action in the environment.[53]

Note that in each of these examples the agency's explicit acknowledgement of uncertainty, whether about the scope or intensity of the effects in question, helped the court to hold that a full EIS—and not an abbreviated substitute—was required. Now, on the one hand, this is somewhat curious. The CEQ rules provide that a complete EIS, where such uncertainties characterize the decision, must make explicit any uncertainty that is material to the decisions being made and that the duty to do so is central to the EIS itself.[54] Yet these same rules make no such provision for EAs or FONSIs, stating only that an EA shall "[b]riefly provide sufficient evidence and analysis for determining whether to prepare an [EIS] or a [FONSI]."[55] The Ninth Circuit derived this duty entirely from its own precedents requiring NEPA agencies to take a 'hard look' at their actions and, derivatively, to take that same hard look at any issues raised by other participants in its NEPA processes.[56] A regional circuit's colloquially verbalized doctrine (a doctrine not adopted, for the most part, in the other courts of appeal), however, is rather more ad hoc and imprecisely tailored to such complex managerial questions as arise in NEPA compliance than is a code that explicitly balances the disparate variables of risk management mentioned above.[57] On the other hand, these outcomes in the Ninth Circuit seem to resonate with NEPA's deeper concerns for overall rationality and, in particular, the assessment, disclosure, and mitigation of environmental risks.[58] The puzzle, of course, is who must or may—and according to what standards they shall—resolve the priority questions inherent in quantifying and/or qualifying such risks. If the core problem is as I have argued—matching bite-size in NEPA compliance routines with the information that is or may become available—then it seems that one regional circuit's precedents are a weak catalyst at most.

It is fairly self-evident that broader NEPA bites like PEISs represent a *possibility* of NEPA compliance for whole streams of agency activity because, in at least some contexts, an action agency will be able to 'tier' its more defined and/or geographically delimited activities from a completed PEIS and proceed operationally with a quite abbreviated NEPA process for any covered 'actions' so tiered.[59] Indeed, some agencies like the Department of Interior have taken CEQ's encouragements to tier to heart.[60] NEPA compliance by this route is only a possibility, though, not a guarantee. The unpredictable nature of NEPA litigation and the fact that some judges are less deferential than others as to the agency's scale/type selections render any cost-benefit compliance calculus herein inherently tentative.[61] In some cases the courts defer to obviously imperfect tiering decisions,[62] while in other cases they reject them outright.[63] A certain ambiguity affecting the rules enters this calculus as well in that agencies seem to have a duty,

in even a standard EIS, to analyze all 'connected,' 'cumulative,' and 'similar' proposed actions.[64] Agencies have found great variability in what courts believe to
be connected, cumulative, and/or similar proposed actions within the meaning
of this particular CEQ rule.[65] And because most agencies have the authority to
structure and denominate their 'actions' and the decisions leading thereto
according to their own (bureaucratic) priorities, the suspicion in scaling cases
of many different kinds is always that the NEPA bites are being sized in order to
minimize the deliberative costs thereof.[66]

Whether it is a more *cause*-centric or more *action*-centric set of boundaries in
scoping, choosing the right boundaries for analysis will never be a formulaic exercise. But it could be more deliberately engineered if NEPA compliance were
managed more like a network. As after-the-fact scrutiny of the Department of
Interior's deepwater drilling NEPA processes revealed following the Deepwater
Horizon tragedy, agencies can expertly 'tier' their NEPA processes from very
general PEISs to more specific EISs to still more specific EA/FONSIs and still more
specific CATEXs—and yet completely miss the threats that, in retrospect, are
serious and unavoidable.[67] Telescopic approaches to risk, like the Department of
Interior's, tend to emphasize the manageability of hazards at broad scales and the
unpredictability of the same hazards at smaller scales. In each, the result is the
same: the risk is minimized or occluded. Of course, whatever the scale of the
chosen NEPA document/process—whether it is a programmatic EIS or an EA/
FONSI—NEPA veterans know that the timing and scope of the NEPA routine
must match the needs of the decision maker or else NEPA processes will be beside
the point at best.[68] Those who are engineering the NEPA network must scale the
bite sizes to fit both the agency's preexisting organizational scheme *and* the informational landscape that might be used to enhance the agency's awareness of risks.
Quite simply, the needle to be threaded is leaving agencies enough discretion to
manage their limited resources for adequate risk analyses, while at the same time
checking those same agencies off of their likely tendencies to align risk assessment
and risk management with their existing priorities.[69] If simple deference to agencies would do, this would not be that big of an issue. But agencies are subject to
many kinds of failure, some of which ought to be remedied by judicial review.

In *Kern v. Bureau of Land Management*,[70] the agency had completed an EIS with
only a cursory discussion of a particular risk (the spread of an invasive fungus
lethal to an ecologically important tree species) that was regional in scope and had
also begun to use that EIS to confine the analyses it was preparing in connected
EAs.[71] In the EIS, the agency referenced a set of guidelines for controlling the

fungus, guidelines that had never been the subject of any NEPA or other external process but that, through such reference, confined the scope of analysis to the set of proposed actions (timber sales) then being considered.[72] In the EA completed by the time of litigation, the tiering had rendered the analysis of the problem even more cursory—excluding, in essence, most of the problem and the variables comprising it from view even within the project area for which the EA had been prepared.[73] The causes of *phytophthora lateralis*'s spread are much more complex than the Bureau of Land Management (BLM) apparently had the resources to capture.[74] The panel's opinion was remorselessly critical, though, holding that BLM had failed to prepare an adequately inclusive EIS, an adequately focused EA, or an adequately considered risk mitigation plan as a basis for tiering.[75] The agency had, in short, completely neglected the risk problem it was facing—in large part because it had decided that the bite it was taking was relatively insignificant, notwithstanding several contrary indications in the record.[76]

This problem of scaling in NEPA analyses seems to be growing in significance, especially as we better understand the limits of our own predictive tools. Environmental modeling—where data gaps force us to substitute predictive probabilities—has progressed significantly, but still disappointingly, over the last several decades.[77] Even an agency that routinely and diligently engages in NEPA 'scoping' will remain the captive of the information to which it has real time access.[78] And as a NEPA task force concluded in 2003, while more agencies are utilizing programmatic NEPA documents, as the "scope expands, cumulative effects become more complex, solutions to problems affect multiple agencies, and information sharing becomes essential."[79] Expanding the scope and scale of risk analysis inevitably means expanding the circle of agents whose (own) information must be tapped and incorporated into the enterprise. Thus, communications and communication failures will become increasingly important to NEPA. Yet the capacity to gather, process, integrate, and timely deliver decisionally relevant information inevitably turns on the agency's technological and managerial acumen—matters on which courts are increasingly reluctant to review or revise agency judgments.[80] The courts, in short, lack the means to force agencies to improve their NEPA deliberations; all they can do is set aside flawed judgments that happen to reach them for review. And in doing so courts may well be mistaking their familiarity with the case before them in all of its living color with a wider or deeper understanding of NEPA compliance as a whole.[81] Courts like the Ninth Circuit may hear a disproportionate share of NEPA cases, but theirs is still an unquestionably biased and unrepresentative selection of the wider NEPA universe affected by their decisions.

This mismatch between our NEPA ambitions and legal institutions stems directly from NEPA as an expression of both our *consequentialist* hopes for environmental protection and our path-dependent laws that aim for environmental quality today. Oddly, as our chosen means to more collective rationality in governance of ourselves and our environment, NEPA demands what our legal system is not adapted to deliver: a network, rather than a series of mostly disconnected and episodic parts. NEPA speaks of a single 'Federal Government,'[82] which is actually comprised of hundreds of distinct, managerially isolated 'agencies.'[83] It speaks of a single 'environment'[84] when what we actually experience are particular places, times, resources, events, etc. We ache to understand how society's myriad and interconnected *interventions* in nature will ramify throughout its complex systems, even as we deny ourselves the necessary tools for doing so by misapprehending the structure of causality in nature's complex systems.

Quite simply, our 'common sense' notions of causation still dominate our legal system (including NEPA and its jurisprudence) and distract us from the very endeavors to which they are supposed to yield for our own good: the more 'scientific,' systemic study of cause and effect.[85] These common sense notions conceive of cause-to-effect in terms of discrete interventions in nature, intentional or at least volitional, that thereafter produce a recognizable 'effect' or outcome in such 'constant conjunction' and 'regular sequence' that the two may be rightly joined by some univocal generalization.[86] Yet, as John Stuart Mill argued in his *A System of Logic*, if we are to have any hope of deriving such generalizations that withstand scrutiny, it can only come from carefully designed and executed processes of observation, hypothesis, and experimentation.[87] Mill wrote of the kinds of strictures that our practices of science now observe—not of the ways in which our legal traditions establish causation and fault.[88] And for most decisions in the real world, Mill's strictures are simply not practical.[89]

With NEPA's purposes and goals in mind, the question is thus: how is NEPA affecting the 'Federal Government'[90] as a whole? NEPA might be imagined as a pyramid with a vast universe of excluded agency actions (consciously or otherwise) at the base (the CATEX layer), a far smaller collection of agency actions processed as some form of EA—be it an EA/FONSI or some other EA (the EA layer)—with a top layer of EISs and perhaps even a tiny peak of programmatic EISs. This pyramid metaphor captures the relative numerosity of each NEPA action type in today's world, but it would be highly misleading to take the metaphor much further. Pyramids are highly structured, consciously and tightly engineered, and can last for millennia. The NEPA pyramid is nothing of the sort.

While CEQ has played a big role in this pyramid's construction, it could hardly be called the architect or engineer.[91] For the courts and NEPA plaintiffs have played at least as much of a role in constructing our NEPA pyramid and that pyramid is, by most accounts, aging poorly.

In fact, if the metric is its utility to our environmental decision making more generally, NEPA seems to be declining rather than improving in its overall performance over time. Collectively, our deliberations with one another are increasingly anchored to *compliance* with a set of legal norms that NEPA itself did not specify and that, instead, are under constant but uneven development by our processes of common law adjudication and quasi-administration by CEQ. What they should instead be anchored to is our collective (in)capacities to recognize and appropriately appreciate cause and effect in the environment. These capacities, we have learned, are at their sharpest when we find well-ordered means for collaborating and jointly producing the predictions and projections that can, if done well, reduce overall uncertainties about natural systems and how they will respond to our stresses.[92] For all its preset routines and standardized outputs, though, NEPA has yet to gel as an overall system or network that better enables such forms of collaboration across our institutional and habitual boundaries. This is in part because NEPA routines are themselves so atomizing: an action agency has an artificially particular 'action,' a defined 'purpose,' a set of 'alternatives' from which it must choose,[93] a set sequence of NEPA process steps,[94] and a familiar gauntlet to run in the event the NEPA compliance routine is challenged in federal court. But it is also partly because CEQ has not been able to cultivate a sense among NEPA agents that they are all constituents of the same system. NEPA is an emergent, engineered system meant to improve government decision making and not just to disclose the risks inherent therein.[95] Part II advances an issue-oriented interpretation of NEPA that would nudge its elements toward that end.

II. RECOGNIZING A SYSTEM: AN ISSUE-ORIENTED INTERPRETATION OF NEPA PRACTICE TODAY

Ecosystem management has remained on the periphery for so long, in part because it requires constant adaptation, structured learning, and a host of cross-organizational relationships that most administrative agencies simply cannot sustain.[96] In theory, NEPA should be a means to the end of ecosystem management; NEPA should be prompting federal natural resource managers to plan and operate modestly, adaptively, and systemically.[97] In practice, NEPA seems to be prompting such choices much too infrequently. Part II argues that this breakdown consists in each of the major recurring issues in NEPA practice today. There is a short list of such issues—of which CEQ personnel and other NEPA veterans are acutely aware and by which they are constantly confounded. They break down into the following three general categories.

First, NEPA practitioners are constantly having to gauge the *availability* (specifically, the *relative* availability) of usable information. NEPA compliance is ultimately a matter of searching out, assembling, and then assessing *available information* as the CEQ rules and the NEPA precedents define it. Indeed, whether a required NEPA routine has been as robust as necessary is mostly a matter of what information was deemed available to the process at the time of execution. Second, NEPA analysts are constantly confronting (1) the causal indeterminacies of our world and the ubiquity of multicausal phenomena, and (2) the practical impossibility of establishing very many causal necessities in the world as we know it. Our consequentialism in the pursuit of environmental quality requires us to seek out, isolate, and apprehend what is practically impossible to apprehend most of the time: the necessary and/or sufficient causes of environmental disturbance, damage, and depletion. Indeed, in the language of statutes like NEPA, the Endangered Species Act (ESA), and others, our naïve notions of 'cumulative' and/or 'indirect' effects help us mostly to ignore these epistemic challenges and sweep all possible causes—great and small—up into the same legal bins. Bounding the analyses that are productively carried out under such laws and separating them from those that are not worth what they entail is a constant struggle. Finally, NEPA practitioners know that each of the issues arising out of the foregoing basic challenges is under constant development by and within two separate, largely antagonistic institutions: courts and administrative agencies. Their institutional rivalry is old, entrenched, and unchangeable, except perhaps at the margins. Navigating this rivalry, therefore, is a big part of any strategy for updating and enhancing NEPA. Part II takes each of these issue sets in turn.

Uncertainty is a pervasive, all-encompassing condition of environmental protection, whatever form the protective effort(s) takes.[98] Uncertainty has many sources, only some of which are eliminable. Perversely, though, NEPA and other similar statutes (such as the ESA) have put a premium on the sources of uncertainty that are practically ineradicable, because of how often they allow some predicate action to be stalled or even permanently enjoined while some incorrigible uncertainty preoccupies the would-be actors. Indeed, if environmental protection takes the (indirect) form of data gathering, processing, dissemination, and analysis—as in NEPA—then pervasive uncertainty itself becomes the foundational, permanent condition that cannot be ignored or overcome, but rather must be understood and managed. NEPA law now treats information availability in a curious way, though.[99]

Consider the following context within the Ninth Circuit. In *Lands Council v. Powell*,[100] the Forest Service had proposed a logging project, coupled with road reconstruction and other restorative activities, in order to carry out a wider watershed protection effort in the Idaho Panhandle National Forest.[101] The real issue was the information the service used to conclude that its quid-pro-quo would actually improve aquatic habitat. When questioned about its study of past logging in the project area and the cumulative effects thereof, for example, the Forest Service simply allowed that past "timber harvests have contributed to the environmental problems in the Project area," but did not include any more specific or detailed analysis of past logging practices in its final EIS.[102] Indeed, among the several defects the court noted in the service's NEPA process, the lack of timely and specific information connecting the proposed actions with their probable future consequences explains virtually the entire case.[103] The panel remanded the EIS to the service with instructions to cure the informational deficits.[104]

Conversely, in *League of Wilderness Defenders v. Allen*,[105] a different panel of the Ninth Circuit—a mere five years later—found most of the same 'defects' to be harmless NEPA error.[106] The court in *Allen* confronted a Forest Service proposal to log an area of 'late successional' forest (an area with large, mature trees) in the hopes of reducing the risk of catastrophic wildfire.[107] The service could not quantify the risks at issue, nor could it quantify the likelihood that its proposed actions would in fact mitigate those risks. The *Allen* court found this particular data gap unavoidable and immaterial to the service's overall NEPA compliance. The totality of Ninth Circuit precedent, in short, offers little guidance to the agency official who must decide whether s/he anticipates the 'highest deference' a court can pay to an agency in its "technical analyses and judgments within its area of expertise"[108] or instead the court's own 'de novo' review of the risks being weighed and the informational bases being employed.[109] A single circuit, that is, has turned data (un)availability—or, at the least, an agency's judgments thereon—into an exercise in hair-splitting with essentially conflicting signals.

A reviewing court that faults an agency's NEPA process for perceived data deficits must essentially ignore its own inability to judge the (relative) availability of information.[110] This is a general problem in environmental—if not in all of administrative—law, properly defined. The availability of relevant information is precisely what expertise empowers one to judge.[111] Courts might even be accused of encouraging informational excesses in their routine nudges of agencies toward ever greater informational awareness.[112] The majority in *Allen* took a deferential view of the service's confessed failure to utilize one particular study on point because that study had been completed after the administrative record had been closed.[113] In other similar cases, the very same circuit has been much less forgiving.[114] CEQ, for its part, does nothing to define with particularity or purpose what constitutes 'available' as opposed to 'unavailable' information for NEPA's sake.[115] Of course, a great deal of critically important information is obtainable in some sense; it is normally a function of how much one is willing to do to get it. Indeed, risk managers are constantly having to value information (without having it) in order to invest rationally in their searches.[116] The public law literature about these dynamics tends to focus on certain principal/agent problems,[117] but a deeper trouble is afoot. If relevant information is somewhere possessed but cannot be timely found, society as a whole seems worse off. No matter who is the principal and who is the agent, better, faster, cheaper information helps all parts of a society. And NEPA law does little more than accentuate our informational failures by amplifying the legal consequences thereof. What NEPA should instead do is transition its own basic units into the raw material of a wider, more intentionally engineered cognitive network of risk managers.[118] If NEPA law were doing the latter, it would reward the conscious and purposeful completion of NEPA routines as if they were also subroutines within a wider, more continuous system. But NEPA law does nothing of the sort today. NEPA law does not even conceive of NEPA itself as what could be called a *joint cognitive system*, but it should—as Parts III and IV will argue.

One particularly troubling type of NEPA information, though, has always been the causal necessities and/or sufficiencies connecting a proposed agency action with its expected effects or consequences. A cause is *necessary* just in case it is one of a set of conditions jointly sufficient for the production of the consequence.[119] A cause is *sufficient* just in case its consequence follows from it without variation.[120] Yet science and philosophy both tell us that establishing generalizations about necessary and sufficient causes is and will remain an extremely costly, provisional, and permanent endeavor wherein the generalizations we do establish remain subjective to their core.[121] And in the absence of such shifting scientific achievements, the legal system is left with incomplete, suggestive insights that constitute evidence of causal relationships and nothing more. It is just this sort of evidence that NEPA veterans must use in choosing their NEPA compliance routines, though. And with a variety of different types of evidence, different preferred defaults, and usually more conflicting evidence than can be sorted out, NEPA actors often make inferences about likely outcomes in the world that appear implausible (or worse) to others.

The CEQ rules provide that the 'effects' of an action include "[d]irect effects, which are caused by the action and occur at the same time and place," as well as "[i]ndirect effects, which are caused by the action and are later in time or farther removed in distance, but are still reasonably foreseeable."[122] Add to this a duty to weigh and consider the 'cumulative impacts' of an action, *i.e.*, "the incremental impact of the action when added to other past, present, and reasonably foreseeable future actions regardless of what agency (Federal or non-Federal) or person undertakes such other actions,"[123] and a NEPA analysis can expand very quickly in depth and breadth. The puzzling thing is, however, that the CEQ rules nowhere specify *when* or *to what degree* cumulative impacts must enter into particular NEPA bites. Indeed, the only context in which cumulative impacts *must by law* be raised in a NEPA process are in those EISs that consider other, related (proposed) actions, the consequences of which are, in some sense, additive with those of the actions under consideration.[124] In 1997, CEQ released a handbook titled "Considering Cumulative Effects Under the National Environmental Policy Act,"[125] and yet never specified the contours of this duty in that 'guidance' either. Today, as climate change enters into more NEPA routines, the relationships between discrete agency actions and globally-scaled processes of environmental disturbance become a matter urgently in need of legal development.[126]

The Supreme Court has twice held that effects bearing no "reasonably close causal relationship" with the proposed action in question need not enter into a NEPA analysis.[127] But what effects bear this 'reasonably' close causal connection? This determination necessarily arises in *every* NEPA compliance routine; even if it is not the 'direct' but rather only the 'indirect' or 'cumulative' consequences of actions which are potentially 'significant' within the meaning of NEPA § 102(2)(C), it still raises the possibility that the proposed action should not be categorically excluded nor found to have no significant impact.[128] Should the severity or gravity of the possible effect(s) matter or should it be gauged solely on the basis of the expected frequency thereof? NEPA precedents have mushroomed in their complexity surrounding questions of this kind and the answers are beginning to vary significantly.

The CEQ rules state that EISs must consider "reasonably foreseeable adverse impacts."[129] But these rules are gradually receding in their influence behind a vast mountain of federal court precedents applying what the courts call the 'rule of reason' on the possibilities that are 'reasonably foreseeable.' These holdings vary subtly, but significantly, from circuit to circuit and sometimes even district to district.[130] Perhaps most importantly for our purposes is the fact that one bench in particular, the Ninth Circuit, has regularly stretched the rule of reason to demand the further analysis of risks that action agencies (and other courts) routinely ignore.[131] Nowhere does the so-called rule of reason in the Ninth Circuit contrast more vividly with the law CEQ has enacted than in the procedural requirements for agencies applying a CATEX as their NEPA routine.

In *California v. Norton*,[132] the Department of Interior had applied an established CATEX for certain oil leasing and permitting decisions in suspending a class of leases (so that they might be prolonged in effect).[133] The State of California sued and argued that an EIS was necessary or, at the very least an EA was, given the gravity of the possible effects of drilling off its coasts.[134] The district court concluded that the agency had not provided an adequately reasoned explanation of its CATEX application and the Ninth Circuit affirmed.[135] The panel held that the agency, in applying its CATEX, failed adequately to *document* its CATEX decision.[136] "It is difficult for a reviewing court to determine if the application of an exclusion is arbitrary and capricious where there is no contemporaneous documentation to show that the agency considered the environmental consequences of its action and decided to apply a categorical exclusion to the facts of a particular decision."[137] With that, the Ninth Circuit established the practical equivalent of a legal rule—at least within the nine western states comprising the Ninth Circuit—requiring agencies to document and substantiate their CATEX decisions as they render them.[138] While this reasoning is rather familiar in standard of review doctrine, the court was walking a fine line given the utter lack of any CATEX documentation requirement in the CEQ rules.[139] It is not the courts of appeal, after all, that budget an agency's NEPA compliance appropriations. And a holding like that in *Norton* can easily up-end whatever planning has been done thereon.[140]

As this part has argued, the issues that now characterize NEPA practice are imbricated in a distinctly antagonistic institutional setting: the entrenched rivalry of courts and administrative agencies. Pleasant metaphors of dialogue and the like do not capture the dynamics at work between the repeat-players of NEPA practice within action agencies and the plaintiffs who bring them to federal court. Indeed, the Ninth Circuit's NEPA jurisprudence has long exemplified the challenges that agencies face in the form of our pluralistic, noncentralized judiciary. Part III offers some modest steps to CEQ in the hopes that it can become more of a conscious and intentional NEPA engineer.

III. NEPA AND THE COGNITIVE ECOSYSTEM:
PIECES INTO WHOLES (AND VICE VERSA)

Not long after it became evident that species imperiled by global climate disruption would be listed pursuant to the ESA and thus necessitate findings that federal agency actions did not further 'jeopardize' their continued existence, the ESA's administrators moved to 'clarify' the rules linking cause-to-effect in such findings.[141] The amended rules severed causal contact between agency action and any effects that were "manifested through global processes" that "[c]annot be reliably predicted or measured at the scale of a listed species' current range."[142] While ESA consultations should probably expand in scope from what they encompass today, this mismatch of broadly-scaled impacts to small-scale spans of control is not being solved in the ESA context either. An agency taking an 'action' that may or may not augment a global threat by some infinitesimally small increment will not care at all about global cause and effect management of such risks. Still, a first principle of adaptive, ecosystem-based management is a commitment to and, thus, the authority to engage in, *iterated decision making*.[143] Iterated decision making requires that "learning is both plausible and valuable" and that those in command have "the ability to change management direction in response to learning."[144] With the authority to govern natural systems like watersheds or species assemblages as radically dispersed as it has become, though, coherent iterated decision making has become the core political and managerial challenge. Quite simply, preliminary coordinative steps are too hard to take and too likely to become locked into place long after their flaws are manifest. Subsequent reconsideration and correction become practically impossible.[145] In this context, a systemic advantage goes to those actors who can avoid sinking too much into any particular choice and, less obviously, to those who can over-communicate the grounds of their choices today in hopes of better informing subsequent decision making.

CEQ's recent guidance on establishing, applying, and revising CATEXs holds real promise here. In the document, CEQ 'clarifies' its rules on the establishment, application, preparation of documentation for, and periodic review of CATEXs.[146] The document even recommended for the first time that the public involvement and transparency for which other NEPA routines are so famous be extended to this most cursory of NEPA routines, precisely because it is so ubiquitous.[147] Overall, what is most striking about the CEQ's CATEX guidance, though, is its deliberate conceptualization of CATEXs as a continuous routine wherein past subroutines are used to improve subsequent similar steps in the future. CEQ emphasizes that "documentation prepared when categorically excluding an action should be as concise as possible to avoid unnecessary delays and administrative burdens,"[148] while at the very same time declaring that "[past] FONSIs cannot be relied on as a basis for establishing a categorical exclusion unless the absence of significant environmental effects has been verified through credible monitoring of the implemented activity or other sources of corroborating information."[149] Finally, CEQ for the first time explicitly encouraged agencies to harness the power latent within already-completed NEPA routines, noting that past NEPA predictions constitute an invaluable source of information for current NEPA actors.[150] It urged NEPA actors to "consider information and records from other private and public entities, including other Federal agencies that experience with the actions covered in a proposed categorical exclusion" when they are substantiating and documenting the environmental effects of the subject category of actions.[151]

The Supreme Court has held that consequences which a NEPA actor cannot control—because, for example, it has insufficient authority to do so—need not enter into that particular agent's NEPA routines.[152] Rules of that sort make CATEXs that much more likely perhaps, but they do nothing to address the problems of scaling NEPA bites or of adequately detailing the norms of information availability in NEPA routines.[153] A CATEX that is adequately substantiated and widely communicated invites retrospective ground-truthing from a variety of actors possessed of a variety of perspectives and information. Over time, a series of such decisions could enable a certain kind of benchmarking.[154] When they bear on risks or causal relationships of wider and/or deeper significance, official 'findings' of the sort attract exactly the kind of attention that can lead to enhanced societal awareness of risk over time. If that happened, more risk judgments would be recognized and reviewed by more actors—actors who enjoy better access to information in the aggregate than any subset of them acting alone. Increasing the sheer sample size of these events is perhaps even more likely to improve societal awareness of risk than simply enlarging a handful of micro-managed events where risks are taken more seriously than anywhere else.[155] Doing so normalizes risk analysis by fitting it to the decisions most agencies make most often. In short, systematizing CATEX practice could improve our overall awareness of environmental risk more effectively than just expanding the scope of a short list of EISs or EA/FONSIs.

CEQ's recent guidance on CATEX procedures seems to envision a more *deliberately engineered* use of CATEXs—in that these (mostly cursory) NEPA analyses will (1) be more proximate to the risks actually being born, while (2) enabling more learning-by-monitoring—than the status quo.[156] CEQ even seems prepared to trade the appearance of panoptic awareness and considered judgment, i.e., the PEIS or EIS that soars in cost, girth, and publicity, for more modest reforms of the rough-and-ready tools that are in widest circulation already. Still, CEQ needs to choose its own tools more carefully. The reforms of CATEX practice mentioned have thus far come only in the same weak forms of 'law'—an informally adopted, not-to-be-codified, agency interpretation—which courts and others have now long struggled to recognize as a type of (legal) reason for action.[157] CEQ 'guidance' aiming to systematize CATEX procedures or 'cumulative impact analysis' across agencies and to eventually reap the benefits of common mode operations, *e.g.*, economics of scale, benchmarking, *etc.*, is a poor substitute for specific and definite legal rules thereon. Indeed, CEQ's guidance on CATEXs is arguably amending prior, long-standing guidance to the opposite effect: just the kind of agency action courts have struggled with the most.[158] CEQ must do more than 'clarify' the norms on NEPA's bite sizes; it must specify and order them.[159] The courts are no more likely to recognize or be appropriately guided by CEQ's informal interpretations than are the other myriad agencies that must interpret and apply NEPA in their own operations. This does not mean that these other actors will not notice or respond to such outputs; rather, it means that their responses will vary too much and that the systemic effects will be even less uniform than they might be under the system as it presently exists. To manage NEPA deliberately as a system, one must start from the right expectations of uniformity and standardization. CEQ must eventually reach a decision about its aging NEPA regulations and the amendments they demand for a more systemic approach to NEPA's cognitive ecosystem; it must implement that decision with legally binding rules that optimally standardize NEPA's bites.

IV. RECOMMENDATIONS

If we are to improve our assessment, communication, and management of ecosystem risks by way of NEPA processes, it will begin with our notions of risk themselves. At least since the philosopher David Hume wrote we have known that "there is no justification for regarding what has been observed to happen in the past as any sort of reliable guide to the future."[160] We have known, that is, that there are no *logical* foundations in inductive reasoning—reasoning from something specific to something general or from something in the past to something in the future. Hume argued that this was most assuredly true of cause and effect, specifically of our inferring a law or generalization from a mere sequence of events or occurrences that seemed regular in nature.[161] Critically and shrewdly, Hume also argued in his most mature philosophy that nothing about the world determines our notions of cause and effect so much as do our 'habits of mind,' 'sentiments,' and 'impressions.'[162] That is to say, our notions of causation ultimately reduce to our means of perceiving and conceptualizing the supposedly 'constant conjunction' and 'regular sequence' of causes and effects.[163] And that means that they are just as much the product of our own biases, information, and flawed judgments as they are of the environment we observe.

Of course, the real puzzle has been the practice of science. Our scientific knowledge vastly exceeds its observational bases, leading the philosopher C.D. Broad to have once called induction "the glory of science and the scandal of philosophy."[164] This is perhaps even a good and sufficient reason for philosophy to relax its attitude toward induction.[165] Because of our sciences' successes in prediction, we generally (happily) trade logical validity for reliability and/or predictive accuracy—in a word, for *probability*. It is highly probable that the force of gravity will continue to work later today even though it is not logically proven, whether from a lifetime of observations, Newtonian mechanics, or the general theory of relativity. Many of the issues natural resources professionals encounter have this same structure, too. For example, will logging in an infested area increase the probability of *phytophthora lateralis*'s spread to other, uninfested stands of Port Orford cedar? If recreational vehicles and other uncontrolled travel are likely to continue the spread of the pathogen in any event (if spread is causally overdetermined), what possible reason would a multiple-use agency have to ban logging in the areas concerned? A series of CATEXs on the question which were at least *explicit* about these factors of judgment would almost certainly be a better means for synthesizing a probabilistic approach to something like spread dynamics than a voluminous EIS consisting of the inconclusive information the science offers at any given moment.[166] Besides rendering the decision more transparent at its origin it is also likely to record for posterity what too often goes unrecorded—the real-time guesswork that substitutes for logically rigorous risk assessments.

Because real probabilities are unknowable without the assembly of dispersed information, our practical trouble is almost always that, for too many real-time decisions, we simply lack the necessary information—and rushed efforts to gather it can be a waste or worse.[167] Consequently, even our best probabilistic reasoning is often deeply flawed and it is usually never more so than when we must make predictions about other humans' behaviors.[168] Those who engineer NEPA compliance routines cannot change the fact that human behavior is unpredictable. But they can turn the necessity of flawed predictions into less of a failing by reimagining these predictions as a stream of data from which later decision makers may benefit. Those who try to fit data and automated data systems together with humans, for whom cognition is always a limiting factor, think of these systems of as joint cognitive systems. A noted principle of joint cognitive systems engineering (JCS) is known as the *law of demands*: whatever "makes work difficult, to a first approximation, is likely to make work hard for any JCS regardless of the composition of human and/or machine agents."[169] Cause and effect relationships are hard to apprehend no matter the scales of cognition we achieve. Thus, even richly staffed and sophisticated teams of personnel aided by the best digital technology money can buy will face insuperable obstacles when it comes to predicting broad-scale phenomena like markets, environmental harm, and the like. But that is exactly what a typical NEPA routine entails. And human judgment and behavior are notoriously interactive; people form preferences, make their decisions and plans, and generally conduct their lives in large part based on their own assessments of those around them and how they will react to behavior of various kinds. In NEPA's cognitive economy, the JCS engineer must turn to the problem of bite sizes and help the NEPA analyst revisit risks encountered at different scales—without simply repeating findings or conclusions that may have been justified at a different scale or an earlier time—but which, in the moment, ignore relevant differences of timing, scale, or knowledge.[170]

The human sciences' inability to provide us with much that is informative about these dynamics—much that is prospective and predictive—stems from a number of distinct failures. For example, 'common sense' foils social scientists in how they formulate their hypotheses about human behavior, how they gather and sort the evidence supposedly confirming or denying the hypothesis under review, and how they reformulate any subsequent hypotheses.[171] Common sense deals with every situation in its particulars and is always on hand. But common sense often tricks us into thinking we know more than we do. It helps us commit basic errors of inference about social systems, for example, when what we are really basing our inferences upon is our knowledge of individuals.[172] Social systems actually exhibit what social scientists call 'emergence,' an independent pattern and existence that cannot be predicted solely by recourse to its constituents.[173] NEPA's objectives implore us to replace our ignorance of such emergent phenomena with predictive knowledge, even while the questions NEPA raises can normally be answered in the present only by guesswork. Changing that basic reality will require leaps forward, which can only be taken with the aid of much more sophisticated techniques of studying, recording, and monitoring our social and natural systems over time.[174]

Yet if, as even Hume conceded, *similarity* is the key to successful prediction, then what NEPA routines need most of all are discriminating accounts of the relevantly similar parts of our recorded past to the possible futures which their interventions in the world are creating.[175] For example, as the sociologist Charles Perrow argued in his classic *Normal Accidents*, the coupling of complex technological systems can actually increase their risk of failure, precisely because of those systems' interactivities and the possibilities of cascading (or 'common mode') failures therein.[176] Perrow's tight coupling and the characteristic risks of bundling complex systems that are susceptible to common mode failures is a path in risk assessment that agencies should almost certainly pursue more often. The limitations of Perrow's thesis, of course, are that it does not specify the various 'modes' of complex systems *ex ante*. And that is what enhanced collaboration and inter-institutional exchange can bring to the table. The more communication among different teams that NEPA processes foster, the more likely NEPA is to improve our overall awareness of environmental risk. A NEPA engineer should seek to facilitate a cognitive economy wherein the threats to environmental quality, which our best available information alerts us to, become the focus of the decision makers who are most able to avoid or mitigate those threats. And that will require a more structured approach to 'tiering,' 'scoping' and the analysis of 'cumulative' impacts by interagency teams than that engendered by CEQ's permissive, highly-deferential rules that essentially ignore the problems of bite sizes today.[177]

Neither CEQ nor NEPA's veterans have done much to solve problems like this, but they are problems from which we can no longer afford to shrink. The ESA faces similar problems, as do many other environmental programs. Individually discrete 'actions' are the prompts by which we consider our environmentally disturbing, depleting, and destructive social fabric—a fabric from which discrete actions can only be disentangled through the most artificial of means, but which itself is not what our environmental laws or agents were designed to reorder. Thus, what we must do is transition these prompts and cues into the constant reminders that we desperately need capacities to identify and mitigate overall—ecosystemic—risks. NEPA § 102 also directs covered agencies to "utilize a systematic, interdisciplinary approach which will insure the integrated use of the natural and social sciences and the environmental design arts in planning and in decisionmaking which may have an impact on man's environment."[178] Reconceiving of our NEPA routines as pieces of a longer-term routine—as subroutines that animate a larger system of decision making on the basis of information that is always evolving—could go a long way toward improving both their individual and aggregate utilities. The Council on Environmental Quality is the only institutional actor that seems capable of doing any of this, though. Until it does so with more dispatch and purpose, NEPA will lag behind where we need it to be.

ACKNOWLEDGMENTS

Alison Glunt and Anna Leonenko provided outstanding research assistance.

NOTES

1. 40 C.F.R. § 1500.1(a) (2010).

2. *See, e.g.*, Natural Resources Def. Council v. Vermont Yankee Nuclear Power Corp., 435 U.S. 519, 558 (1978).

3. 42 U.S.C. § 4332(2)(C) (2006).

4. *Id.* § 4332(2)(C)(i)–(iv).

5. In 1977, President Carter ordered CEQ to transition what had been a set of loosely drafted 'guidelines' implementing NEPA § 102 into governing 'regulations' with which covered 'Federal agencies' were ordered to 'comply.' Exec. Order No. 11991, 40 C.F.R. § 1500.3 (1977). While the regulations explicitly provide that they are binding "except where compliance would be inconsistent with other statutory requirements," 40 C.F.R. § 1500.3, CEQ's authority to issue binding rules at all is nowhere provided in the text of NEPA and neither is the President's. Originally this might have meant that such rules lacked the force of law, *see* Thomas W. Merrill & Kathryn Tongue Watts, *Agency Rules with the Force of Law: The Original Convention*, 116 Harv. L. Rev. 467, 503–28 (2002), but the modern Supreme Court has put the matter in considerable doubt. *Id.* at 528–75. And ever since Andrus v. Sierra Club, 442 U.S. 347, 358 (1979), where the Court stated that "CEQ's interpretation of NEPA is entitled to substantial deference," the force of the NEPA rules themselves has rarely been questioned.

6. *See* National Environmental Policy Act, 43 Fed. Reg. 55978, 55979–80 (1978).

7. *See* Erik Hollnagel & David D. Woods, Joint Cognitive Systems 1 (2005).

8. *See id.* at 5.

9. *See* Gary Klein et al., Macrocognition, 18 Intelligent Sys. 81, 81–82 (2003).

10. The Federal Highway Administration, a leader in EIS production, once found that its average EIS took some 3.6 years to complete and that some took as long as twelve! U.S. Dep't of Transp., Federal Highway Admin., Evaluating the Performance of Environmental Streamlining: Development of a NEPA Baseline for Measuring Continuous Performance 4.1.1 (2001), *available at* http://www.environment.fhwa.dot.gov/strmlng/baseline/index.asp. Where it governs, "[o]ne of the undisputed strengths of NEPA, and the EIS process specifically, is that it provides the principal avenue for public information and consequent involve-

ment in the planning and decision-making processes of the federal government." H. Welles, *The CEQ NEPA Effectiveness Study: Learning from Our Past and Shaping Our Future*, in Environmental Policy and NEPA 193, 205 (Ray Clark & Larry Canter eds., 1997).

11. *See, e.g.*, Bradley C. Karkkainen, *Toward a Smarter NEPA: Monitoring and Managing Government Environmental Performance*, 102 Colum. L. Rev. 903, 918–22 (2002) [hereinafter Karkkainen, *Smarter NEPA*] (calling this the incentive to 'overstuff' an EIS with more and more of whatever information is on-hand). *Id.* at 922.

12. *See, e.g.*, Robert W. Adler, *In Defense of NEPA: The Case of the Legacy Parkway*, 26 J. Land Resources & Envtl. L. 297 (2006). When surveyed, most of the traditional groups from which NEPA plaintiffs emerge respond that NEPA's successes have come in "opening the decision-making process for public input...." Welles, *supra* note 10, at 206.

13. *Cf.* Bradley C. Karkkainen, *Bottlenecks and Baselines: Tackling Information Deficits in Environmental Regulation*, 86 Tex. L. Rev. 1409, 1431 (2008) [hereinafter Karkkainen, *Bottlenecks and Baselines*].

14. Karkkainen, *Smarter NEPA, supra* note 11, at 924.

15. In the finalized 1978 rules, CEQ maintained that its rules had "the threefold objective of less paperwork, less delay, and better decisions." National Environmental Policy Act-Regulations: Implementation of Procedural Provisions, 43 Fed. Reg. 55978–55979 (Nov. 29, 1978) "Ultimately, of course, it is not better documents but better decisions that count." 40 C.F.R. § 1500.1(c).

16. 40 C.F.R. § 1501.2 (2010).

17. *See generally*, Council on Env'tl Quality, Exec. Office of the President, Report Regarding the Minerals Management Service's: National Environmental Policy Act Policies, Practices, and Procedures as They Relate to Outer Continental Shelf Oil and Gas Exploration and Development (Aug. 16, 2010), *available at* http://ceq.hss.doe.gov/current_developments/docs/CEQ_Report_Reviewing_MMS_OCS_NEPA_Implementation.pdf. Moreover, courts are at their most deferential when allocative choices of the sort arise. *See also*, Eric Biber, *The Importance of Resource Allocation in Administrative Law*, 60 Admin. L. Rev. 1 (2008). *But cf. id.* at 23 ("Any time a court reviews an agency decision, the court is in some way interfering with agency resource allocation, and not just where a court compels an agency to take a particular action.").

18. In 1978, as CEQ was finalizing its rules, it introduced what is now known as the 'scoping procedure;' "an early and open process for determining the scope of issues to be addressed and for identifying the significant issues related to a proposed action." 40 C.F.R. § 1501.7. The concept of scoping, the Council noted, "was one of the innovations in the proposed regulations most uniformly praised by members of the public..." National Environmental Policy Act-Regulations: Implementation of Procedural Provisions, 43 Fed. Reg. 55978, 55982. Out of deference to action agencies, however, the Council ultimately decided to leave "important elements of scoping to agency discretion" by not formally specifying what a scoping procedure *must* consist in. *Id.*

19. *Id.*

20. *See* George J. Mannina, Jr., *NEPA at 40*, 39 Envtl. L. Rep. News & An. 10660, 10660–61 (2009) (noting that too many parties, especially action agencies themselves, do not taking the scoping process seriously enough).

21. *See* Forty Most Asked Questions Concerning CEQ's National Environmental Policy Act Regulations, 46 Fed. Reg. 18026, 18033 (March 23, 1981) (noting that tiering is appropriate where general discussions of risk can cover more specific ones and giving as an example plans for wide 'geographic areas' that will be implemented stepwise within that area); *see also* Guidance Regarding NEPA Regulations, 48 Fed. Reg. 34263, 34267 (July 28, 1983) [hereinafter Hill Memorandum]. In explaining tiering as a practical means by which to solve the problem of bites, CEQ observed that "where a Federal agency adopts a formal plan which will be executed *throughout a particular region*, and later proposes a specific activity to implement that plan in the same region, both actions need to be analyzed under NEPA...." *Id.*

22. *See infra* Part III.

23. *See, e.g.,* Carol Borgstrom, Integrating NEPA Into Long-Term Planning at DOE, 39 Envtl. L. Rep. News & An. 10642, 10642 (2009) (describing a small range of Dep't of Energy activities that typically merit a programmatic EIS and observing that "[p]reparing a PEIS takes considerable time, effort, and resources" and that some have cost $30 million and more).

24. NEPA rather naïvely assumes in its notion of agency 'action,' *see* 42 U.S.C. 4332(2)(C), that some *a priori* (or at least some externally given) definition of agency action will precede a covered agency's (stated) intentions, plans, or structured decision. That the agency itself must give form thereto, however, is the subject of many of the totemic early NEPA precedents—prior to the Supreme Court's decision in Strycker's Bay Neighborhood Ass'n v. Karlen, 444 U.S. 223 (1980). *See, e.g.,* Calvert Cliffs Coordinating Comm. v. United States Atomic Energy Comm. Inc., 449 F.2d 1109 (D.C. Cir. 1971); Hanley v. Kleindienst, 471 F.2d 823 (2d Cir. 1972); Natural Res. Defense Council, Inc. v. Morton, 458 F.2d 827 (D.C. Cir. 1972); Environmental Defense Fund v. Corps of Eng'rs of the United States Army, 492 F.2d 1123 (5th Cir. 1974). Even after CEQ's guidelines were in place and eventually became CEQ's *rules* the nature of the administrative action driving the NEPA process remained a key driver of NEPA disputes. *See e.g.,* Aeschliman v. United States Nuclear Regulatory Comm'n. 547 F.2d 622 (D.C. Cir. 1976), *rev'd,* Vermont Yankee Nuclear Power Corp. v. Natural Resources Defense Council Inc., 435 U.S. 519 (1978).

25. *See generally,* William D. Araiza, *In Praise of a Skeletal APA:* Norton v. Southern Utah Wilderness Alliance, *Judicial Remedies for Agency Inaction, and the Questionable Value of Amending the APA,* 56 Admin. L. Rev. 979 (2004); Lisa Shultz Bressman, *Judicial Review of Agency Inaction: An Arbitrariness Approach,* 79 N.Y.U. L. Rev. 1657 (2004).

26. Borgstrom, *supra* note 23, at 10642.

27. Lynton Keith Caldwell, The National Environmental Policy Act 13 (1998).

28. *See* Glen W. Suter II et al., Ecological Risk Assessment 3–4 (2d ed. 2007).

29. *See* Tim Snell & Richard Cowell, *Scoping In Environmental Impact Assessment: Balancing Efficiency with Precaution?,* 26 Envtl. Imp. Assessment Rev. 359, 365–66 (2006). How an individual or team assesses risks depends on its compartmentalization of expertise versus leadership/accountability, its grasp of uncertainty versus variability, and its default settings in the absence of convincing proof. *See* National Research Council, Science and Decisions: Advancing Risk Assessment 5–10 (2009).

30. National Research Council, *supra* note 29, at 19.

31. 40 C.F.R. § 1508.4.

32. *See* Ted Boling, *Making the Connection: NEPA Processes for National Environmental Policy*, 32 Wash. U. J. L. & Pol'y 313 (2010). For example, the Dep't of Interior's former Minerals Management Service, in permitting deepwater offshore oil drilling, applied a series of categorical exclusions by which the risks of a catastrophic blowout at depth in the western Gulf of Mexico were barely understood or calibrated. *See* National Commission on the BP Deepwater Horizon Oil Spill and Offshore Drilling: The National Environmental Policy Act and Outer Continental Shelf Oil and Gas Activities 29 (2010) (Staff Working Paper No. 12), *available at* http://www. oilspillcommission.gov/sites/default/files/documents/The%20National %20Environmental%20Policy%20Act%20and%20Outer%20Continental%20Shelf%20Oil %20and%20Gas%20Activities.pdf ("Because the idea that drilling a deepwater well on the Outer Continental Shelf is the sort of action that 'does not have a significant effect on the human environment' appears, with hindsight, to be inherently illogical, the question arises whether BLM properly applied the CE procedures."). Likewise, an agency that little acknowledges the environmental risks which it encounters is unlikely to weigh those risks appropriately. *See* Suter II et al., *supra* note 28, at 5 (noting that risk assessment can be valued only as a tool for decision making and that decision makers must inevitably prioritize the hazards they balance).

33. *See* Boling, *supra* note 32, at 322–25.

34. *See* National Commission on the BP Deepwater Horizon Oil Spill and Offshore Drilling, Report to the President: Deep Water: The Gulf Oil Disaster and the Future of Offshore Drilling 55–85 (2011) [hereinafter Report to the President]. The original categorical exclusion for deep water drilling in the Gulf was Congress's doing in 1978. *Id.* at 62.

35. *See* Jamison E. Colburn, *Necessarily Unpredictable?: Oil Spill Risks Beyond the Horizon*, 30 Miss. C. L. Rev. 307 (2011).

36. 40 C.F.R. 1508.9.

37. *See, e.g.*, Sierra Club v. Watkins, 808 F. Supp. 852 (D.D.C. 1991).

38. 161 F.3d 1208 (9th Cir. 1998), *cert.* denied, 527 U.S. 1003 (1999).

39. *Id.* at 1212.

40. *Id.* at 1213–14. "An EIS is required of an agency in order that it explore, more thoroughly than an EA, the environmental consequences of a proposed action whenever 'substantial questions are raised as to whether a project *may* cause significant [environmental] degradation.'" *Id.* at 1216 (quoting Idaho Sporting Cong. v. Thomas, 137 F.3d 1146, 1149 (9th Cir. 1998), *overruled* by Land Council v. McNair, 537 F.3d 981 (9th Cir. 2008). *Idaho Sporting Congress* is arguably the source of the 'substantial questions' doctrine in the Ninth Circuit. There, the court concluded that the agency's decision to rely on its own chosen expert to the exclusion of plaintiffs' proffered evidence as to the significance of effects from its action showed that it had not taken the requisite 'hard look' at its action required by NEPA (Ninth Circuit precedent on NEPA, in particular). *Idaho Sporting Cong.*, 137. F.3d at 1154.

41. *Blue Mountains Biodiversity Project*, 161 F.3d at 1210–12.

42. Between the time the appeal was heard and the district court's denial of injunctive relief, the Forest Service allowed more than half of the project area in question to be logged. *Id.* at 1215. The appeal and the resulting remand forcing the Forest Service to replace its EA with an EIS was, then, aimed at a fraction of the original action in question and the 'searching environmental review' of which the panel spoke, *Id.* at 1216, was surely not intended for just that particular project area.

43. *But see* David B. Lindenmayer et al., *Salvage Logging and its Ecological Consequences* (2008) (collecting evidence that salvage logging is generally very costly and of wide ecological significance).

44. 241 F.3d 722 (9th Cir. 2001).

45. *Id.* at 731–36.

46. *Id.* at 737. "Preparation of an EIS is mandated where uncertainty may be resolved by further collection of data or where the collection of such data may prevent 'speculation on potential . . . effects. The purpose of an EIS is to obviate the need for speculation. . . .'" *Id.* at 732.

47. *Id.* at 733.

48. This is an aspect of environmental decision making that has been obvious from virtually the inception of modern environmental/natural resources law: additional search may simply be the compounding of decisional costs with little or no real improvement of one's decisional position. *See, e.g.*, Marcia R. Gelpe & A. Dan Tarlock, *The Uses of Scientific Information in Environmental Decisionmaking*, 48 S. Cal. L. Rev. 371 (1973).

49. *Cf. Babbitt*, 241 F.3d at 735 (quoting discussion in EA that "[f]ollow up research and monitoring w[ould] be essential to define humpback whale use patterns in Glacier Bay resulting from [the chosen] alternative"). In a footnote, the court rather argumentatively (and naïvely) suggested that the service would need only to "determine the current effects of vessel traffic, and extrapolate or project from that data the effects of increased traffic," *id.* at 733 n.12, as if the agency was certain to have the necessary baseline data from which to determine the 'effects' of then-current vessel traffic levels, as well as the resources needed to construct useful extrapolative models.

50. 402 F.3d 846 (9th Cir. 2005).

51. *Id.* at 855.

52. *Id.* at 855–57.

53. *Id.* at 855–56. There were, moreover, credible assertions by the facility operator and the agency that throughput at the facility would increase whether the newer dock was constructed or not. *Id.* at 857.

54. *See Id.* at 867–68.

55. The CEQ rule requires that:
[i]f the information relevant to reasonably foreseeable significant adverse impacts cannot be obtained because the overall costs of obtaining it are exorbitant or the means to obtain it are not known, the agency shall include within the environmental impact statement: (1) a statement that such information is incomplete or unavailable; (2) a statement of the relevance of the incomplete or unavailable information to evaluating reasonably foreseeable significant adverse impacts on the human environment; (3) a summary of existing credible scientific evidence which is relevant to evaluating the reasonably foreseeable significant adverse impacts on the human environment, and (4) the agency's evaluation of such impacts based upon theoretical approaches or research methods generally accepted in the scientific community. *For the purposes of this section, reasonably foreseeable includes impacts which have catastrophic consequences, even if their probability of occurrence is low, provided that the analysis of the impacts is supported by credible scientific evidence, is not based on pure conjecture, and is within the rule of reason.*

56. 40 C.F.R. § 1502.22(b) (2009) (emphasis added). No effort has ever been mounted by CEQ to define 'exorbitant' information costs.

57. 40 C.F.R. § 1508.9(a)(1).

58. *Cf.* Sierra Club v. Bosworth, 510 F.3d 1016, 1018 (9th Cir. 2007) (stating that an agency must take a 'hard look' at its actions and quoting the CEQ rules to the effect that, if the action "*may* have a significant effect upon the . . . environment, an EIS must be prepared," by adding the emphasis and citing *Babbitt*, 241 F.3d at 730). This particular incantation of the 'hard look' that the Ninth Circuit expects of NEPA agencies is not without irony: the Supreme Court itself has said in substantially the same contexts that the standard of review in NEPA cases is a narrow and constrained. *See, e.g.*, Kleppe v. Sierra Club, 427 U.S. 390, 413–15 (1976) (concluding that the agency's decision to forego a broader-scale EIS, though not the only plausible decision, was not arbitrary); Marsh v. Oregon Natural Res. Council, 490 U.S. 360, 370 (1989) (upholding agency decision to forego supplementing completed EIS with newly acquired information by applying a 'rule of reason' in reviewing agency's judgment).

59. *Cf.* National Research Council, *supra* note 29, at 240–56 (2009) (concluding that a systematic decision-making framework must structure risk assessment by federal agencies, propounding that framework, and arguing that the biggest challenge in such a project is adequately pursuing the multiple objectives agencies always have in performing and using risk assessments); Frederick Schauer, *Do Cases Make Bad Law?*, 73 U. Chi. L. Rev. 883 (2006) (arguing that predictable cognitive errors are common in judges' use of individualized cases to make general rules). This is not to argue, however, that adjudication of particularized cases has no role to play in NEPA law. *See* Jeffrey J. Rachlinski, *Bottom-Up Versus Top-Down Lawmaking*, 73 U. Chi. L. Rev. 933 (2006); *see infra* Part III.

60. *See e.g.*, Wendy B. Davis, *The Fox is Guarding the Henhouse: Enhancing the Role of EPA in FONSI Determinations Pursuant to NEPA*, 39 Akron L. Rev. 35, 41–52 (2006) (arguing that findings of no significant impact are often arbitrary and that several Ninth Circuit decisions like those reviewed here aim to attack that basic shortcoming).

61. *See, e.g.*, Ryan M. Seidemann & James G. Wilkins, Blanco v. Burton: *What Did We Learn from Louisiana's Recent OCS Challenge?*, 25 Pace Envtl. L. Rev. 393 (2008) (arguing that the Dep't of Interior's 'tiering' practices rewarded the unwarranted dismissal of certain risks in outer continental shelf oil and gas development programs); *see* Nevada v. Dep't of Energy, 457 F.3d 78, 91–92 (D.C. Cir. 2006) ("The decision whether to prepare a programmatic EIS is committed to the agency's discretion."); Fund for Animals v. Kempthorne, 538 F.3d 124, 137–39 (2d Cir. 2008) (upholding programmatic impact statement on wildlife extermination plan against challenge that site-specific exterminations were not going to be assessed individually); Northern Alaska Envtl. Ctr. v. Kempthorne, 457 F.3d 969 (9th Cir. 2006) (upholding programmatic impact statement for whole oil and gas leasing program over challenges that agency's commitment to analyze site-specific actions in the future left risks understudied); North Alaska Envtl. Ctr. v. Lujan, 961 F.2d 886, 891 (9th Cir. 1992); *cf.* Minnesota Public Interest Research Group v. Butz, 498 F.2d 1314 (8th Cir. 1974) (confirming the general propriety of tiering prior to CEQ's NEPA rules).

62. *See* Report to the President, *supra* note 34, at 81–85.

63. *See* The NEPA Task Force Report, Modernizing NEPA Implementation: Report to the Council on Environmental Quality 38–39 (2003). "[T]he courts have not developed a specific test to determine the specificity required in programmatic EISs." *Id.* at 38.

64. *See, e.g.,* Nevada v. Dep't of Energy, 457 F.3d 78, 91–92 (D.C. Cir. 2006); Churchill County v. Norton, 276 F.3d 1060, 1073–79 (9th Cir. 2001). Relatedly, agencies accuse of segmenting their 'actions' in order to avoid analyzing the environmental consequences thereof in the aggregate can face searching scrutiny—especially in the Ninth Circuit. *See, e.g.,* Thomas v. Peterson, 753 F.2d 754, 757–61 (9th Cir. 1985).

65. *See, e.g.,* Kern v. BLM, 284 F.3d 1062 (9th Cir. 2002); Hall v. Norton, 266 F.3d 969 (9th Cir. 2001). "Some agencies...have abandoned the concept of tiering concluding that it is ineffective and inefficient." NEPA Task Force Report, *supra* note 62, at 38.

66. The rules state that "[p]roposals or parts of proposals which are related to each other closely enough to be, in effect, *a single course of action* shall be evaluated in a single impact statement," 40 C.F.R. § 1502.4(a) (emphasis added), and that 'scoping' the NEPA process at its outset must sort out the "issues to be addressed" and "the significant issues to be analyzed in depth in the [EIS]." *Id.* at § 1501.7(a)(2). Finally, the rules define the 'scope' of an EIS to include "cumulative actions, which *when viewed with other proposed actions* have cumulatively significant impacts" should be "discussed in the same impact statement," *id.* at § 1508.25(a)(2) (emphasis added), as well as 'connected' and 'similar' actions. *Id.* at § 1508.25(a)(1), (3). Thus, because there is no other text within the rules explicitly incorporating 'cumulative impact' analysis, and because this text only requires cumulative impact analysis in connection with duly connected, currently pending proposals for action, this is the only applicable requirement thereof—notwithstanding several Ninth Circuit precedents that purport to find an independent duty to analyze 'cumulative impacts' in the text of § 1502.16. *See* Muckleshoot Indian Tribe v. U.S. Forest Serv., 177 F.3d 800, 809–10 (9th Cir. 1999); City of Carmel-By-The-Sea v. U.S. Dep't of Transp., 123 F.3d 1142, 1160 (9th Cir. 1996).

67. Cases prominently rejecting an agency's narrowing determination(s) include Klamath-Siskiyou Wildlands Ctr. v. BLM, 387 F.3d 989 (9th Cir. 2004); Save the Yaak Committee v. Block, 840 F.2d 714 (9th Cir. 1988); Barnes v. Babbitt, 329 F. Supp. 2d 1141 (D. Ariz. 2004); Sierra Club v. United States, 23 F. Supp. 2d 1132 (N.D. Cal. 1998); Shoshone Paiute Tribe v. United States, 889 F. Supp. 1297 (D. Idaho 1994); Alpine Lakes Protective Soc'y v. United States Forest Serv., 686 F. Supp. 256 (D. Mont. 1988). Cases prominently upholding an agency's narrowing determination(s) include Coalition on West Valley Nuclear Wastes v. Chu, 592 F.3d 306 (2d Cir. 2009); Wilderness Workshop v. BLM, 531 F.3d 1220 (10th Cir. 2008); Basin Mine Watch v. Hankins, 456 F.3d 955 (9th Cir. 2006); Citizens Comm. to Save Our Forests v. United States Forest Serv., 297 F.3d 1012 (10th Cir. 2002); Society Hills Towers Owners Ass'n v. Rendell, 210 F.3d 168 (3d Cir. 2000); Inland Empire Public Lands Council v. Schultz, 992 F.2d 977 (9th Cir. 1993).

68. *See* Daniel R. Mandelker, NEPA Law and Litigation §§ 7:10–7:12 (2d ed. 1996).

69. *See generally* National Commission on the BP Deepwater Horizon Oil Spill and Offshore Drilling, *supra* note 32.

70. *Cf.* NEPA Task Force Report, *supra* note 62, at 36 ("Agency definitions [of programmatic versus operational analyses] are strongly oriented toward their mission and/or culture. . . . Some Federal agencies use the term programmatic analysis to describe analyses that directly support decisionmaking . . . [while o]thers use the term for data gathering and analyses covering a vast area where no decisions to take or change agency actions are being made."); National Research Council, Science and Decisions, *supra* note 29, ("Well-designed risk-assessment processes create products that serve the needs of a community of consumers. . . .").

71. *See* NEPA Task Force Report, *supra* note 62, at 39 ("Reliance on programmatic NEPA documents has resulted in public and regulatory agency concern that programmatic NEPA documents often play a 'shell game' of when and where deferred issues will be addressed, undermining agency credibility and public trust."); *cf.* Daniel R. Mandelker, *supra* note 67 (gathering the many recent attempts to reform 'categorical exclusion' practices).

72. 284 F.3d 1062 (9th Cir. 2002).

73. *Id.* at 1067–68.

74. *Id.* BLM also indicated that it would undertake EA/FONSIs and/or EISs in connection with particular timber sales as appropriate. *See id.* at 1070–72.

75. *Id.* at 1074.

76. Predicting the spread of an invasive species has long required the integration of both observational and theoretical work, turning on detailed knowledge of the species, increasingly sophisticated spread models, and of the possible long-distance dispersal events. Even being one of the most studied problems in biology, though, spread is still simply not very well understood. *See* Alan Hastings et al., *The Spatial Spread of Invasions: New Developments in Theory and Practice*, 8 Ecology Letters 91 (2005). Controlling spread, especially of an organism like *phytophthora lateralis* in an environment like the public lands of the West, is "complicated by mixed ownership, varied control of road maintenance, and public use of forests for a wide range of recreational and economic activities." Everett M. Hansen et al., *Managing Port-Orford-Cedar and the Introduced Pathogen Phytophthora lateralis*, 84 Plant Disease 4, 8 (2000).

77. *Cf.* Kern, 284 F.3d at 1072 ("NEPA is not designed to postpone analysis of environmental consequence to the last possible moment. Rather, it is designed to require such analysis as soon as it can reasonably be done."). The court's opinion in *Kern* does not differentiate between 'related actions' within the meaning of § 1508.25(a) and 'cumulative impacts' within the meaning of § 1508.27(b), at least for purposes of assessing the sufficiency of BLM's Environmental Assessment. *See Kern*, 284 F.3d at 1074–75.

78. *Cf. Kern*, 284 F.3d at 1078 (holding that BLM should have analyzed the likely future timber sales in the action area because of the similarity of the risks arising therefrom and because the agency itself had acknowledged a need for some form of cumulative impacts analysis).

79. *See* L.W. Canter, *Cumulative Effects and Other Analytical Challenges of NEPA, in* Environmental Policy and NEPA 115–37 (Ray Clark & Larry Canter eds., 1997).

80. *Cf.* Lands Council v. Powell, 395 F.3d 1019, 1025 n.2 & 1027 (9th Cir. 2005) (noting that the Forest Service "conducts scoping on all proposed actions" but then finding the scale of risk analysis deeply flawed in the Forest Service's NEPA process). In *Lands Council*, the plaintiffs even attempted to depose an official from another agency, the United States Geological Survey, in an effort to supplement the record which they alleged omitted critical information tending to undermine the Forest Service's predictions about its proposed action. *Id.* at 1025.

81. NEPA Task Force, *supra* note 62, at 39.

82. *See* Biber, *supra* note 17, at 20–25. This reluctance is hardly uniform across courts, though. In Lands Council v. Powell, for example, the Ninth Circuit first held that the EIS the Forest Service prepared was insufficiently detailed in light of the related actions under considerations

and then held that the risks growing out of the cumulative impacts (as described by the plaintiffs) has been insufficiently investigated. The court found that "the lack of up-to-date evidence [on the cumulative impacts of forestry within a particularly stressed watershed] prevented the Forest Service from making an accurate cumulative impact assessment...." *Lands Council*, 395 F.3d at 1031.

83. Rachlinski, *supra* note 58, at 938–51. Cognitive psychology's prominent refrain on 'probability neglect' and the clear disparities between 'expert' and lay risk assessments, *see, e.g.*, Cass R. Sunstein, Risk and Reason (2002), would seem at least *prima facie* to undermine the epistemic case for judicial—as opposed to agency—authority on risk management. It is far from clear, however, that the decentralized structure of our judiciary as a whole does not more than offset the cognitive shortcomings of individual judges. *See* Rachlinski, *supra* note 58, at 951–63.

84. 42 U.S.C. § 4321.

85. *Cf.* 5 U.S.C.A §§ 551(1), 706 (2011) (defining 'agency' to mean "each authority of the Government of the United States, whether or not it is within or subject to review by another agency," excluding several such authorities explicitly, and rendering every agency 'action' reviewable according to defined criteria or 'standards of review').

86. 42 U.S.C. § 4321.

87. On this divergence between scientific conceptions of cause and effect and our more 'common sense' notions that still dominate the legal system, *see* H.L.A. Hart & Tony Honoré, Causation in the Law 9–32 (2d ed. 1985).

88. *See id.* at 44–61 (attributing this notion of causation to Hume and Mill).

89. *See id.* at 16–22.

90. *See id.* at 44–51, 65–68.

91. This is as true in a work-a-day life as it is within most federal agencies. *Cf.* National Research Council, Science and Decisions, *supra* note 29, at 93–94 (finding that administrative agencies lack clear guidance on the appropriate degree of detail, rigor and sophistication needed in assessing the variability and uncertainty for any given risk assessment).

92. 42 U.S.C. § 4332(2).

93. Indeed, whatever CEQ's role since 1978 when it formalized NEPA's procedural routines throughout its regulations (and there is good reason to discount its influence overall), the original CEQ guidelines themselves were structured and self-evidently influenced by a relatively small collection of judicial precedents that happened to interpret the statute before the agency did so. *See* Herbert F. Stevens, *The Council on Environmental Quality's Guidelines and Their Influence on the National Environmental Policy Act*, 23 Cath. U. L. Rev. 547, 556–73 (1974); Frederick R. Anderson, NEPA in the Courts (1973).

94. *See* Orrin H. Pilkey & Linda Pilkey-Jarvis, Useless Arithmetic (2007); National Research Council, Science and Decisions, *supra* note 29. Of course, confronted with serious information deficits, many environmental impact assessors simply resort to imprecise and vagueness. *See* Paul J. Culhane, *The Precision and Accuracy of U.S. Environmental Impact Statements*, 8 Envtl. Mon. & Ass. 217, 235 (1987) ("EIS forecasts are often confoundingly vague.").

95. *See* 40 C.F.R. § 1502.14 (requiring a rigorous exploration of all 'reasonable alternatives' to the proposed action, including an 'alternative of no action').

96. *Cf.* 40 C.F.R. § 1502.1–1503.4, 1507.3 (describing the compliance routines entailed in the production of an EIS or other NEPA document).

97. CEQ's rules have always made clear that NEPA documents are "more than a disclosure document" and that they must also "be used by Federal officials in conjunction with other relevant material to plan actions and made decisions." 40 C.F.R. § 1502.1. But this sort of injunction is atomizing as well; it fails to instill the proper respect for the NEPA routine as but an iteration, a part of a larger whole.

98. *See generally* Judith A. Layzer, Natural Experiments (2008).

99. *See* Robert B. Keiter, *Beyond the Boundary Line: Constructing a Law of Ecosystem Management*, 65 U. Colo. L. Rev. 293, 312–14 (1994); Julie Thrower, *Adaptive Management and NEPA: How A Nonequilibrium View of Ecosystems Mandates Flexible Regulation*, 33 Ecology L.Q. 871 (2004). In the minority of instances where ecosystem-based management has been instituted coincident with sustained success in resource management in Layzer's study sample, she found that it was by institutions that also "circumscribed the planning process by articulating a strong, pro-environmental goal and employing regulatory leverage" in ways "likely to conserve biodiversity or restore damaged ecosystems." Layzer, *supra* note 97, at 289.

100. *Cf.* Daniel C. Esty, *Environmental Protection in the Information Age*, 79 N.Y.U. L. Rev. 115, 119 (2004) (arguing that 'information failures' are at the heart of most environmental and natural resource management failures).

101. The NEPA regulations specify that EISs should be "analytic rather than encyclopedic," 40 C.F.R. § 1502.2(a), that EISs should employ "an interdisciplinary approach which will insure the integrated use of the natural and social sciences and the environmental design arts," *id.* at § 1502.6, that agencies "shall insure the professional integrity, including scientific integrity, of the discussions and analyses in [EISs]" *id.* at § 1502.24, and that agencies using information submitted to them by "applicants" must "independently evaluate the information submitted" and take "responsibility for its accuracy." *Id.* at § 1506.5(a). They also require that the information used in a NEPA process "be of high quality," *id.* at § 1500.1(a), and that analyses be "[a]ccurate." *Id.* The regulations nowhere specify how to gauge the relative (or absolute) availability of information, however, nor what makes for an adequately robust NEPA process.

102. 395 F.3d 1019 (9th Cir. 2005).

103. *Id.* at 1025. "The Project is designed to improve the aquatic, vegetative, and wildlife habitat in the Project area." *Id.* The 'design' mentioned by the court included so-called fuels-treatment to reduce the risk of catastrophic fire, removal of obsolete roads and stream-channel crossings, and the replacement of nonnative with native, more disease-tolerant tree species. Each of these project objectives was selected in step with the Interior Columbia Basin Ecosystem Management Plan. *See* U.S.D.A. Forest Service, Idaho Panhandle Forests Coeur d'Alene River Ranger District, Iron Honey Resource Area Final Environmental Impact Statement I-1 (Nov. 2001) (copy on file with author) [hereinafter Iron Honey FEIS].

104. 395 F.3d at 1027; *see* Iron Honey FEIS, *supra* note 102, at III-1 to III-27. Using the logging projects to fund/underwrite specified aquatic restoration measures justified 'all action alternatives' to greater or lesser extent. *Id.* at III-27 to III-29. The service's "vague discussion of the general impact of prior timber harvesting" without "discussion of the environmental impact from past projects on an individual basis" was, in the court's words, "inadequate." 395 F.3d at 1027.

105. *See id.* at 1027–28 (calling service's failure to inventory and specifically describe past timber harvests and their effects "inadequate"); *id.* at 1030–31 (rejecting service's use of fish count data that were thirteen years old); *id.* at 1032 (rejecting the service's incomplete disclosures and characterizations of its modeling program's limits).

106. *See id.* at 1037.

107. 615 F.3d 1122 (9th Cir. 2010).

108. *Id.* at 1135–38 (reversing district court's finding that service's cumulative impact analysis was deficient because it failed to specify the time, place and scale of prior effects and holding that service was entitled to 'aggregate' the past effects as CEQ guidance had suggested it could).

109. *See id.* at 1125–26. A 'commercial' benefit from the logging projects was also documented in the record of decision. *Id.* at 1126 n.1. But the panel was careful to note that the agency's "primary objective" was protective and that "it spent approximately three years doing the analysis necessary" under the applicable legal requirements. *Id.*

110. *Id.* at 1131.

111. *See Powell*, 395 F.3d at 1030.

112. On the perils of judicial overreaching in this dynamic, *see* Jerry L. Mashaw & David L. Harfst, The Struggle for Auto Safety (1990).

113. The literature on this factor in principal/agent problems is vast. *See* Nicola Persico, *Committee Design with Endogenous Information*, 71 Rev. Econ. Stud. 165 (2004); Sean Gailmard, *Expertise, Subversion, and Bureaucratic Decision*, 18 J.L. Econ. & Org. 536 (2002); Timur Kuran & Cass R. Sunstein, *Availability Cascades and Risk Regulation*, 51 Stan. L. Rev. 683 (1999); Philippe Aghion & Jean Tirole, *Formal and Real Authority in Organizations*, 105 J. Pol. Econ. 1 (1997); *see also* Thomas W. Gilligan & Keith Krehbiel, *Organization of Informative Committees by a Rational Legislature*, 34 Am. J. Pol. Sci. 531 (1990).

114. *See* Wendy E. Wagner, *Administrative Law, Filter Failure, and Information Capture*, 59 Duke L.J. 1321 (2009).

115. *Allen*, 615 F.3d at 1133–34. This deference was in marked contrast with the panel's opinion in *Powell* which flirted with opening the administrative record notwithstanding strong precedent ruling out that very step. *See Powell*, 395 F.3d at 1029 (citing Florida Power & Light Co. v. Lorion, 470 U.S. 729, 743–44 (1985)).

116. *See Powell*, 395 F.3d at 1030–32; Klamath-Siskiyou Wildlands Ctr. v. BLM, 387 F.3d 989, 993–96 (9th Cir. 2004) (denouncing BLM's NEPA process for failing to include quantitative data on cumulative effects within the project area and faulting the EAs under challenge for a lack of necessary information). The two sides of the Ninth Circuit occasionally meet up in en banc proceedings where such issues are presented. In Lands Council v. McNair, 537 F.3d 981 (9th Cir. 2008) [hereinafter Lands Council II], the court took en banc a case very similar to *Powell* and *Allen* in "order to clarify some of [its] environmental jurisprudence" that had arisen in the review of Forest Service actions. *Id.* at 984. Much of the opinion is concerned with the availability of preliminary injunctions and was, in effect, overruled by Winter v. Natural Resources Defense Council, Inc., 129 S.Ct. 365 (2008). The opinion in Lands Council II, however, bristles with attention to the relative institutional competence of courts and administrative agencies. *See, e.g.*, Lands Council II, 537 F.3d at 988 ("In essence,

Lands Council asks this court to act as a panel of scientists that instructs the Forest Service how to validate its hypotheses . . . chooses among scientific studies . . . and orders the agency to explain every possible scientific uncertainty.").

117. CEQ issued a 'memorandum' to heads of federal agencies in 2005. *Memorandum from James L.* Connaughton, to Heads of Federal Agencies, Guidance on the Consideration of Past Actions in Cumulative Effects Analysis (June 24, 2005), *available at* http://ceq.hss.doe.gov/nepa/regs/Guidance_on_CE.pdf.

118. *See, e.g.*, Adrian Vermeule, Law and the Limits of Reason 57–60 (2008).

119. *See, e.g.*, Matthew C. Stephenson, *Information Acquisition and Institutional Design*, 124 Harv. L. Rev. 1422 (2011).

120. *Compare* Daniel Mach, *Rules Without Reasons: The Diminishing Role of Statutory Policy and Equitable Discretion in the Law of NEPA Remedies*, 35 Harv. Envtl. L. Rev. 205, 219–25 (2011) (describing the discretionary provision of injunctive relief to NEPA plaintiffs under established doctrines and analyzing the condition of 'irreparable harm' in the NEPA context), *with* Michael B. Gerrard & Michael Hertz, *Harnessing Information Technology to Improve the Environmental Impact Review Process*, 12 N.Y.U. Envtl. L. Rev. 18 (2003) (describing the primitive state of NEPA information management law and the fact that most NEPA documents are effectively lost upon completion).

121. *See* Hart & Honoré, *supra* note 86, at 112.

122. *See id.* at 122–23.

123. *Cf.* Colin Howson, Hume's Problem 168–78 (2000) (describing the orthodox model of inductive inference that is rooted in probabilism); *see also* W.V.O. Quine, *Two Dogmas of Empiricism*, 60 Phil. Rev. 20 (1951).

124. 40 C.F.R. §§ 1508.8(a)–(b).

125. 40 C.F.R. § 1508.7.

126. *See* 40 C.F.R. §§ 1501.7, 1508.25 (mandating a 'scoping' process for EISs in which the necessary 'issues' are identified and specifying the definition of connected, cumulative, and similar 'actions' that should be analyzed in a single EIS).

127. Council on Environmental Quality, Exec. Office of the President, Considering Cumulative Effects Under the National Environmental Policy Act (1997) [hereinafter Considering Cumulative Effects] (copy on file with author). Note that the CEQ rules state that "[e]ffects and impacts as used in these regulations are synonymous." 40 C.F.R. § 1508.8. The rules maintained by the Dep'ts of Interior and Commerce implementing ESA § 7 take a subtly different approach to the terminology, though, in step with that statutory provision's language. *See* 50 C.F.R. § 402.02 (2010) (defining 'effects of the action').

128. *See, e.g.*, Amy L. Stein, *Climate Change Under NEPA: Avoiding Cursory Consideration of Greenhouse Gases*, 81 U. Colo. L. Rev. 473 (2010).

129. Dep't of Transp. v. Public Citizen, 541 U.S. 752, 767 (2004) (quoting Metropolitan Edison Co. v. People Against Nuclear Energy, 460 U.S. 766, 774 (1983)).

130. *See, e.g.*, Center for Food Safety v. Johanns, 451 F. Supp. 2d 1165, 1183–86 (D. Hawai'l 2006). This is assuming cumulative impacts must be considered at these other, more preliminary NEPA stages. CEQ's rules, as mentioned above, do not make that manifest, although

CEQ's 'handbook,' Considering Cumulative Effects, states that "adequate consideration of cumulative effects requires that EAs address them fully." Considering Cumulative Effects, *supra* note 126, at 4; *see also* Kern v. BLM, 284 F.3d 1062, 1075–76 (9th Cir. 2002) (requiring cumulative effects analysis in EA).

131. 40 C.F.R. § 1502.22(b) (2010).

132. *See* Sierra Club v. Marsh, 976 F.2d 763 (1st Cir. 1992) (requiring that an agency consider a risk under NEPA if it is "sufficiently likely to occur that a person of ordinary prudence would take it into account"); City of New York v. U.S. Dep't of Transp., 715 F.2d 732 (2d Cir. 1983) (articulating a concept of 'overall risk' in assessing agency compliance with 40 C.F.R. Part 1500); New Jersey Dep't of Envtl. Prot. v. U.S. Nuclear Regulatory Comm'n, 561 F.3d 132 (3d Cir. 2009) (requiring agency to analyze a risk under NEPA only where it is not "too far removed from the natural or expected consequences" of the agency action); Webb v. Gorsuch, 699 F.2d 157 (4th Cir. 1983) (requiring agency to analyze a risk under NEPA only if that risk is a "significant aspect of the probable environmental consequences" of the agency action); City of Shoreacres v. Waterworth, 420 F.3d 440 (5th Cir. 2005) (applying the *Marsh* standard of risk's being "sufficiently likely to occur that a person of ordinary prudence would take it into account in reaching a decision" to conclude that a risk not caused or at all enhanced by the agency action is not "reasonably foreseeable"); Save Our Cumberland Mountains v. Kempthorne, 453 F.3d 334 (6th Cir. 2006) (holding that NEPA requires a 'hard look' at the potential environmental impacts of a proposed action and that this entails examination of normal accidents like automobile collisions); Porter County Chapter of Izaak Walton League of America v. Atomic Energy Comm'n, 533 F.2d 1011 (7th Cir. 1976) (holding that an impact statement prepared in the licensing of a nuclear installation need not consider accidents that are 'remote' possibilities); Arkansas Wildlife Fed'n v. U.S. Army Corps of Eng'rs, 431 F.3d 1096 (8th Cir. 2005) (holding that the selection of risks that are "sufficiently likely to occur that a person of ordinary prudence would take [them] into account" is the agency's judgment to make and should not be disturbed by a court barring obvious errors); Ground Zero Ctr. for Non-Violent Action v. Dep't of the Navy, 383 F.3d 1082 (9th Cir. 2004) (holding that risks calculated as between one in a hundred million and one in a trillion need not enter into NEPA process at all); Lee v. United States Air Force, 354 F.3d 1229 (10th Cir. 2004) (holding that concept of 'overall risk' could be used to exclude risks with expected frequencies of one in 43,500 from NEPA process); Carolina Environmental Study Group v. United States, 510 F.2d 796 (D.C. Cir. 1975) (holding that severity of a consequence does not necessarily require a risk's inclusion in NEPA process).

133. *See, e.g.,* Greenpeace Action v. Franklin, 14 F.3d 1324, 1332 (9th Cir. 1992) (holding that wherever 'substantial questions' are raised by a commenter or by information that is brought to light in a NEPA process, a further 'hard look' by the agency is required in that NEPA process); Idaho Sporting Cong. v. Thomas, 137 F.3d 1146 (9th Cir. 1998) (applying 'substantial question' doctrine); Blue Mountains Biodiversity Proj. v. Blackwood, 161 F.3d 1208 (9th Cir. 1998) (same); National Parks & Conservation Ass'n v. Babbitt, 241 F.3d 722 (9th Cir. 2001) (same); San Luis Obispo Mothers for Peace v. Nuclear Regulatory Comm., 449 F.3d 1016 (9th Cir. 2006) (same).

134. 311 F.3d 1162 (9th Cir. 2002).

135. *Id.* at 1164–65.

136. *Id.* at 1165–66.

137. *Id.* at 1169–70.

138. *Id.* at 1175–77.

139. *Id.* at 1176. "Post hoc invocation of a categorical exclusion does not provide assurance that the agency actually considered the environmental effects of its action before the decision was made." *Id.*

140. Until California v. Norton is overruled by the Ninth Circuit en banc or by the Supreme Court, it remains the law in effect in the Ninth Circuit. *But see* Miller v. Gammie, 335 F.3d 889, 893 (9th Cir. 2003) (en banc) *abrogated by* Fossen v. Blue Cross and Blue Shield of Montana, Inc., 2011 WL 4926006 (9th Cir. 2011) (holding that panels and district courts are not bound by prior panel or en banc opinions that are "clearly irreconcilable with the reasoning or theory of intervening higher authority"). Because the Ninth Circuit is so large geographically and so disproportionately responsible for NEPA's jurisprudence, this is an especially important 'rule' for NEPA practitioners.

141. The Supreme Court in Citizens to Preserve Overton Park v. Volpe, 401 U.S. 402, 416 (1971), held that an agency making a decision unstructured by any formal rules of process still must record that decision contemporaneously if it expects to be judged upon review as having acted for its stated reasons and nothing more. One of the issues with CATEXs going back to their creation is the perception that they can easily become, if they must be excessively documented, more trouble than they are worth. *See, e.g.,* Hill Memorandum, *supra* note 26, at 34265.

142. Subsequent events—CEQ's release in 201 of a 'guidance' interpreting the CATEX rules—only confirm the troubles with holdings like that in *Norton.* In the guidance, CEQ is abundantly clear that its newly recommended CATEX substantiation practices be implemented consistent with other agency NEPA policies and priorities. *See* Council on Environmental Quality, Final Guidance for Federal Departments and Agencies on Establishing, Applying, and Revising Categorical Exclusions Under the National Environmental Policy Act, 75 Fed. Reg. 75628, 75636 (Dec. 6, 2010).

143. *See* Interagency Cooperation Under the Endangered Species Act, 73 Fed. Reg. 76272 (Dec. 16, 2008).

144. 73 Fed. Reg. at 76287 (finalizing amended 50 C.F.R. § 402.03(b)(2)) (Dec. 16, 2008). After the 2008 presidential election and enactment of the 2009 Omnibus Appropriations Act, Pub. L. No. 111-8 (2009), these amendments to Part 402 were abrogated in their entirety and the preexisting consultation rules were reinstated. *See* Dep't of the Interior & Dep't of Commerce, Final Rule, Interagency Cooperation Under the Endangered Species Act, 74 Fed. Reg. 20421 (May 4, 2009).

145. *See, e.g.,* Holly Doremus et al., *Making Good Use of Adaptive Management* 2 (Ctr. for Progressive Reform Paper # 1104, 2011) *available at* http://papers.ssrn.com/sol3/papers.cfm?abstract_id=1808106at 2.

146. *Id.* at 7.

147. *See, e.g., Id.* ("Adaptive management cannot help when there is no way to correct an initial mistake, as when the decision in question is to allow irreversible alteration of the environment.").

148. *See* 75 Fed. Reg. at 75628–75631.

149. 75 Fed. Reg. at 75636 (endorsing and recommending to others the Dep't of Energy's practice of posting key CATEX decisions and documents to its website).

150. *See* 75 Fed. Reg. at 75630.

151. *See Id.*

152. *See* 75 Fed. Reg. at 75628–01("Agencies can obtain useful substantiating information by monitoring and/or otherwise evaluating the effects of implemented actions that were analyzed in EAs that consistently supported Findings of No Significant Impact."). "Implemented actions analyzed in an EIS can also be a useful source of substantiating information if the implemented action has independent utility to the agency, separate and apart from the broader action analyzed in the EIS." *Id.*

153. *See* 75 Fed. Reg. at 75629.

154. *See* Dep't of Transp. v. Public Citizen, 541 U.S. 752, 766 (2004).

155. *See* Colburn, *supra* note 35, at 313–25.

156. *See* 75 Fed. Reg. at 75634.

157. *See* Vermeule, *supra* note 117, at 57–96 (comparing different institutional arrangements and arguing that those which best control for 'epistemic freeriding' or rational ignorance and those that best randomize predictable biases are most likely to perform best over time).

158. On the concept of a deliberative polyarchy and the importance of learning-by-monitoring thereto, *see* Charles F. Sabel & Jonathan Zeitlin, *Learning from Difference: The New Architecture of Experimentalist Governance in the EU*, 14 Eur. L.J. 271 (2008); Charles F. Sabel, *A Real-Time Revolution in Routines, in* The Firm as Collaborative Community: Reconstructing Trust in the Knowledge Economy 106 (Charles Heckscher & Paul Adler eds., 2006); Michael C. Dorf & Charles F. Sabel, *A Constitution of Democratic Experimentalism*, 98 Colum. L. Rev. 267, 446 (1998).

159. I have written about this problem with agency rules at length. *See* Jamison E. Colburn, *Agency Interpretations*, 82 Temp. L. Rev. 657 (2009).

160. *See, e.g.,* Jon Connolly, Note, *Alaska Hunters and the D.C. Circuit: A Defense of Flexible Interpretive Rulemaking*, 101 Colum. L. Rev. 155 (2001).

161. The so-called 1983 Hill Memorandum stated CEQ's belief that CATEXs were being underutilized and that the problem was one of unrealistic expectations about their substantiation. *See* Hill Memorandum, 48 Fed. Reg. at 34265 ("Accordingly, the Council strongly discourages procedures that would require the preparation of additional paperwork to document that an activity has been categorically excluded."). Twenty-seven years later CEQ "clarifie[d]" that "documentation may be appropriate to demonstrate that the proposed action comports with any limitations identified in prior NEPA analysis and that there are no potentially significant impacts expected as a result of extraordinary circumstances" and that documenting a CATEX "provides the agency the opportunity to demonstrate why its decision to use the categorical exclusion is entitled to deference." 75 Fed. Reg. at 75628, 75636.

162. Howson, *supra* note 122.

163. *See* David Hume, A Treatise of Human Nature 139 (L.A. Selby-Bigge & P.H. Nidditch eds., 2d ed., 1978) (1738) ("That even after the observation of the frequent or constant conjunction of objects, we have no reason to draw any inference concerning any object beyond those of which we have had experience."); *see also* David Hume, Enquiries Concerning Human Understanding and of the Principles of Morals 76–77 (L.A. Selby-Bigge & P.H. Nidditch eds., 3d ed., 1975) (1777) (arguing that cause, in its natural aspect, is said to be "[a]n object

followed by another, and where all the objects similar to the first are followed by objects similar to the second.").

164. Hume, *supra* note 160, at 75. Hume's discussion in his *Enquiries*—a later, clearer exposition of many of the ideas in his *Treatise*—simplified but retained his basic argument that inductive statements were essentially probabilistic statements by implication. *See* A.J. Ayer, Hume 89–91 (2000).

165. Howson, *supra* note 122, at 10–12. Hume's argument was especially attuned to the fact that "the course of nature may change" and that, therefore, what we take to be unchanging norms in natural phenomena are simply heretofore uncontradicted regularities. *Id.* at 11 (*quoting* David Hume, An Enquiry Concerning Human Understanding (1748)).

166. Howson, *supra* note 122, at 10 (*quoting* C.D. Broad, *The Philosophy of Francis Bacon, in* Ethics and the History of Philosophy 143 (1952)).

167. *See, e.g.,* Nelson Goodman, Fact, Fiction, and Forecast (3d ed. 1973).

168. *Cf.* Alan Hastings et al., *The Spatial Spread of Invasions: New Developments in Theory and Evidence*, 8 Ecology Letters 91, 98–99 (2005) (noting increasing evidence that evolution influences the dynamics of invading populations and that unpredictable adaptations must be understood before rates of spread become more predictable and observing that evolutionary dynamics are simply too complex for the present state of the research).

169. *See* Karkkainen, *Bottlenecks and Baselines, supra* note 13, at 1410–13, 1439–44; Wagner, *supra* note 113, at 1351–71 (arguing that administrative law inadvertently reinforces a tendency among factions to artificially augment and/or constrain the flows of information to and from regulatory agencies such that appropriate processing and usage of that information is more challenging for the agencies). These familiar challenges are all known to risk assessors in the major regulatory agencies, *cf.* National Research Council, *supra* note 29, at 245–51 (identifying three critical and distinct phases of appropriate risk assessment and management), but little has been done to remedy the basic predicament. *See* Doremus et al., *supra* note 144.

170. *See* Philip E. Tetlock, Expert Political Judgment (2005) (showing that regression-based algorithms outperform putative experts in predicting human decision making without exception).

171. Hollnagel & Woods, *supra* note 7, at 19.

172. CEQ's NEPA regulations require that narrower impact statements be tiered to broader statements if it will "eliminate repetitive discussions of the same issues." 40 C.F.R. § 1500.4(i) (2010). As was evident in the Deepwater Horizon tragedy, though, tiering of the kind can degenerate into the hiding of specific threats that are made to seem remote and improbable at *both* broad and narrow scales. Colburn, *supra* note 35, at 313–25.

173. *See* Duncan J. Watts, Everything Is Obvious Once You Know the Answer 27–28 (2010) (arguing that 'common sense' often works like mythology by providing explanations for whatever particulars the world yields without enhancing our understanding of the causal forces underlying real world events).

174. *See Id.* at 61–67.

175. *See Id.*

176. *Compare* Hollnagel & Woods, *supra* note 7, at 113–41 (collecting stories of 'surprise' from agents trying to cooperate through automation and finding that collaborative work

that occurs in the process of recognizing and solving a problem will necessarily develop unpredictably and often unsatisfactorily), *with* Bill Tomlinson, Greening Through IT 87 (2010) (arguing that computation and the continued spread of computational power is the only factor of human civilization that can even possibly solve the scale and cognition problems we face).

177. Gilboa and colleagues, for example, have suggested a way of improving frequentist probabilities by a process of what they call *similarity-weighted* frequency estimates wherein more similar cases get higher weights in the computation of frequencies. *See* Itzhak Gilboa et al., *On the Definition of Objective Probabilities by Empirical Similarity*, 172 Synthese 79 (2010). I take this possible solution up in an examination of NEPA routines at the Dep't of Interior's Minerals Management Service and the events leading to the Deepwater Horizon tragedy. *See* Colburn, *supra* note 35, at 325–31.

178. Charles Perrow, Normal Accidents 62–100 (1984). Bhopal, Chernobyl, and the *Challenger* explosion arguably further confirmed his thesis. *See* Charles Perrow, Normal Accidents 353 (2d ed. 1999).

179. Updating the NEPA rules to foster the emergence of such a cognitive economy would entail at least: (1) conditioning the completion of PEISs and other broad-scale NEPA documents on the adequate use of *past* EISs, EA/FONSIs, and/or CATEXs bearing on similar judgments; (2) further specifying the degree(s) of connection(s) that make actions and/or impacts 'cumulative' or 'connected' such that they should be analyzed together; and (3) specifying consequences and/or alternatives for NEPA processes and documents other than the EIS where 'incomplete or unavailable information' prevents an agency from evaluating its actions. *Cf.* 40 C.F.R. § 1502.22 (2010) (providing for the completion of EISs where necessary information is unavailable).

180. 42 U.S.C. § 4332(2)(A). Separately, NEPA § 102(2)(B) requires all agencies to "identify and develop methods and procedures . . . which will insure that presently unquantified environmental amenities and values may be given appropriate consideration in decisionmaking along with economic and technical considerations." *Id.* at § 4332(2)(B). Combined, these are NEPA's 'study and integration' mandates.

8

Restoration and Law in Ecosystem Management
Robert W. Adler, University of Utah

E cosystem management has been defined in many different ways,[1] but most of those definitions share several key attributes. All involve managing at the scale of whole ecosystems rather than individual species or resources. All focus on ecosystem structure, function, and processes as opposed to individual components, and on the dynamic nature of ecosystems rather than a single state. All define a process to determine and articulate ecological and social goals to be attained and monitored through the management process, often by way of adaptive management. Depending on the goal-setting process used, however, some of those definitions include the idea of *improving* or *restoring* as well as maintaining ecosystems,[2] but others do not, at least not explicitly.[3] Accordingly, some—but not all—definitions of ecosystem management incorporate the concept of ecosystem restoration.

This key difference can be explained by either physical differences between ecosystems or differences in societal goals. Although relatively rare on twenty-first century Earth, some ecosystems may be in close to a pristine or 'natural' condition—that is, comparatively unimpaired by past or ongoing human activities.[4] In those places, the goal of an ecosystem management process may be largely to protect and maintain ecosystem structure, function, and services. Other places, including many that are the subject of major ecosystem management programs (such as the Everglades or old growth forests in the Pacific Northwest), have been impaired to varying degrees by a wide range of human activities, from resource extraction (logging, mining, oil and gas drilling) to development (road construction, urbanization, dams, and water diversions) to waste contamination. Depending on the goals established by any given decision-making process, restoration of past harm may be a key component of an ecosystem management program. Even where the program goal is maintenance—as opposed to improvement of ecosystem function or services—restoration may be an important program component as compensatory mitigation for planned resource development activities within the ecosystem, so that ecosystem health is maintained rather than degraded over time.

Depending on which of the above situations is involved, restoration decisions present a different balance between scientific and social or economic choices, and the law, in turn, plays an important role in guiding those choices and how they are made. At least in theory, if a goal is established to 'restore' an ecosystem to as close to a pristine (or unimpaired) state as possible, the subsequent decision-making process should be largely scientific. This avoids value choices balancing ecosystem and human (societal) needs, or viewed differently, indicates that the choice has already been made in favor of ecosystem needs, as in the case of national parks or designated wilderness areas. Even in those cases, however, determining 'on the ground' restoration goals is challenging. Ecosystems are not static, which means that there is no single 'correct' or desired state to which an ecosystem—or one or more of its components or attributes—should be restored.[5]

Increasingly, although the concept of restoration is characterized by a similar variety of definitions as ecosystem management,[6] ecologists define restoration in terms of ecosystem functions and processes, rather than end goals.[7] This definition recognizes the reality that ecosystems, like the species that inhabit them, naturally evolve over time. Moreover, even when restoring national parks or other ecosystems where the applicable statutory or other legally defined goal may be an 'unimpaired' condition,[8] it will be a rare case in which restoration decisions involve no balance between ecological and social or economic needs.[9]

In areas that have been degraded sufficiently so that full restoration of ecosystem functions and processes is no longer possible, or in areas in which ecosystem managers must balance ecosystem and societal needs due to applicable legal standards or competing societal uses, restoration decisions arguably become even more complex, because they involve at least two interrelated sets of decisions. In those cases, rather than determining restoration goals solely from a scientific perspective, managers first must determine the appropriate balance between societal and ecosystem needs, in conjunction with the scientific assessment of which restoration goals or methods are best suited to the balance chosen.[10] Those kinds of choices must be made through democratic process in addition to scientific analysis.[11] And given the significant uncertainties inherent in that scientific assessment, the process is likely to be iterative, often through the process of adaptive management (addressed elsewhere in this volume).[12] Sometimes statutes or other legal standards govern or guide those decisions. In other cases, they establish procedures through which such decisions must be made.

Finally, in the context of future decisions, we also need to come to grips with the equally thorny question of how much reliance to place on restoration as opposed to protection or conservation. Indeed, some critics argue that undue reliance on the prospect of future restoration can dilute preservation or conservation efforts, and suggests hubris about our relationship to the natural world that generates a mentality of "if you destroy it, we can build it again."[13] Others, however, view restoration as a "positive counterbalance to the disruptive effects of modern human activities."[14] Elsewhere, I have suggested that restoration conceptually constitutes a logical third prong of the modern environmental era, in which mitigation was designed to minimize harm from ongoing human activities, prevention was designed to reduce or eliminate harm from future actions, and restoration is a necessary third strategy to undo the harm from centuries of past degradation.[15] Another key role of law, then, is to guide decisions about when it is appropriate to rely on restoration as an acceptable alternative to prevention.

A number of different sources of law can guide or sometimes limit ecosystem restoration and management. These include state and local law, common law doctrines, and even constitutional principles governing property rights and use. However, federal environmental, natural resources, and land management statutes predominate in governing or requiring a large percentage of major ecosystem restoration efforts.[16] This chapter evaluates the most significant federal statutes affecting restoration decisions and implementation, with particular attention to the difficult conceptual issues outlined above. Although other sources of law and relevant cases will be discussed incidentally, analysis of all 'law' governing ecosystem restoration would require a much longer discussion.

I. EXISTING FEDERAL STATUTES GOVERNING OR INFLUENCING ECOSYSTEM RESTORATION

Federal statutory provisions that govern or influence environmental restoration decisions fall within at least four broad conceptual categories: (1) statutes establishing legal standards dictating or influencing the applicable ecological end goal, by articulating either legislative goals or criteria for particular public lands, or broader ecological objectives or environmental standards regardless of location; (2) statutes mandating particular restoration actions, such as remediation requirements for hazardous waste contamination, reclamation of lands damaged by mining or other resource extraction activities, or compensatory mitigation requirements for anticipated development; (3) statutes dictating or affecting the *process* of making restoration decisions by requiring consideration of alternatives to proposed actions, and the comparative consequences of those options, or by establishing processes to involve stakeholders directly in a public participation or collaborative process; and (4) place-specific statutes establishing, guiding, or supporting particular ecosystem-based restoration programs.

A. *Federal Statutes Establishing Ecological Goals*

1. *ECOLOGICAL AND PRESERVATION MANDATES*

Some federal statutes establish or articulate standards that govern or guide ecological goals, either generally or for particular geographic areas. Although applicable only to the designated public land units, for example, some of the statutes governing federal land management articulate goals or establish enforceable standards that may affect ecological restoration decisions and practices. Some of those statutes appear to have chosen restoration to relatively pristine conditions, while others explicitly require a balance between ecological restoration and other goals.

In the National Park System Organic Act (National Park Act),[17] Congress directed the Park Service to manage national parks "to conserve the scenery and the natural and historic objects and the wild life therein and to provide for the enjoyment of the same in such manner and by such means as will leave them *unimpaired* for the enjoyment of future generations."[18] Use of the term 'conserve' suggests maintenance of existing resources and conditions as opposed to restoration. Moreover, the statute's general management planning mandate requires consideration of 'preservation' but not, at least not explicitly, restoration of national parks.[19] However, that focus probably reflects the fact that Congress established many national parks (excluding areas established mainly to protect historical resources or to provide for public recreation) in areas that were largely unimpaired by human activities. The 'unimpairment' standard in this provision, however, implies that where park resources have been degraded, restoration to a condition equivalent to what would exist absent human activities is required. Although some courts have conferred significant discretion on the Park Service in interpreting and implementing the unimpairment standard,[20] others have imposed a duty on the service, at a minimum, to quantify and analyze adverse impacts before determining them acceptable.[21] Moreover, Professor Keiter has noted that many existing national parks, especially in the eastern states, were created from previously damaged lands that the Park Services later restored, and argues that a broader program of restoring damaged ecosystems might serve as one strategy to expand the national park system.[22]

Thus, arguably Congress has already struck the balance in national park restoration efforts in favor of restoration to a condition as close to its 'natural,' or unimpaired, state as possible. That view is probably overstated, however, because some development activities are allowed in national parks to facilitate their use and enjoyment by the public,[23] and because parks can also be degraded by external threats. Restoration might be required to address those impacts, but clearly restoration would still have to be balanced against opportunities for future park use and enjoyment, and for activities that occur entirely outside of park boundaries. Again, whether correctly or not, this may reflect a congressional presumption that national parks are largely in an unimpaired condition, and therefore require preservation rather than restoration as their main management focus. For example, managers of Zion National Park propose to conduct riparian restoration activities to remedy the impacts of a road built too close to the Virgin River, which runs through the heart of the park's tourist activity.[24] Moreover, Congress has adopted specific restoration mandates for some national parks degraded by human activities within or outside the park, as it did in the Grand Canyon Protection Act.[25] In the case of marine sanctuaries, Congress expressly included a goal to "where appropriate, restore and enhance natural habitats, populations, and ecological processes,"[26] and required management plans to identify restoration as well as protection efforts.[27]

National Wildlife Refuges administered by the U.S. Fish and Wildlife Service (FWS) under the National Wildlife Refuge System Administration Act (Refuge System Act)[28] are a second example of federal land units in which Congress appears to have chosen a largely unimpaired or 'natural' state as a restoration and management goal. Congress articulated the mission for the refuge system as "the conservation, management, and where appropriate, *restoration* of the fish, wildlife, and plant resources and their habitats within the United States for the benefit of present and future generations of Americans."[29] Moreover, Congress defined the terms 'conserving,' 'conservation,' 'manage,' 'managing,' and 'management' as "to sustain and, where appropriate, *restore and enhance*, healthy populations of fish, wildlife, and plants. . . ."[30]

Consistent with the conservation-oriented mission of the refuge system, Congress directed FWS to "provide for the conservation of fish, wildlife, and plants, and their habitats within the System,"[31] to "ensure that the biological integrity, diversity, and environmental health of the system are maintained for the benefit of present and future generations of Americans,"[32] and to expand the refuge system "in a manner that is best designed...to contribute to the conservation of the ecosystems of the United States."[33] Congress also reached value judgments about human—as well as ecological—uses of National Wildlife Refuges, but established as a general rule that only "compatible wildlife-dependent recreation is a legitimate and appropriate general public use of the System,"[34] and that such uses "shall receive priority consideration in refuge planning and management."[35] Moreover, Congress prohibited the Secretary of the Interior from allowing, expanding, or renewing any use on refuge lands that is not 'compatible with' the purposes of the refuge.[36] The compatibility standard for the refuge system establishes a presumption that development activities may occur, but that fish and wildlife and related ecological values take priority for purposes of both restoration and management goals.

Two other federal preservation statutes, the Wilderness Act[37] and the Wild and Scenic Rivers Act,[38] set aside specific areas within other federal land management units (such as national parks and national wildlife refuges) for special protection, and thus also suggest a Congressional presumption in favor of ecological restoration over other uses and goals. Congress established the National Wilderness Preservation System to ensure that ongoing growth and development would not leave the nation with "no lands designated for preservation and protection in their natural condition," and with a mandate to manage those lands "in such manner as will leave them unimpaired for future use and enjoyment as wilderness, and so as to provide for the protection of these areas, the preservation of their wilderness character."[39] Congress further defined 'wilderness' as:

> an area where the earth and its community of life are untrammeled by man, where man himself is a visitor who does not remain [and] an area of undeveloped Federal land retaining its primeval character and influence, without permanent improvements or human habitation, which is protected and managed so as to preserve its natural conditions and which (1) generally appears to have been affected primarily by the forces of nature, with the imprint of man's work substantially unnoticeable; (2) has outstanding opportunities for solitude or a primitive and unconfined type of recreation; (3) has at least five thousand acres of land or is of sufficient size as to make practicable its preservation and use in an unimpaired condition; and (4) may also contain ecological, geological, or other features of scientific, educational, scenic, or historical value.[40]

However, because of the statutory presumption that lands designated as wilderness are already unimpaired by human factors as a condition of designation, the primary mandate of the statute is preservation of those features as opposed to restoration.[41] Thus, the main substantive provisions of the statute prohibit additional human activities that might impair wilderness character, although with a number of significant exceptions that allow ongoing and additional human uses.[42] As with national parks, however, these strict preservation standards suggest that any degradation that does occur should be remedied by restoration efforts governed by largely scientific objectives. In addition, in some recent wilderness bills, Congress has included restoration incentives for adjacent private lands as part of a comprehensive ecosystem management and protection program.[43]

Likewise, the goal of the Wild and Scenic Rivers Act is to preserve certain stretches of river "in their free-flowing condition to protect the water quality of such rivers and to fulfill other vital national conservation purposes."[44] Like wilderness areas, the basic presumption is that river reaches are already in a largely unimpaired condition as a prerequisite to designation,[45] and the basic substantive requirements in the act prohibit additional dams, other water projects, and other developments on adjacent federal land that would impair wild and scenic river values.[46] However, unlike the Wilderness Act, the Wild and Scenic Rivers Act envisions that some river reaches might become eligible for designation "upon restoration to this condition."[47] Moreover, Congress directed that "[e]ach component of the national wild and scenic rivers system shall be administered in such manner as to protect *and enhance* the values which caused it to be included in said system without, insofar as is consistent therewith, limiting other uses that do not substantially interfere with public use and enjoyment of these values," with an emphasis on protecting "esthetic, scenic, historic, archeologic, and scientific features."[48] Congress also directed agencies administering system components to cooperate with EPA and state water pollution control agencies to eliminate or reduce pollution in those river reaches,[49] suggesting both preservation and restoration goals for the Wild and Scenic Rivers system.

2. MULTIPLE USE MANDATES

By contrast, Congress expressly provided for multiple uses in other categories of federal lands, including resource use and extraction, as well as ecological conservation and public recreation. Those designations suggest an even more difficult set of decisions about the role of restoration in ecosystem management, because restoration goals will likely be tempered by other uses and goals, and 'full' ecological restoration will most often be unlikely or impossible in order to accommodate those other uses, except in distinct segments of those land areas set aside for wilderness or other special protection.

In the Forest and Rangeland Renewable Resources Planning Act of 1974 (as later amended and re-designated in the National Forest Management Act of 1976),[50] Congress instructed that national forest lands must be managed on a 'multiple use-sustained yield' basis,[51] which reflects a balance between extractable resources for human use and preservation of ecological, recreational, and other nonextractive forest values. Congress defined 'multiple use' to allow for consumption of extractable resources "[w]ithout impairment of the productivity of the land" and "not necessarily the combination of uses that will generate the greatest dollar return or the greatest unit output."[52] Likewise, Congress defined 'sustained yield' to mean "high-level ... output of the various renewable resources of the national forests without impairment of the productivity of the land."[53] The main focus of national forest planning is maintenance of forest conditions to support a sustained yield of timber, water supply, and other economic resources.[54] Even under the multiple use and sustained yield mandate of these statutes, however, the national forest planning and management process provides for restoration in some circumstances and for some purposes. This reflects the difference discussed above between a value choice to 'restore' an ecosystem to its natural conditions as closely as possible, and a decision to 'restore' selected ecosystem values, services, or conditions within the context of a balance between human and ecosystem needs. In some cases the statute is designed to 'restore' the forest's capacity to produce goods and materials, and in others to restore ecosystem components. For example, the act requires the National Forest Service, as part of its inventory of national forest resources, to evaluate "opportunities for improving their yield of tangible and intangible goods and services"[55] (thus reflecting both objectives), and to reforest areas that have been cut or otherwise denuded as well as areas "that are not growing at their best potential rate of growth."[56] Similarly, the act directs the service to "protect and, where appropriate, improve the quality of soil, water, and air resources" in national forests,[57] and to provide, where 'appropriate' and 'practicable,' for a "diversity of tree species similar to that existing in the region controlled by the plan."[58]

General public domain lands managed by the Bureau of Land Management (BLM) are subject to a similar balance under the Federal Land Policy and Management Act of 1976 (FLPMA).[59] Like the Forest Planning Act, FLPMA establishes a multiple use and sustained yield mandate for federal lands that have not been reserved for specific purposes.[60] Unlike the National Forest Management Act, however, FLPMA does not appear to include specific restoration tools, objectives, or mandates. This may reflect that Congress chose an even more utilitarian set of goals for these general public domain lands, but that conclusion appears contradicted by the fact that one of the statute's policies is to protect ecological and environmental values (among others), to "preserve and protect certain public lands in their natural condition," and to "provide food and habitat for fish and wildlife,"[61] as well as the statutory mandate to identify and to "give priority to the designation and protection of areas of critical environmental concern."[62] It is more likely, then, that Congress simply conferred even broader discretion on BLM to determine what management tools are appropriate for particular kinds of lands.

Under the Sikes Act, even military lands are subject to a range of conservation and restoration programs, reflecting the military as well as other uses and values of those areas. This is surprisingly important from an ecological perspective because the Department of Defense manages a large amount of acreage that is often not subject to significant development pressures or other wildlife impacts, precisely because it is generally limited to military uses.[63] The act directs the Secretary of Defense to "carry out a program for the conservation *and rehabilitation* of natural resources on military installations,"[64] supported by "an integrated natural resources management plan" for each military installation with significant natural resources.[65] With respect to military lands, Congress made the rather obvious policy choice that military needs take priority, directing that conservation activities be "[c]onsistent with the use of military installations to ensure the preparedness of the armed forces."[66] Within that constraint, however, Congress required the secretary, "to the extent appropriate and applicable," to provide for a wide range of conservation activities, including fish and wildlife habitat enhancement or modifications and wetland protection, enhancement, and restoration,[67] as well as restoration and management of migratory game birds.[68]

3. ENVIRONMENTAL PROTECTION STANDARDS AND REQUIREMENTS

The Clean Water Act (CWA)[69] and the Clean Air Act (CAA)[70] are the two most general environmental statutes affecting ecological restoration, in the sense that they apply nationwide and require the establishment and attainment of ambient environmental standards for water and air quality (as opposed to individual source pollution control obligations alone). Although the applicability of these pollution control statutes to ecosystem restoration and management efforts may not seem obvious, at least with respect to water pollution, Congress established this linkage in both the text and the legislative history of the CWA. Indeed, in some cases, such as the Everglades and the California Bay-Delta, CWA compliance has been one key factor in stimulating major ecosystem restoration programs.[71]

The CWA begins with an overarching ecological restoration objective, "to *restore* and maintain the chemical, physical, and biological integrity of the Nation's waters" (emphasis added).[72] On its face, the term 'restore' applies to physical, biological, and chemical integrity of the nation's waters, suggesting ecological restoration, as well as reduction or elimination of future chemical pollution, with which the CWA is most closely associated.[73] Moreover, additional CWA provisions reinforce that the scope of the law extends beyond chemical pollution to ecological restoration and protection.[74] For example, the CWA distinguishes between discharges of 'pollutants,' which consist of discrete releases of contaminants into waterways from point sources,[75] and 'pollution,' which means "the man-made or man-induced alteration of the chemical, physical, biological, and radiological integrity of water."[76]

In the legislative history of the 1972 statute, which added this overarching statutory objective to the existing Federal Water Pollution Control Act, as well as later amendments, the House and Senate committees confirmed that the statutory term 'chemical, physical, and biological integrity' established a goal of ecological restoration, rather than simply pollution reduction or elimination. Although the legislative history uses some terms that are not necessarily useful in establishing ecological restoration objectives, such as 'pristine state,' other aspects of the explanations are more useful, and in many ways consistent with concepts in current restoration ecology. For example, the 1972 Senate Report calls for a return to ecological conditions that are "functionally identical to the original."[77] The 1972 House Report defines 'integrity' to mean "a condition in which the natural structure and function of ecosystems is maintained," and a state that is not entirely free of human alteration, but alteration limited to perturbations within the system's natural ability to respond.[78] Both the 1977 and 1985 Senate Reports indicate that the target state for restored ecosystems should support a "balanced, indigenous population," meaning the mix and balance of species that would occur naturally, or absent human alterations to the ecosystem.[79]

Several provisions of the CWA help to effectuate these ecosystem restoration goals. First, the act requires either states or EPA to promulgate water quality standards that define the beneficial uses for which individual waters or categories of waters must be protected, and water quality criteria sufficient to do so.[80] Those standards essentially define restoration goals for the nation's aquatic ecosystems and, 'wherever attainable,' must include water quality sufficient to support the "protection and propagation of fish, shellfish, and wildlife and . . . recreation in and on the water."[81] Thus, although Congress authorized individual states to identify more specific restoration goals for individual water bodies, it preempted those decisions to some degree by proclaiming that all waters must be restored sufficiently, wherever possible, to protect recreation and fish and aquatic life at a minimum. From an ecosystem restoration and management perspective, it is interesting that many states, with EPA's active encouragement, have adopted water quality criteria that extend beyond individual chemical pollutant standards (e.g., no more than x mg/L of a particular contaminant) to broadly-defined biological criteria based on the abundance and diversity of multiple trophic levels of indigenous biota compared to relatively unimpaired reference systems.[82]

Other provisions of the act require, although with varying degrees of stringency, that those water quality goals be implemented through permit limits on point source discharges,[83] best management practices for nonpoint sources of pollution,[84] and comprehensive pollutant load allocations and reduction plans for aggregate sources of pollution within watersheds.[85] The act is deficient, however, in specific mechanisms or mandates to restore ecosystems from the wide range of insults to aquatic ecosystem health other than discharges of chemical or biological pollutants, such as dams, water diversions, levees and channelization, or exotic species.[86] However, where water quality standard (WQS) violations have provided the necessary impetus to develop comprehensive watershed restoration and management programs, as has occurred in places like the Everglades, Chesapeake Bay, and San Francisco Bay-Delta, broader restoration efforts are often identified as a more cost-effective and successful means of ecosystem restoration. Moreover, as discussed further below,[87] restoration efforts in the form of compensatory mitigation are required as a condition of wetlands loss and degradation incurred through activities permitted under section 404 of the act for discharges of dredged or fill material into waters of the United States.

Although it does not use the word 'restore,' one of the declared purposes of the CAA is similarly "to protect and *enhance* the quality of the Nation's air resources so as to promote the public health and welfare and the productive capacity of its population" (emphasis added).[88] The linkage between air quality and ecosystem restoration may be less clear in both the text of the CAA and its legislative history, and the practical connection may be less obvious as well. However, at least some CAA operative provisions have affected ecosystem restoration and management efforts in the past, and other components of the CAA may simply be untapped as ecosystem restoration drivers or tools. The prevention of significant deterioration (PSD) program in the CAA includes a process for designation and protection of 'Class I' airsheds, to remedy "any existing impairment of visibility" in those areas,[89] which include national parks, national memorial parks, national wilderness areas, and international parks of specified sizes.[90] Those provisions are directly applicable if the concept of restoration is viewed broadly enough to include aesthetic, as well as ecological restoration, as a component of ecosystem management, and air pollution in Class I areas may also have ecological impacts. Congress adopted the acid rain reduction program enacted in Title IV-A of the CAA[91] in response to threats to natural resources and ecosystems, as well as property and public health.[92] Moreover, the national ambient air quality standards (NAAQS) adopted by EPA are designed to protect both public health and welfare, which includes impacts on such environmental resources as farms, forests, and other ecosystem attributes.[93] Thus, like WQS, the NAAQS establish ecological restoration goals to the extent that attainment efforts are designed to protect those resources.

B. Federal Statutes Mandating Restoration, Reclamation, or Mitigation

Whereas the statutes discussed above establish ecological and other goals that may guide or directly govern restoration decisions on federal lands, other federal statutes more expressly mandate or provide for affirmative restoration efforts. Some apply only to federal lands, but others either require or facilitate restoration on private lands, or on lands within the jurisdiction of other levels of government. Some provide open-ended authority for whatever restoration actions are appropriate under the circumstances, while others mandate specific kinds of restoration or remediation under particular circumstances.

1. GENERAL FEDERAL LANDS RESTORATION MANDATE

In addition to the statutes governing programs on particular land units, a more generic federal statute governs conservation and restoration programs on federal lands other than military reservations, national parks, national wildlife reservations, or Indian reservations.[94] The Sikes Act requires the Secretaries of Interior and Agriculture to "plan, develop, maintain and coordinate programs for the conservation and rehabilitation of wildlife, fish, and game," including "specific habitat improvement projects,"[95] and pursuant to comprehensive plans developed by those federal departments in consultation with the states.[96] The act defines 'conservation and rehabilitation programs' as "those methods and procedures which are necessary to protect, conserve, and enhance wildlife, fish, and game resources to the maximum extent practicable on public lands . . . consistent with any overall land use and management plans for the lands involved," including restoration activities such as propagation, transplantation, and regulated taking.[97] Thus, this planning, conservation, and rehabilitation program is designed to supplement, rather than replace, other federal land management planning processes—such as those mandated by FLPMA and the NFMA—and comprehensive fish and wildlife conservation and rehabilitation plans must be "consistent with" those overall land use plans.[98] The act also authorizes state agencies to require and sell stamps to allow hunting and fishing on public lands, with the proceeds to be used to implement conservation and rehabilitation programs, including purchase of lands or access rights.[99]

2. WILDLIFE PROTECTION AND RESTORATION REQUIREMENTS

A second category of federal restoration mandates is included in statutes designed to protect and restore—or 'recover'—various species of threatened, endangered, or otherwise sensitive species of plants and animals. Although the most prominent example is the Endangered Species Act,[100] others include the Bald and Golden Eagle Protection Act,[101] the Migratory Bird Treaty Act,[102] and the Marine Mammal Protection Act.[103] In addition, Congress has adopted a set of related statutes providing specifically for fish and wildlife restoration and protection, including the Fish and Wildlife Coordination Act[104] and the Federal Aid in Wildlife Restoration Act (also known as the Pittman-Robertson Wildlife Restoration Act).

The nominal focus of the Endangered Species Act (ESA) is individual species. However, Congress also articulated a broader goal of providing "a means whereby the ecosystems upon which endangered species and threatened species depend may be conserved."[105] Moreover, while the term 'conserve' appears to suggest protection and management of current populations, the statutory definition of "'conserve,' 'conserving,' and 'conservation'" is much broader, including "all methods and procedures which are necessary to bring any endangered species or threatened species to the point at which the measures provided pursuant to this chapter are no longer necessary."[106] Thus, the ESA should more properly be viewed as not only a species restoration statute, but at least for ecosystems that support threatened and endangered species, as an ecosystem restoration statute. Although limited or qualified in a number of ways, several operative provisions of the ESA reinforce and implement this intent.

First, several provisions of the ESA articulate or require the Secretary of Interior[107] to establish more specific restoration goals. For example, the act defines "'critical habitat' for a threatened or endangered species" to include "those physical or biological features ... essential to the conservation of the species,"[108] suggesting a focus on restoration of ecosystem attributes necessary to support listed species. In the recovery planning process, the secretary must establish "objective, measurable criteria" and time estimates for species recovery, which has been implemented by reference to such factors as minimum viable population size goals and targets for the number of discrete population groups within a species meta-population.[109]

Second, a number of ESA provisions require restoration as an affirmative strategy for species recovery. On identification of a species as threatened or endangered, the secretary is concurrently required, with some exceptions,[110] to designate critical habitat for that species.[111] Critical habitat designations can be used to focus restoration and other related components of ecosystem management efforts. For each listed species, the ESA also requires the secretary to adopt recovery plans for species conservation and survival, including "site-specific management actions" to achieve species recovery,[112] and which can also be incorporated into larger ecosystem management plans. For collaborative ecosystem management efforts that also involve state governments, recovery plans can be integrated with state management programs and policies through cooperative agreements.[113]

Finally, although the ESA is also a strict regulatory statute with respect to proposed activities that might harm threatened or endangered species, restoration tools are built into its enforcement provisions as well. Indeed, the regulatory prohibitions in the act serve as much as a stimulus for restoration and recovery actions as they do as bans on development activities *per se*. Section 7 of the ESA requires all federal agencies whose actions might jeopardize the continued existence of a listed species or destroy or adversely modify critical habitat for such species to consult with the secretary before proceeding.[114] Although the secretary has the power to issue an unadorned jeopardy finding, far more often the project is allowed to proceed based on "reasonable and prudent alternatives" the secretary deems sufficient to prevent jeopardy,[115] which typically involve measures reasonably characterized as designed to promote species recovery through various restoration strategies. Likewise, section 9 of the ESA imposes a presumptive ban on the 'taking' of endangered species by any person or entity,[116] and the term 'take' is defined to include a range of detrimental activities beyond overtly killing.[117] Section 10, however, authorizes the secretary to issue incidental take permits if the applicant prepares and implements a conservation plan designed to minimize and mitigate the effects of the taking.

Arguably, neither the requirement in section 7 for reasonable and prudent alternatives, nor in section 10 for conservation plans as a condition of incidental take permits, however, require mitigation sufficient to 'improve' species status. Section 7 requires only alternatives sufficient to prevent a violation of section 7(a) (2), meaning that the action will not jeopardize the continued existence of the species. Section 10 contains an arguably weaker standard, requiring that the permitted taking and conservation plan, considered together, will not "*appreciably reduce* the likelihood of the survival and recovery of the species in the wild."[118] From the perspective of species recovery, therefore, these provisions might be construed to constitute mitigation, rather than true restoration. However, they also promote actions typical of restoration in order to comply with the statute and successful efforts under these provisions might still promote species recovery, given uncertainties in predicting how much harm will result from the permitted take and how much effective mitigation (or restoration) results from the implemented alternatives or conservation plans, respectively. Moreover, in *Gifford Pinchot Task Force v. U.S. Fish & Wildlife Service*,[119] the Ninth Circuit invalidated FWS regulations that defined critical habitat as requiring only species *survival*, because that would ignore the more ambitious recovery goals of the statute.

The Migratory Bird Treaty Act predated the ESA by more than half a century and was adopted expressly to implement migratory bird treaties with Great Britain and later with Mexico, Japan, and the Soviet Union.[120] It was added to legislation Congress had passed at the turn of the twentieth century, providing that the Interior Department's duties and powers include "the preservation, distribution, introduction, and *restoration* of game birds and other wild birds," and in particular articulated a statutory goal "to aid in the restoration of such birds in those parts of the United States adapted thereto where the same have become scarce or extinct, and also to regulate the introduction of American or foreign birds or animals in localities where they have not heretofore existed."[121] By contrast, the Bald and Golden Eagle Protection Act[122] includes both criminal and civil penalties for the unlawful taking of eagles,[123] and allows for exceptions pursuant to federally issued permits for scientific, exhibition, or religious purposes,[124] but incorporates no provisions for mitigation or restoration to compensate for those takings.

Likewise, one goal of the Marine Mammal Protection Act (MMPA) is to "replenish any species or population stock" that has fallen below its "optimum sustainable population," in addition to more traditional conservation and protection measures.[125] As such, the statutory definition of "'conservation' and 'management'" includes efforts to increase—as well as maintain—population levels and habitat acquisition, improvement, and protection.[126] Like the ESA, the MMPA includes various prohibitions on the unlawful taking of species covered by the act; but unlike the ESA, various permits for incidental taking of marine mammals are tied only to findings of minimal impacts and efforts to reduce those effects, but not to affirmative restoration or replacement efforts.[127] However, some permits may be issued expressly for the purpose of "enhancing the survival or recovery" of a marine mammal species,[128] and for each species listed as 'depleted' under the act, the Secretary of Commerce is required to promulgate a conservation plan with "the purpose of conserving and restoring the species or stock to its optimum sustainable population."[129]

In addition to these statutes aimed at particular protected species, at least two other federal statutes promote fish and wildlife restoration efforts generally. The Fish and Wildlife Coordination Act (FWCA) authorizes the Secretary of the Interior to cooperate with and to provide assistance to other federal, state, and private entities "in the development, protection, rearing, and stocking of all species of wildlife, resources thereof, and their habitat, in controlling losses of the same from disease or other causes, [and] in minimizing damages from overabundant species,"[130] among other purposes. Subject to several exemptions,[131] the FWCA establishes a process under which any federal department or agency that impounds, diverts, deepens, or otherwise controls or modifies water bodies, or allows such activities by permit or license, must consult with the FWS and relevant state wildlife agency "with a view to the conservation of wildlife resources by preventing loss of and damage to such resources as well as providing for the development and improvement thereof in connection with such water-resource development."[132] That process must include reports of recommendations from the participating agencies with specific estimates of potential damage to wildlife and specific recommendations for mitigation and compensation for those damages.[133] Although not expressly requiring any particular level of wildlife restoration, the statute requires "such justifiable means and measures for wildlife purposes as the reporting agency finds should be adopted to obtain maximum overall project benefits."[134] When a federal department or agency conducts the project, it must include "adequate provision" for wildlife conservation, maintenance, and management "consistent with the primary purposes" of the project.[135]

In the Pittman-Robertson Act and subsequent amendments, Congress provided for a cooperative federal-state program of wildlife restoration efforts, approved jointly by federal and state agencies and funded by a combination of state hunting and fishing license fees and federal matching grants.[136] The act defines a 'wildlife-restoration project' as "the selection, restoration, rehabilitation, and improvement of areas of land or water adaptable as feeding, resting, or breeding places for wildlife," including habitat acquisition and construction efforts designed to facilitate their use for wildlife restoration.[137] As such, it focuses heavily on habitat restoration and acquisition in order to improve the health of wildlife populations. In order to receive federal funding, states must submit a wildlife conservation strategy, which the Secretary of Interior must approve, that focuses on low population and declining species of wildlife "that are indicative of the diversity and health of wildlife of the State," and that identifies adaptive research, conservation, and restoration strategies, and provides for monitoring designed to focus on such species.[138] However, because of the sources of funding and resulting political pressures on state agencies, there is some reason to suspect bias in favor of game fish and animals at the expense of other species or ecological habitat in general. The act strongly supports the prevailing model of fish and wildlife management in the United States, in which fees from hunting and fishing licenses are obligated entirely to state fish and game departments and used exclusively for those purposes,[139] and, in fact, federal funding is allocated in part based on the number of licensed hunters in each state.[140] Although that requirement prevents the diversion of hunting and fishing fees to state expenditures that are not related to wildlife at all, the hunters and fishers who pay those fees understandably prefer them to be used to benefit the resources they use, potentially at the expense of nongame species that are valued for nonconsumptive uses.

3. SITE RECLAMATION, REMEDIATION, AND MITIGATION REQUIREMENTS

Other federal statutes, such as the Surface Mining Control and Reclamation Act (SMCRA),[141] the Comprehensive Environmental Response, Compensation, and Liability Act (CERCLA, or 'Superfund'),[142] the Oil Pollution Act of 1990 (OPA),[143] and the Resource Conservation and Recovery Act (RCRA),[144] require site-specific restoration of individual sites, in the form of either reclamation or remediation, that have been damaged by ongoing or past development activities. Similarly, section 404 of the CWA[145] requires compensatory mitigation as a condition of issuance of permits to discharge material in ways that destroy or degrade wetlands or other aquatic ecosystem values and functions. Because many of these laws focus on past, ongoing, and future activity, they must address the issue of *whether* restoration is appropriate, as well as which restoration goals and standards should apply.

Although each statute applies in specific circumstances or to specific activities, they can also be incorporated, where applicable, as restoration strategies within broader ecosystem management or restoration programs. However, from a definitional perspective there is some question about the relationship between 'reclamation,' 'remediation,' and 'mitigation' one the one hand, and ecological 'restoration' on the other. In some of the statutes, as shown below, Congress uses these terms somewhat interchangeably. Restoration experts, however, note that there can be subtle but important differences. For example, Eric Higgs notes that 'reclamation' is closely allied with, but not identical to, restoration; "[t]o reclaim something means to rescue land from an undesirable state."[146] Although that typically means converting damaged land to a productive use, the extent to which that constitutes restoration in an ecological sense depends on the intended future land use.[147] Similarly, Higgs defines 'remediation' as a process of "remedying environmental insults," but without necessarily focusing on either a return to historical conditions or restoring ecological integrity.[148] Thus, although the statutes discussed in this section focus on narrow efforts to address particular kinds of environmental harm, their end goals nevertheless depend on the same kinds of value judgments—whether made by statute or through agency discretion—involved in more broadly focused restoration efforts.

SMCRA requires affirmative steps to restore both abandoned mines and active mines once mining is completed,[149] but is also designed to prevent future mining on lands "where reclamation . . . is not feasible" or where existing ecological or other resources are deemed too important to risk through the designation of "lands unsuitable for mining."[150] Thus, in SMCRA, Congress made the judgment that postmining restoration is an appropriate trade-off for the future development of coal resources through surface mining, but only where successful restoration is possible and where existing resources are not too valuable to risk. Congress designated some of those areas as unsuitable in the statute itself, including national parks, national wildlife refuges, national trails, wilderness, wild and scenic river areas, and certain national forest lands.[151] Both the secretary and individual states can designate other lands as unsuitable for mining, as part of their surface mining programs or other land use and ecosystem management planning processes, because of their sensitivity or importance for ecological, historic, cultural, scientific, or aesthetic values, or because mining could adversely affect long-range ecological or other natural resources.[152]

Title IV of SMCRA[153] establishes an Abandoned Mine Reclamation Fund[154] financed through fees on active surface coal mining operations.[155] Those funds are administered by the Secretary of the Interior for reclamation activities on federal lands, and to issue grants to state programs[156] for the "reclamation and restoration of land and water resources adversely affected by past coal mining."[157] Where previously mined lands exist within areas subject to larger ecosystem management and restoration programs, this program can be used as one tool to restore land and water resources and either economic activities (such as agriculture or silviculture) or ecological services on which they depend.

Title V of SMCRA applies to proposed or active surface coal mines. For areas in which mining is allowed, the act serves the dual purpose of ensuring appropriate protection during mining through environmental performance standards,[158] and requiring appropriate reclamation pursuant to standards articulated in the statute and in regulations promulgated by the secretary and by states with delegated programs. The most fundamental reclamation standard requires the operator to "restore the land affected to a condition capable of supporting the uses which it was capable of supporting prior to any mining, or higher or better uses of which there is a reasonable likelihood."[159] Among the other detailed reclamation standards are requirements to restore the mined area to its approximate original premining contour,[160] to segregate and remove topsoil and to protect it with appropriate cover crops or otherwise,[161] and to replace soil conditions necessary to restore soils and other growing conditions on prime farmlands.[162] With respect to revegetation, the act establishes one of the more detailed restoration mandates in federal law, i.e., to "establish on the regraded areas, and all other lands affected, a diverse, effective, and permanent vegetative cover of the same seasonal variety native to the area of land to be affected and capable of self-regeneration and plant succession at least equal in extent to the natural vegetation of the area; except, that introduced species may be used in the revegetation process where desirable and necessary to achieve the approved post-mining land use."[163] In short, the standards that apply to reclamation or restoration activities depend first on a value judgment about appropriate postmining land uses, and then on a determination about what specific reclamation requirements are necessary to support those uses.[164]

Section 404 of the CWA,[165] which governs the discharge of dredged or fill material into wetlands and other waters of the United States, similarly must address the dual issue of whether restoration is acceptable as a condition of prospective degradation of aquatic resources, and if so, what nature and degree of restoration suffices to offset that anticipated harm. In this program, restoration is known as 'compensatory mitigation' in recognition of the fact that regulators intentionally allow mitigation in return for expected future activities and the harm they are expected to cause. Notably, this concept stands in sharp contrast to other provisions in the CWA, which prohibit restoration as an alternative to prevention through the implementation of strict, technology-based standards.[166] Unlike surface water discharges of pollutants, however, for which strict treatment requirements can minimize—if not eliminate—environmental harm through prevention-oriented requirements, discharges that involve the permanent filling of wetlands and other waters necessarily destroy or degrade aquatic ecosystems and resources. Moreover, experience shows the extreme difficulty of restoring or replacing wetland resources and values once they are lost.[167] Therefore, an important threshold question is when mitigation is permissible as compensation for anticipated future harm.

The section 404 program answers this question through a hierarchy of requirements. Compensatory mitigation is not even considered until all other available measures are taken, to avoid and then to minimize any resulting harm to wetlands and other aquatic ecosystems,[168] although some have questioned the effectiveness of this 'avoidance first' approach in actual practice.[169] Compensatory mitigation is allowed, therefore, only for unavoidable adverse impacts, and then is designed to result in 'no net loss' of wetland values and functions.[170] Regulators then must grapple with a difficult set of scientific issues, regarding what kinds of compensatory mitigation, and at what locations, are sufficient to meet that goal. Although the resulting regulations are complex,[171] as is true for the SMCRA reclamation regulations discussed above, in general they prefer wetland restoration over wetland enhancement or establishment, restoration within the same watershed in which wetland resources are lost or degraded, and a preferred focus on measures of lost ecological values and functions rather than raw acreage.[172] Ultimately, with compensatory mitigation and similar programs designed to replace lost ecological resources by restoring others, the issue is whether the nature and value of the replacements matches or exceeds what has been lost.[173]

Another group of statutes, principally CERCLA, OPA, and RCRA, as well as their predecessor in section 311 of the CWA (which governs oil and hazardous substance spills),[174] mandates remediation of harm from past activities, typically in the form of contamination by petroleum products and hazardous materials, due to activities predating modern environmental regulation, unintentional spills or other releases, or violations of current requirements. Because of the context of those kinds of events and circumstances, these statutes do not address the issue of whether restoration should occur; instead, they all presume that remediation of harm is necessary, and should be conducted or financed by the parties responsible for the contamination. This presumption and liability approach is consistent with the 'polluter pays' principle of environmental law.[175] As a result, with respect to restoration these statutes focus primarily on remedy selection, i.e., what cleanup or remediation levels are necessary and sufficient under various circumstances.

  CERCLA imposes strict, joint and several, and retroactive liability on parties responsible for hazardous substance releases (including site owners and operators, past owners and operators at the time of disposal, transporters that play a significant role in choosing disposal sites, and generators of released hazardous substances).[176] More importantly for purposes of this analysis, CERCLA requires site remediation to standards established by EPA on a site-specific basis, after an opportunity for public participation,[177] based on a combination of cleanup standards in the statute itself,[178] the National Contingency Plan (NCP) promulgated by EPA,[179] and any other 'legally applicable' or 'relevant and appropriate' federal or state environmental or land use statutes.[180] Remedy selection is necessarily a difficult and complex process, which must account for significant differences among sites in terms of the nature and degree of contamination, the populations and resources at risk, available remediation methods, and other factors. Certain key principles govern the remedy selection process, however. For example, remedies must be cost-effective, account for both short-term and long-term costs and impacts, and prefer permanent reduction of waste volume, toxicity, and mobility over nontreatment alternatives.[181] Ultimately, however, CERCLA cleanups must "assure[] protection of human health and the environment."[182] Similar site cleanup standards apply to the cleanups at existing hazardous waste treatment, storage, and disposal facilities under RCRA (known as 'corrective action')[183] and to spills of oil and other petroleum products under OPA, the cleanup of which are also governed by the NCP.[184]

  The site cleanup requirements explained above primarily require remediation of contaminated areas to protect human health and to avoid additional environmental harm. Other provisions of CERCLA and OPA, however, allow government trustees of public natural resources to collect natural resource damages from responsible parties, and to use those funds to restore or replace those lost or damaged resources. Those provisions potentially authorize remedies that can focus more broadly on ecosystem restoration goals. In 1990, Congress enacted a similar statute, the Park System Resources Protection Act (PSRPA),[185] which adopts a similar natural resource damages remedy for damages to national park resources caused by a wider range of actions (not only the release of hazardous substances or oil), and similar requirements apply to marine sanctuaries.[186]

CERCLA, OPA, and the PSRPA, respectively, allow federal, state, and tribal government natural resource trustees to recover response costs and damages from any person who destroys, causes the loss of, or injures natural resources covered by the respective statutes.[187] The statutory definitions include compensation for the cost of restoring, replacing, or acquiring the equivalent of those resources, and lost use values pending such restoration, replacement, or acquisition.[188] More important, Congress mandated that recovered amounts can only be used to reimburse response costs and damage assessments, and to "restore, replace, or acquire the equivalent of" lost or damaged resources.[189] Those requirements focus far more heavily on ecosystem restoration, but raise interesting issues about the degree to which various natural resources can be restored or replaced, and about the appropriate methods for assessing economic damages for lost use values.[190] They also signal a congressional policy of ecosystem restoration in the face of unintentional releases ranging from major disasters, such as the *Exxon Valdez* and Deepwater Horizon oil spills, to smaller, more localized spills. Unfortunately, however, evidence suggests that these natural resource damage provisions are actually used by federal and state trustees far less frequently than may be appropriate.[191]

C. Federal Statutes Affecting Restoration Decision-making Processes

Several of the federal statutes discussed above include procedures that can help agencies and others involved in restoration and ecosystem management programs decide appropriate restoration goals and strategies. For example, the comprehensive management planning processes in FLPMA, the National Forest Management Act, and the National Park Service Organic Act discussed above require the agencies responsible for those federal land units to consider alternative management policies, and to consider the views of other agencies, various stakeholders and the general public in deciding the appropriate balance between restoration and other objectives. The Coastal Zone Management Act (CZMA) similarly requires state coastal zone management plans and programs to consider restoration among other coastal ecosystem management strategies,[192] with appropriate opportunities for participation by other federal, state, and local agencies, as well as the general public.[193] Likewise, agency consultation required under the ESA and the FWCA,[194] and the mitigation hierarchy discussed above under section 404 of the CWA,[195] involves both alternatives analysis and public participation in ways that facilitate evaluation of appropriate restoration goals and options, with room for consideration of competing value judgments about restoration choices.

The National Environmental Policy Act (NEPA)[196] provides more generic authority and requirements for federal agencies to involve all affected agencies, other levels of government, affected stakeholders, and members of the public in ecosystem restoration decisions. This is true in particular for "every recommendation or report on proposals for legislation and other major Federal actions significantly affecting the quality of the human environment," which requires the preparation of an environmental impact statement (EIS).[197] When an EIS is required (and to some extent, even when an agency prepares only a less detailed environmental assessment (EA) for restoration projects deemed not to cause significant impacts), federal agencies must address a range of issues relevant to the kinds of restoration choices discussed generally above, and for the specific substantive statutes evaluated. The analysis must include, *inter alia*, the environmental impact of the proposed action, any unavoidable adverse impacts from the proposal, alternatives to the proposed action (including alternative restoration strategies), an evaluation of the relationship between short-term uses and long-term environmental productivity, and any irreversible and irretrievable commitments of resources caused by the proposed action.[198]

An initial question might be why the EIS requirement—which Congress adopted mainly to address proposed actions with adverse environmental impacts—should apply to restoration programs designed to *improve* environmental health or ecological integrity?[199] Textually, however, the statute applies to projects that "significantly affect the human environment," not only to projects with adverse effects. Moreover, many of the restoration choices described above do not involve a simple decision about whether something should be restored or not. Often they involve difficult tradeoffs between actions that might be beneficial to some resources and detrimental to others; or that involve choices among different restoration options, with different ranges of costs and benefits.[200] Evaluation of impacts and alternatives in a NEPA process, with participation by all affected interests, can facilitate those decisions.[201]

A related issue surrounding NEPA and other participatory decision processes is how well they can accommodate the concept of adaptive management, which is often critical to successful restoration efforts, given the many uncertainties in how ecosystems respond to particular restoration strategies or other management actions. A complete, thoughtful, and inclusive process is desirable, both as a way to achieve consensus on the many value choices inherent in restoration efforts and as a way to choose among alternative technical means of attaining the selected goals. Adaptive management, however, relies on a subsequent set of considered management experiments, with subsequent management (and related restoration) decisions informed by the results of those experiments.[202] Especially where subsequent restoration efforts are time-dependent, the need to continue lengthy NEPA and other processes can serve as an impediment, rather than an aid to successful restoration programs.[203] Agencies can minimize this impact through a 'tiered' NEPA process in which more abbreviated EAs are used in subsequent restoration stages, following a more complete initial EIS process.[204] Even an EA, however, can consume considerable time and resources that might impede successful restoration programs. A recent lawsuit challenging the adequacy of the Glen Canyon Adaptive Management Program under NEPA, the ESA, and the Grand Canyon Protection Act resulted in four lengthy federal district court opinions over a period of several years.[205] This illustrates how ensuing phases in restoration program decisions, subject to tiered NEPA and ESA analysis or not, can present frequent litigation targets that might help to improve agency decision making, but that also might divert agency resources and delay program decisions and implementation.

1. SPECIFIC FEDERAL ECOSYSTEM RESTORATION STATUTES

In addition to (or at times as part of) these general federal statutes, Congress has also adopted a number of specific statutes or statutory provisions establishing ecosystem restoration and management efforts for particular areas. Some of the many examples include the Comprehensive Everglades Restoration Plan facilitated by the Water Resources Development Act of 2000 and several other pieces of federal legislation,[206] the Grand Canyon Protection Act,[207] the CALFED Bay-Delta Authorization Act,[208] and watershed restoration programs for a series of large, inter-jurisdictional water bodies under the CWA.[209] Although the varied details of these place-based restoration statutes and programs are beyond the scope of this analysis, the very number and diversity of those laws, in addition to the wide array of more general statutes discussed above, suggests that the 'law' of ecosystem restoration may be *too* varied, as discussed in the following, concluding section.

II. CONCLUSION: TOWARD A MORE COMPREHENSIVE LAW OF RESTORATION

This survey of federal statutes requiring or guiding ecological restoration activities indicates that Congress has provided ample authority and, in many cases, significant resources for ecological restoration on federal lands and elsewhere. This very breadth of applicable statutes, however, suggests several related challenges.

First, because some of these laws are designed primarily to address a wide range of issues other than ecological restoration, they do not always 'fit' the restoration process very well. For example, the requirements of NEPA, even with the flexibility of tiered EIS processes, can lead to troublesome delay in restoration programs pursued through adaptive management strategies. A separate federal environmental restoration statute, which establishes a process that requires agencies to weigh critical restoration decisions with appropriate public input, but that is more consistent with the timing of adaptive management programs, is worthy of consideration. The ESA, despite its focus on ecosystems as well as species, is driven by particular recovery goals for individual species rather than general ecosystem restoration. The CWA includes a comprehensive aquatic ecosystem restoration goal, but lacks effective mechanisms to redress many sources of aquatic ecosystem degradation. Those limitations also suggest that a patchwork of laws with some restoration components is insufficient to establish and implement an effective national program of ecosystem restoration.

Second, in some cases restoration efforts are governed by multiple, potentially inconsistent substantive or procedural requirements, even considering federal statutes alone. For example, the comprehensive Everglades restoration program involves, among other matters, the impacts of water pollution on endangered species within both a national park and several national wildlife refuges, and major federal actions to address them. Those factors alone require simultaneous compliance with NEPA, the CWA, the ESA, and the statutes governing management of national parks and wildlife refuges, in addition to federal and state legislation enacted specifically to facilitate Everglades restoration. A uniform ecosystem restoration statute alone would not reconcile all possible inconsistencies among those laws, but could, for example, exempt a restoration from NEPA if a functionally equivalent process is employed.

Third, the existence of a large number of ecosystem-specific restoration statutes and provisions suggests mixed messages about the state of the federal 'law' of ecosystem restoration. Where bioregionalism or other sociopolitical factors generate sufficient support, Congress has been willing to support them, with significant amounts of federal funding in some cases.[210] And the diversity of restoration approaches included in these place-specific statutes, and the fact that they were needed to support the respective restoration efforts, possibly suggests that generic legislation alone may not account adequately for the varying ecological, political, economic, social, and other factors involved in restoration programs and decisions. On the other hand, the number and diversity of place-specific restoration statutes might suggest that the existing menagerie of generic federal statutes outlined above governing various restoration program goals, methods, and sources of authority is too fragmented and incomplete to account for the full range of necessary or desirable ecosystem restoration efforts. Otherwise, the more specific statutes would not be necessary. Moreover, the fact that some regions have generated sufficient political support for massive amounts of federal funding and agency efforts for ecosystem restoration may not result in an entirely rational distribution of available resources among the many possible ecosystem restoration targets around the country. Other ecosystems may be equally worthy candidates for attention, but unable to attract similar attention, due to geography, representation in Congress, or other factors. That also suggests that a more comprehensive federal approach to ecosystem restoration might be appropriate.

NOTES

1. *See* Richard O. Brooks, Ross Jones & Ross A. Virginia, Law and Ecology 267–68 (2002) (collecting definitions of ecosystem management from diverse sources).

2. *See id.* (quoting definitions by the Environmental Protection Agency, the Fish and Wildlife Service, the Bureau of Land Management, and the Interagency Ecosystem Management Task Force).

3. *See id.* (quoting definitions by Clark & Zaunbrecher, Grumbine, Noss & Cooperrider, the Ecological Society of America, the Forest Ecosystem Management and Assessment Team, the Keystone National Policy Dialogue on Ecosystem Management, and the Society of American Foresters).

4. For example, although oil development has been proposed for many years in the coastal plain of the Arctic National Wildlife Refuge, currently it is devoid of significant human development.

5. *See* Dave Egan & Evelyn A. Howell, *Introduction, in* The Historical Ecology Handbook 7 (Dave Egan & Evelyn A. Howell eds., 2005); Eric Higgs, Nature by Design 38, 119 (2003).

6. *See* Higgs, *supra* note 5, at 93–110.

7. *See* Egan & Howell, *supra* note 5, at 7; Higgs, *supra* note 5, at 39; Margaret A. Palmer, Donald A. Falk & Joy B. Zedler, *Ecological Theory and Restoration Ecology, in* Foundations of Restoration Ecology 1, 1 (Donald A. Falk, Margaret A. Palmer & Joy B. Zedler eds., 2006).

8. *See infra* notes 17–48 and accompanying text.

9. *See* Egan & Howell, *supra* note 5, at 10; Higgs, *supra* note 5, at 13, 106.

10. *See* Higgs, *supra* note 5, at 41 (noting that setting restoration goals involves both "scientific and cultural knowledge").

11. *See id.* at 211, 226, 256; Robert W. Adler, *Restoring the Environment and Restoring Democracy: Lessons from the Colorado River*, 25 Va. Envtl. L.J. 55 (2007).

12. *See* Mary Doyle, *Introduction: The Watershed-Wide, Science-Based Approach to Ecosystem Restoration, in* Large-Scale Ecosystem Restoration, at ix, xiii (Mary Doyle & Cynthia A. Drew eds., 2008) (identifying federal statutes such as the Clean Water Act and Endangered Species Act as principle drivers of five large ecosystem restoration programs studied).

13. Higgs, *supra* note 5, at 2–4.

14. Egan & Howell, *supra* note 5, at 3.

15. *See* Robert W. Adler, Restoring Colorado River Ecosystems 7–10 (2007).

16. *See id.* at x.

17. Act of Aug. 25, 1916, ch. 408, 39 Stat. 535 (codified as amended at 16 U.S.C. §§ 1–4 (1997)).

18. 16 U.S.C. § 1 (2006) (emphasis added).

19. *Id.* § 1a–7.

20. *See, e.g.,* Sierra Club v. Andrus, 487 F. Supp. 443 (D.D.C. 1980), aff'd *sub. nom.* Sierra Club v. Watt, 659 F.2d 203 (D.C. Cir. 1981).

21. *See, e.g.,* Greater Yellowstone Coal. v. Kempthorne, 577 F. Supp. 2d 183 (D.D.C. 2008).

22. Robert B. Keiter, *The National Park System: Visions for Tomorrow*, 50 Nat. Resources J. 71, 96–98 (2010).

23. *See* 16 U.S.C. § 1a–2.

24. *See* U.S. Dep't of the Interior, Nat'l Park Service, Zion National Park General Management Plan 7 (2001); Grand Canyon Trust, The Potential for Restoration Along the Virgin River in Zion National Park (2001).

25. Grand Canyon Protection Act of 1992, Pub. L. 102-575, 106 Stat. 4600, § 1802(a) (requiring operation of Glen Canyon Dam "in such a manner as to protect, mitigate adverse impacts to, and improve the values for which Grand Canyon National Park and Glen Canyon National Recreation Area were established...."). *See infra* notes 207 to 208 and accompanying text.

26. 16 U.S.C. § 1431(b)(3) (2006).

27. *Id.* § 1434(a)(2)(C).

28. 16 U.S.C. §§ 668dd–ee (2006).

29. *Id.* § 668dd(a)(2) (emphasis added).

30. *Id.* § 668ee(4) (emphasis added).

31. *Id.* § 668dd(a)(4)(A).

32. *Id.* § 668dd(a)(4)(B).

33. *Id.* § 668dd(a)(4)(C).

34. *Id.* § 668dd(a)(3)(B).

35. *Id.* § 668dd(a)(3)(C).

36. *Id.* § 668dd(c)(3)(A)(i).

37. Wilderness Act, Pub. L. 88-577, 78 Stat. 890 (1964) (codified as amended at 16 U.S.C. §§ 1131–1136 (2006)).

38. Wild and Scenic Rivers Act, Pub. L. No. 90-542, 82 Stat. 906 (1968) (codified as amended at 16 U.S.C. §§ 1271–1287 (2006)).

39. 16 U.S.C. § 1131(a).

40. *Id.* § 1131(c).

41. *See id.* § 1133(b).

42. *See id.* § 1133(c) & (d).

43. *See, e.g.,* Steens Mountain Cooperative Management and Protection Act of 2000, § 122(c), Pub. L. No. 106-399, 144 State. 1655, 1664 (2000).

44. 16 U.S.C. § 1271.

45. *See id.* § 1273(b).

46. *Id.* §§ 1278, 1279.

47. *See id.* § 1273(b).

48. *Id.* § 1281(a) (emphasis added).

49. *Id.* § 1283(c).

50. Rangeland Renewable Resources Planning Act of 1974 Pub. L. No. 93-378, 88 Stat. 476 (codified as amended at 16 U.S.C. §§ 1600–1687 (2006)).

51. *See* Multiple-Use Sustained-Yield Act of 1960, Pub. L. No. 86-517, 74 Stat. 215 (codified as amended at 16 U.S.C. §§ 528–531 (2006)).

52. *Id.* § 531(a).

53. *Id.* § 531(b).

54. *See, e.g.,* 16 U.S.C. §§ 1600(3), 1601(a)(2), 1601(d)(1), 1604(e).

55. *Id.* § 1601(a)(2).

56. *Id.* § 1601(d).

57. *Id.* § 1602(5)(C).

58. *Id.* § 1604(g)(3)(B).

59. Federal Land Management and Policy Act of 1976, Pub. L. No. 94-579, 90 Stat. 2743 (codified as amended at 43 U.S.C. §§ 1701–1785 (2006)).

60. *See* 43 U.S.C. §§ 1701(a)(7), 1702(c), 1702(h), 1712(c)(1).

61. *Id.* § 1701(a)(8).

62. *Id.* §§ 1701(a)(11), 1702(a), 1712(c)(3).

63. *See generally*, Robert B. Keiter & Robert W. Adler, *NEPA and Ecological Management: An Analysis with Reference to Military Base Lands, in* Environmental Methods Review (Porter & Fittipaldi eds., 1998).

64. 16 U.S.C. § 670a(a)(1)(A)(2006) (emphasis added).

65. *Id.* § 670a(a)(1)(B).

66. *Id.* §§ 670a(a)(3), (b).

67. *Id.* § 670a(b)(1).

68. *Id.* § 670b.

69. 33 U.S.C. § 1251–1387.

70. 42 U.S.C. § 7401–7671q.

71. *See* Doyle, *supra* note 11, at x.

72. 33 U.S.C. § 1251(a).

73. *See* Robert W. Adler, *The Two Lost Books in the Water Quality Trilogy: The Elusive Objectives of Physical and Biological Integrity*, 33 Envtl. L. 29 (2003).

74. For a more detailed analysis of these provisions than is possible here, *see id.* at 37–43.

75. *See* 33 U.S.C.A. § 1362(14) (West 2008) (defining "point source" as "any discernable, confined and discrete conveyance. . . .").

76. *Id.* § 1362(19).

77. S. Rep. No. 92-414, at 76 (1972), *reprinted in* 1972 U.S.C.C.A.N. 3742.

78. H.R. Rep. No. 92-911, at 76–77 (1972).

79. S. Rep. No. 95-370, at 51 (1977), *reprinted in* 1977 U.S.C.C.A.N. 4376; S. Rep. No. 99-50, at 15 (1985).

80. 33 U.S.C. § 1313(c) (2006).

81. *Id.* § 1251(a)(2).

82. *See* Adler, *supra* note 73, at 70–75.

83. 33 U.S.C.A. § 1342.

84. *Id.* § 1329.

85. *Id.* § 1313(d).

86. *See* Adler, *supra* note 73.

87. *See infra* notes 165 to 173 and accompanying text.

88. 42 U.S.C. § 7401(b)(1) (2006).

89. *Id.* § 7491(a)(1).

90. *Id.* § 7472(a).

91. *Id.* §§ 7651–76510.

92. *See id* § 7651(1).

93. *See* 42 U.S.C. §§ 7408, 7409.

94. 16 U.S.C. § 670k(4) (2006).

95. *Id.* § 670g(a).

96. *See id.* § 670h.

97. *Id.* § 670k(6).

98. *Id.* § 670h(b).

99. *Id.* § 670i.

100. Endangered Species Act, Pub. L. No. 93-205, 81 Stat. 884 (1973) (codified as amended at 16 U.S.C. §§ 1531–1544 (2006)).

101. 16 U.S.C. §§ 668–668d.

102. 16 U.S.C. §§ 703–712.

103. Marine Mammal Protection Act, Pub. L. No. 92-522, 86 Stat. 1027 (1972) (codified as amended at 16 U.S.C. §§ 1361–1423h (2006)).

104. Fish and Wildlife Coordination Act, Pub. L. No. 85-624, 72 Stat. 563 (1958) (codified as amended at 16 U.S.C. §§ 661–667d (2006)).

105. 16 U.S.C. § 1531(b) (2006).

106. *Id.* § 1532(3).

107. For certain marine species the Secretary of Commerce has this authority. *See id.* § 1533(a)(2).

108. *Id.* § 1532(5)(A).

109. *Id.* § 1533(f)(1)(b)(ii).

110. *Id.* §§ 1533(a)(3)(B), (b)(6)(C).

111. *Id.* § 1533(a)(3)(A).

112. *Id.* § 1533(f).

113. *Id.* § 1535(c).

114. *Id.* § 1536(a).

115. *Id.* § 1536 (b)(3).

116. *Id.* § 1538. Although this ban applies only to endangered and not to threatened species, the secretary has the authority to adopt regulations imposing similar requirements for threatened species. *Id.* § 1533(d).

117. *Id.* § 1532(19).

118. *Id.* § 1539(a)(2)(B)(iv) (emphasis added).

119. 378 F.3d 1059, 1070–71 (9th Cir. 2004).

120. *See* 16 U.S.C. §703(a) (2006).

121. *Id.* § 701.

122. 16 U.S.C. §§ 668–668d.

123. *Id.* §§ 668(a), (b).

124. *Id.* § 668a.

125. *Id.* § 1361(2).

126. *Id* § 1362(2).

127. *See id.* §§ 1371–1374.

128. *Id.* § 1374(c)(4).

129. *Id.* § 1383b(b).

130. 16 U.S.C. § 661.

131. *Id.* § 661(h) (exempting projects impounding less than ten acres or activities by federal agencies primarily for federal land management).

132. *Id.* § 662(a).

133. *Id.* § 662(b).

134. *Id.*

135. *Id.* § 663(a).

136. *See* 16 U.S.C. § 669–669k.

137. *Id.* § 669a(8).

138. *Id.* § 669c(d)(1)(D).

139. *See id.* § 669.

140. *Id.* § 669c(b), (c).

141. Surface Mining Reclamation, Conservation and Recovery Act, Pub. L. No. 95-87, 91 Stat. 456 (1977) (codified as amended at scattered sections of 30 U.S.C.).

142. Comprehensive Environmental Response, Compensation and Liability Act, Pub. L. No. 96-510, 94 Stat. 2767 (1980) (codified as amended at 42 U.S.C. §§ 9601–9675 (2006)).

143. Oil Pollution Act of 1990, Pub. L. No. 101-380, 104 Stat. 484 (codified as amended at 33 U.S.C.A. §§ 2701–2762 (West, 2010)).

144. Resource Conservation and Recovery Act, Pub. L. No. 94-580, 90 Stat. 2796 (1976) (codified as amended, as part of the Solid Waste Disposal Act, at 42 U.S.C. §§ 6901–6992k (2006)).

145. 33 U.S.C. § 1344.

146. Higgs, *supra* note 5, at 99.

147. *Id.*

148. *Id.* at 100.

149. *See* 30 U.S.C. §§ 1201–1202.

150. *Id.* §§ 1202(c); 1272.

151. *Id.* § 1272(e).

152. *Id.* § 1272(a), (b).

153. *Id.* §§ 1231–1244.

154. *Id.* § 1231.

155. *Id.* § 1232.

156. *Id.* § 1235.

157. *Id.* §§ 1231(c)(1), 1233(a).

158. *Id.* § 1265.

159. *Id.* § 1265(b)(2).

160. *Id.*

161. *Id.* § 1265(b)(5).

162. *Id.* § 1265(b)(7).

163. *Id.* § 1265(b)(19).

164. *See generally,* Nat'l Wildlife Fed'n v. Hodel, 839 F.2d 694 (D.C. Cir. 1988) (evaluating adequacy of Interior Dep't regulations establishing reclamation requirements for various postmining land uses).

165. 33 U.S.C. § 1344.

166. *See* Riverkeeper, Inc. v. EPA, 475 F.3d 83, 108–10 (2d Cir. 2007), *reversed on other grounds sub nom.* Entergy Corp. v. Riverkeeper, Inc., 129 S. Ct. 1498 (2009).

167. *See* National Research Council, Compensating for Wetland Losses Under the Clean Water Act 22–45 (2001).

168. 40 C.F.R. § 230.10 (2011).

169. *See* Oliver A. Houck, *Hard Choices: The Analysis of Alternatives Under Section 404 of the Clean Water Act and Similar Environmental Laws,* 60 U. Colo. L. Rev. 773 (1989).

170. *See* Compensatory Mitigation for Losses of Aquatic Resources, 73 Fed. Reg. 19594, 19994–95 (Apr. 10, 2008).

171. *See* 40 C.F.R. Part 230, Subpart J (2011).

172. *Id.* § 230.93.

173. *See* James Salzman & J.B. Ruhl, *Currencies and the Commodification of Environmental Law,* 53 Stan. L. Rev. 607, 657–661 (2000).

174. 33 U.S.C. § 1321.

175. *See* Rio Declaration on Environment and Development, U.N. Conference on Environment and Development, U.N. Doc. A/CONF.151/5/Rev.1, 31 I.L.M. 874 (Principle 16) (1992); New York v. Shore Realty, 759 F.2d 1032 (2d Cir. 1985).

176. *Shore Realty,* 759 F.2d at 1042–44; 42 U.S.C. § 9607(a).

177. *Id.* § 9617.

178. *Id.* § 9621.

179. *Id.* § 9605.

180. *Id.* § 9621(d)(2).

181. *Id.* § 9621(a), (b).

182. *Id.* § 9621(d)(1).

183. 42 U.S.C. § 6924(v).

184. 33 U.S.C. § 2712(a).

185. Park System Resources Protection Act, Pub. L. No. 101-337, 104 Stat. 379 (1990) (codified as amended at 16 U.S.C. §§ 19jj–19jj-4 (2006).

186. *See* 16 U.S.C. §§ 1432(6)–(7), 1443 (2006).

187. 42 U.S.C. § 9607(a)(4)(C); 33 U.S.C. § 2702(b)(2)(A); 16 U.S.C. § 19jj.

188. 42 U.S.C. § 9607(a)(4)(C); 33 U.S.C. § 2706(d)(1); 16 U.S.C. § 19jj(b)(1)(A). In the case of national park resources, damages may be based on the value of the resource as a measure of damages only if restoration or replacement is not possible. 16 U.S.C. § 19jj(b)(1)(B).

189. 42 U.S.C. § 9607(f)(1); 33 U.S.C. § 2607(c), (f); 16 U.S.C. § 19jj-3.

190. *See* Ohio v. Dep't of the Interior, 880 F.2d 432, 474–481 (D.C. Cir. 1989).

191. *See* United States General Accounting Office, Status of Selected Federal Natural Resource Damage Settlements (November 1999); United States General Accounting Office, Superfund: Outlook and Experience with Natural Resource Settlements (April 1999).

192. 16 U.S.C. § 1455(d)(2)(I), (d)(9) (2006).

193. *Id.* § 1455(d)(1), (d)(14).

194. *See* 16 U.S.C. §§ 662a, 1536.

195. *See supra* notes 168 to 173 and accompanying text.

196. 42 U.S.C. § 4321–4370f.

197. *Id.* § 4332(2)(C).

198. *Id.*

199. *Cf.* Kootenai Tribe of Idaho v. Veneman, 313 F.3d 1094, 1120 (9th Cir. 2002) (indicating that "[t]he NEPA alternatives requirement must be interpreted less stringently when the proposed agency action has a primary and central purpose to conserve and protect the natural environment, rather than to harm it."), *abrogated by* Wilderness Soc. v. U.S. Forest Service, 630 F.3d 1173 (9th Cir. 2011).

200. *See* Adler, *supra* note 15, at 137–70 (discussing such tradeoffs in Colorado River restoration programs).

201. *See, e.g., id.* at 144 (describing NEPA process for choosing among alternative restoration flow scenarios through the Grand Canyon).

202. *See* Carl J. Waters & C.S. Holling, *Large-Scale Management Experiments and Learning by Doing*, 71 Ecology 2060 (1990) (passim).

203. *See* Doyle, *supra* note 12, at xiii (noting that adaptive management as a restoration tool remains largely untested, but that the process involves "a long, complex, painstaking process of negotiation among government representatives and stakeholders...to reach agreement on a restoration plan.").

204. *See* 40 C.F.R. § 1502.20 (2011).

205. *See* Grand Canyon Trust v. Bureau of Reclamation, No. CV-07-8164 PCT-DGC, 2008 WL 4417227 (D. Ariz. Sept. 26, 2008); Grand Canyon Trust v. Bureau of Reclamation, 623 F. Supp. 2d 1015 (D. Ariz. May 26, 2009); Grand Canyon Trust v. Bureau of Reclamation, No. CV-07-8164-PHX-DGC, 2010 WL 2643537 (D. Ariz. June 29, 2010); Grand Canyon Trust v. Bureau of Reclamation, 2011 WL 1211602 (D. Ariz. Mar. 30, 2011).

206. Water Resources Development Act of 2000, Pub. L. No. 106-541, 114 Stat. 2687; *see* Terrence "Rock" Salt, Stuart Langton & Mary Doyle, *The Challenges of Restoring the Everglades Ecosystem, in* Large-Scale Ecosystem Restoration, *supra* note 12, at 5, 11–12.

207. Grand Canyon Protection Act, Pub. L. No. 102-575, 106 Stat. 4600 (1992).

208. CALFED Bay-Delta Authorization Act, Pub. L. No. 108-361, 118 Stat. 1681 (2004).

209. 33 U.S.C. §§ 1267 (Chesapeake Bay), 1268 (Great Lakes), 1269 (Long Island Sound), 1270 (Lake Champlain), 1330 (National Estuary Program for series of estuaries).

210. *See* Doyle, *supra* note 12, at xii (noting funding levels of $20 billion for Everglades restoration, $19 billion for Chesapeake Bay restoration, $8 billion for the first eight years of San Francisco Bay-Delta restoration, and over $5 billion for Upper Mississippi River restoration).

9 Landscape-scale Conservation and Ecosystem Services
Leveraging Federal Policies
Lynn Scarlett & James Boyd, Resources for the Future[1]

The context of conservation and resource management reveals a confluence of two trends as the twenty-first century unfolds. First is the broadening compass of conservation to landscape-scale evaluation and action. Efforts are telescoping outward to encompass whole watersheds and ecoregions and to undertake actions at a scale that accommodates interconnected and intersecting land, water, and wildlife issues.[2] Second is the deepening recognition that natural systems provide services—water purification, coastal storm surge mitigation, flood protection, temperature regulation, and more.[3] Alongside these emergent trends, federal laws, regulations, and policies are evolving to better accommodate landscape-scale actions and assessment and protection of ecosystem services, but the fit between existing laws and these trends in conservation focus is often imprecise or poorly articulated.

I. RISING INTEREST IN ECOSYSTEM SERVICES

As these two conservation trends unfold, they potentially link in mutual reinforcement. Understanding that potential requires examining the rising trajectory of interest in ecosystem (or natural) services. Several considerations have broadened interest in ecosystem services.

First is the search for new revenue streams for landowners to support sustainable practices. For example, in 2005 Florida initiated a Ranchlands Environmental Services Project to field-test payment for environmental services in the

LYNN SCARLETT & JAMES BOYD

northern Everglades ecosystem with eight participating ranchers.[4] Through a partnership including two state agencies, the South Florida Water Management District, the U.S. Department of Agriculture's Natural Resources Conservation Service, the World Wildlife Fund, and local ranchers, ranchers received payments to undertake measures to store water, reduce phosphorous loadings, and provide other environmental benefits. The project, which tested concept design and development of measurements, is now transitioning from experimentation to ongoing implementation.

Second are potential cost savings for basic community services. Consider Seattle's use of 'green infrastructure,' which reduces stormwater runoff volumes at a cost 25 percent less than the traditional alternative. Other examples and economic analysis of potential cost savings are accumulating. In a study of 27 water suppliers, each 10 percent increase in forest cover in a source water area decreased treatment and chemical costs by around 20 percent.[5] In an assessment of urban tree cover, the American Forest Foundation has calculated that tree cover results in avoided energy costs of nearly $3 million annually.[6]

Third are opportunities to more cost-effectively meet regulatory requirements to achieve environmental performance. The local surface water utility for the Tualatin Basin in Oregon, Clean Water Services (CWS), bundled into a single permitting action the renewals of four wastewater treatment permits and a stormwater permit.[7] Rather than investing $60 million in expensive refrigeration systems, the agency worked with the adjacent farming community to plant 37 miles of shade trees along the river to cool water temperatures to required standards. The cost of the ecosystem service approach to the community was $6 million, a tenth the cost of the mechanical cooling.

Fourth are costs associated with ecosystem service losses. Over the past century, the annual number of natural disasters has increased more than forty-fold: rising from 10 in the first decades of the past century to 400 to 500 in the last decades of the twentieth century. These hazards translated into costs that climbed from less than $1 billion in 1900 to over $200 billion in 2005.[8] Eying these costs, communities are re-examining natural systems and their potential to meet their economic, environmental, and safety needs.

Fifth is the growing interest in enhancing resilience in a context of changing conditions. Climate change may result in increased incidence of high-intensity storm events and more variability in water availability. Investments in the protection or restoration of floodplains and coastal dunes and sea marshes have potential to enhance resilience. For example, evaluation of dune protection in

North Carolina showed marked reductions in buildings destroyed or damaged from storm events, compared to areas without dune protection.[9]

Scientists and economists have significantly advanced our ability to both measure and value these goods and services, even in the absence of market transactions. Alongside these analytical developments regarding ecosystem services, conservation and resource management initiatives are increasingly broadening to encompass larger landscapes, whole watersheds, and even ecoregions.

II. TRENDS IN LANDSCAPE-SCALE CONSERVATION

Such efforts to focus action at a large-landscape scale are not entirely new. The Everglades Restoration commitment in the United States has its roots in legislation promulgated in the 1980s. A voluntary, multiparticipant initiative to restore Chesapeake Bay also formed in the 1980s. Other large-scale initiatives in the Bay-Delta of California, the Louisiana Gulf Coast, and Great Lakes began taking shape a decade later. But significant focus on ecoregions and large-scale conservation did not gain broad traction until the late 1990s and later.

This evolution was modestly acknowledged in the Cooperative Conservation Initiative of the George W. Bush administration, which emphasized cross-jurisdictional, multi-issue partnerships to address environmental challenges.[10] It was the Obama administration, however, that specifically articulated large landscape conservation as an operating framework. Under the Obama administration, federal agencies began organizing information and conservation strategies within twenty-two ecoregions;[11] the Bureau of Land Management broadened its planning focus to an ecoregional scale;[12] and the President's initiative on America's Great Outdoors emphasized large-scale actions to address interconnected environmental issues.[13] This federal framework mirrored developments among conservation organizations.

III. FEDERAL POLICY SETTING

Thus, two trends converge—the focus on landscape-scale conservation and on ecosystem services—and link in mutual reinforcement. This convergence is occurring within an environmental statutory and regulatory context in which the focus has, typically, been on individual (rather than multimedia) issues, single species (rather than ecosystems), and site-specific action. Moreover, while many statutes accommodate or support evaluation of the impacts of actions on biophysical resources, these evaluations have not generally encompassed the services provided by those biophysical resources and their functions.[14]

Despite these limitations, numerous federal statutes, regulations, and practices include requirements, tools, or aspirations that could support measurement of, assessment of damage to, and creation and protection of ecosystem goods and services. Several may hold particular potential to support landscape-scale conservation and natural asset (ecosystem services) markets.

Several general observations about the nexus between landscape-scale conservation and ecosystem services help illuminate this potential. First, meaningful measurement of ecosystem goods and services often requires measurement boundaries large enough to capture functional interrelationships among ecosystem components. Second, the universe of ecosystem service beneficiaries of a particular wetland, floodplain, or vegetative cover may extend well beyond the immediate location of that natural feature. Third, though measurement of discrete ecosystem components, functions, and values has characterized much of the work to date, more recent efforts are trending toward integrative metrics. These three features of ecosystem services and their valuation all support operating within a landscape-scale framework.

Several statutes and policies could strengthen this linkage of ecosystem services with landscape-scale conservation. Broadly, these policies bundle into three categories:

- *Planning and Priority-setting Tools and Guidance,* such as those used in National Environmental Policy Act evaluations, land-use planning by federal agencies, and Water Resources and Development Act principles and guidelines.
- *Regulatory Mitigation and Impact Reimbursement Tools,* including wetlands mitigation, conservation banking, Federal Energy Relicensing Commission mitigation measures, and natural resource damages mitigation.
- *Grants, Loans, and Other Investments,* such as Farm Bill land conservation payments, Safe Drinking Water Act and Clean Water Act Revolving Loan Funds.

IV. PLANNING AND PRIORITY-SETTING TOOLS AND GUIDANCE

National Environmental Policy Act: The National Environmental Policy Act (NEPA) offers an appropriate starting point for reviewing policy opportunities to place ecosystem services evaluation within a landscape-scale context, as it is the nation's premier, overarching environmental policy foundation.[15] The

Council of Environmental Quality (CEQ) NEPA regulations require that environmental impact statements include analysis of cumulative effects. The concept of cumulative effects builds on the recognition that a single, incremental action may have minimal ecological effects. But, combined with other actions in the same geographic area, effects may cumulatively impact resources and ecosystem components. CEQ regulations define these effects as impacts to "the components, structures, and functioning of affected ecosystems."[16]

This language offers a clear nexus with evaluation of ecosystem goods and services. Cumulative effects analysis, through its broadened geographic and temporal focus, allows NEPA analysts to examine the effects of actions that alter general ecological processes, such as changing hydrologic patterns and sediment transport. Currently, CEQ has not established specific criteria for determining the appropriate scope and scale of the cumulative impact analysis.

Provision in CEQ regulations of definitions and methods for evaluating ecosystem services could strengthen both ecosystem services evaluation and the use of a landscape-scale analytic framework. A key challenge is how to extend the boundaries of NEPA evaluation beyond an individual public land unit. One possible platform for such analysis would be through use of NEPA's provisions for cooperating agency status.[17] Another possible platform for evaluation beyond the boundaries of an individual agency's land unit could be the Department of the Interior's NEPA regulation on consensus-based collaborative management options, which allows such options to be identified as the preferred alternative in the NEPA public review process.[18] Using this regulatory tool, the Bureau of Land Management (BLM) has collaborated with state agencies, nonprofit organizations, and private landowners in the Las Cienegas Watershed south of Tucson in an integrating land-use planning effort at a watershed scale.

NEPA regulations are sufficiently flexible to enable agencies to undertake multi-unit environmental impact analyses—e.g., on BLM and Forest Service lands—to broaden the scale of analysis. In assessing energy development impacts, for example, the BLM has implemented NEPA analysis across multiple BLM land-use planning units within a single NEPA process.

Federal agency resource management plans provide another, related context for applying a landscape-scale and ecosystem services framework. Requiring calculation of ecosystem services benefits could support a more landscape-scale vantage point, as this scale is often a more relevant analytic context for ecosystem services evaluations. In 2010, Denver's water authority entered into an agreement with the U.S. Forest Service to fund fuels treatments on forestland to maintain source water supplies and prevent erosion. Agency planning frameworks that look beyond their jurisdictional boundaries can illuminate these kinds of possibilities. BLM's ecoregional assessments offer a possible model.

Water Resources Development Acts—Principles and Guidelines: Like NEPA, the Principles and Guidelines (P & G) that frame federal implementation of flood control, irrigation, and other water projects shape multiagency decisions and have the potential to significantly reorient decision making to incorporate watersheds and ecosystem-scale analysis.[19] The Principles and Guidelines govern how the Army Corps of Engineers plans, constructs, operates, and maintains water resources projects, but could also apply to Bureau of Reclamation and other water resources projects. The enormous size of water project appropriations means that the Corps' ecological evaluations can affect decisions with potentially large ecological effects.

The guidelines describe an analytical framework and set of evaluation practices to forecast and describe natural resource conditions, as well as formulate, evaluate, and compare alternatives. The White House Council on Environmental Quality (CEQ), the Corps, and other federal agencies revised the guidelines in 2011. The updated guidelines elevate the importance of ecosystem protection by stating that water resources planning and development "should both protect and restore the environment and improve the economic well-being of the nation for present and future generations."[20] Fuller accounting for the environmental benefits of water resource projects provides an opportunity to focus the analytic framework at a watershed and landscape-scale.

V. REGULATORY MITIGATION AND IMPACT-REIMBURSEMENT REQUIREMENTS

Beyond federal provisions for land use planning, federal statutes and regulations also set forth mitigation and liability requirements. These, too, have potential to foster ecoregional or landscape scale conservation in priority areas and strengthen inclusion of ecosystem services evaluation.

Among these mitigation requirements, wetlands and habitat conservation mitigation banks are evolving in extent and project scale. Banking concepts provide a potential platform for pooling mitigation into large conserved areas. They also potentially provide a context to develop 'full-service' or multibenefits credits and include an ecosystem services focus.

Wetlands Mitigation Banks: The Clean Water Act prohibits discharge of dredged or fill material into waters of the United States unless a permit under section 404 of the act is provided by the U.S. Army Corps of Engineers. Permittees are required to avoid or minimize adverse impacts to wetlands and other regulated water resources. Where impacts are unavoidable, federal regulations require compensatory mitigation to replace the loss of wetland and aquatic resources and functions.

The Environmental Protection Agency (EPA) describes three wetlands mitigation options, which include mitigation banking, in-lieu feeds, and permittee-responsible mitigation.[21] A key tool for fulfilling compensatory mitigation requirements is the use of wetlands mitigation banks. A wetlands mitigation bank is a wetland area that has been restored, established, enhanced or preserved, which is set aside to compensate for future conversions of wetlands through development activities. The growing emphasis on wetlands functions rather than simply on acres lost and gained provides an important precursor to evaluating and quantifying ecosystem services.

While wetlands mitigation banks offer a potential context for maintaining and restoring ecosystem functions, values, and services, their benefits have traditionally been measured in acres lost and gained rather than on "preservation of service value."[22] Nonetheless, section 404(b)(1) guidelines provide regulatory authority to consider ecosystem services benefits of wetlands, such as water purification, and the banking tool generally sets the stage for a more landscape-scale or ecosystem focus for mitigation than is generally possible with on-site mitigation efforts.

Conservation Banking: Patterned after wetlands mitigation banks, conservation banks refer to parcels of land protected and managed to conserve listed species under the Endangered Species Act.[23] To provide greater consistency in their use, the U.S. Fish and Wildlife Service (FWS) published conservation banking guidance in 2003.[24] Despite this guidance, the use of conservation banks to date is extremely modest given the numbers of listed species, their geographic spread, and ESA regulatory requirements associated with listed species.

In 2010, an estimated 133 conservation banks protected just over 100,000 acres to provide benefits for some 90 listed species. Banks range in size from just 27 acres to over 27,000 acres. But extensive land development, especially for traditional and renewable energy, creates a potential impetus for greater use of conservation banks. Linked to this issue is how to target conservation banks in high-priority habitat at scales sufficient to provide meaningful benefits. Within this context, development of banks with a multispecies, ecosystem-based focus may emerge, providing opportunities to use species conservation banks as a tool from which to build 'full-service' or multiple-benefits (including ecosystem services) banking.

Hydropower Relicensing: Mitigation of impacts associated with hydropower facilities provides another significant opportunity to move beyond piecemeal, small-scale mitigation and for federal policies to emphasize ecosystem services functionality. Most nonfederal hydropower dams require licensing under the provisions of the Federal Power Act.[25] As a part of this process, the Federal Energy Regulatory Commission (FERC) must consider recommendations from state and federal fish and wildlife agencies to mitigate impacts of licensed facilities. The Federal Power Act also authorizes the Secretaries of the Interior and Commerce to prescribe mitigation measures for hydropower facilities.

Certain mitigation requirements pertaining to impacts on federal lands and provisions prescribed by the FWS or National Marine Fisheries Service (NMFS) are mandatory. As in wetlands mitigation and conservation banking, the mitigation provisions under the licensing process offer a potential source of funding for ecosystem services investments and a potential source of market demand. Hence they provide an opportunity for agencies to steer mitigation toward landscape-scale, high-priority, and multibenefits conservation, including maintenance of ecosystem services.

These opportunities could be strengthened by two policy changes: 1) updating the Hydropower Interagency Memorandum of Understanding to reference ecosystem services evaluation within the context of requirements to evaluate environmental impacts of projects; and 2) setting mitigation funding priorities through mitigation guidance to emphasize that enhancement of ecosystem services outcomes could result in more effective, better targeted mitigation efforts.

Natural Resource Damages Assessments: Federal Natural Resource Damages Act (NRDA) requirements present a similar opportunity to target conservation and restoration investments with an ecosystem services focus. Several U.S. environmental statutes establish liability for injury to natural resources.

In economic terms, the goal of federal NRDA liability is to "make the environment and public whole" following a pollution event.[26] Current NRDA emphasis is on restoration rather than a monetized estimate of lost value as the measure of damages.[27]

Several NRDA implementation trends show potential for using NRDA funds to supplement other conservation funding to achieve broader goals for restoring and sustaining ecosystems and their benefits. The first is the increased use of off-site restoration. For example, $3 million in NRDA funds resulting from a settlement regarding harbor contamination in Rhode Island were combined with private-sector and nonprofit funds toward purchase of 1.5 million acres of loon nesting habitat in Maine.[28] The second is the increasing emphasis on collaborative projects. For example, $400,000 in NRDA funds were combined with Coast Guard and other nonprofit funding to protect and monitor common eider nesting habitat. The FWS estimated in 2009 that it leverages its annual NRDA settlement funding allocation by a seven to one ratio.[29]

VI. GRANTS, LOANS, AND OTHER INVESTMENTS

Beyond planning, mitigation, and liability policies, the federal government is a big direct investor—through direct payments, grants, and loans—in ecosystem goods and services, though program goals and measures have seldom focused explicitly on ecosystem services.

Farm Bill Programs: Most significant among these federal investments are Farm Bill conservation programs. These programs vary in their provisions, with some programs supporting conservation easements and long-term withdrawal of land from production and others providing 'rental payments' for various conservation practices on working lands. Total projected five-year spending under the 2008 Farm Bill was $24.3 billion.[30]

Farm Bill conservation programs have faced criticism because the allocation of funds are not always closely tied to high-priority ecosystems or high-value outcomes. Funds have also been highly distributed, often invested in dispersed and unlinked parcels of land.

However, some form of ecological ranking or criteria—and thus ecological evaluation—is associated with each of the Farm Bill programs, and implementation of Farm Bill programs holds significant potential to emphasize landscape-scale investments in protecting and enhancing ecosystem functions and services. The 2008 Farm Bill specifically references ecosystem services and environmental markets and called for the creation of an Office of Environmental Markets within the U.S. Department of Agriculture. Under the Obama administration, Natural Resource Conservation Service (NRCS) Chief Dave White has strengthened the focus on larger-scale strategic Farm Bill program investments using a 'conservation beyond boundaries' framework. Through the initiative, NRCS is targeting over $200 million in Farm Bill program spending to projects in nine ecoregions with high-priority conservation needs. For example, in one focal area—the Chesapeake Bay Watershed—NRCS is using a landscape focus to target funding in high-priority watersheds within the larger area to reduce nitrogen, phosphorus, and sediment from agricultural sources.[31]

Ranking criteria of the various programs that support this cross-cutting initiative directly embrace or imply the relevance of landscape-scale conservation.

The NRCS Conservation Effects Assessment Project (CEAP) report for the Chesapeake Bay, for example, showed the effectiveness of conservation investments in high-priority areas to reduce erosion and manage nutrients. The CEAP report indicates that high-priority areas have nitrogen losses of 53 pounds per acre while those in areas with moderate need for treatment have losses of half that amount and areas with lowest treatment needs have losses of just 2 pounds per acre.[32]

But ranking schemes continue to evolve. The most developed ranking schemes to date are associated with the Wetland Reserve and Conservation Reserve programs. Ranking or targeting factors currently used are primarily biophysical in nature, rather than based on measures of the economic or social benefits of a given biophysical outcome. One exception is the Conservation Reserve Program's Environmental Benefits Index, which includes both biophysical outcome measures (like soil erosion vulnerability) and social indicators (the number of well-water users in proximity to the land).

The general principle that payments should be directed toward conservation that yields the largest environmental benefit is well-established in policy discussions. Key proposals to improve program performance center on: 1) consolidating programs that share common purposes and/or consolidating different payment types (rental payments, easements, incentives) into a single, multipurpose payment system; 2) better targeting programs to high-priority conservation areas to achieve ecosystem benefits; 3) developing better performance indicators; and 4) improving environmental returns on investment through use of landscape-scale approaches, competitive bidding to lower the cost of conservation program contracts, and linking payments more directly to environmental performance.

Clean Water Act—Total Maximum Daily Loads: The Total Maximum Daily Load (TMDL) program under the Clean Water Act focuses on water quality outcomes, with establishment of effluent loads that can be discharged consistent with achieving those outcomes. The TMDL program creates a context that is *potentially* conducive to 'effluent trading' programs, since different dischargers face different costs to reduce their pollution loadings.

The 2003 Water Quality Trading Policy and 2004 EPA Water Quality Trading Assessment Handbook were designed to facilitate water trading to lower compliance costs and improve water quality.[33] Through the end of 2006, EPA had sponsored eleven pilot projects to assess trading opportunities and issues in various regions.[34] These efforts create some opportunity for urban water managers to pursue ecosystem services investments, especially efforts that link to the broader nonurban watershed and ecosystem restoration and conservation initiatives. Long Island Sound, Connecticut, and New York adopted a basin-wide plan (the Long Island Sound Comprehensive Conservation and Management Plan) to reduce nitrogen loads in the sound by 58.5 percent over fifteen years.[35] The TMDL policy's virtue lies in the fact that environmental planning and compliance are assessed on a watershed basis. Aggregate conditions across a watershed's geography are the focus of evaluation and quantification.

The concept of permit bubbles offers related opportunities. As the concept of ecosystem services has gained traction, permit bubbles—particularly in the context of water quality—provide a potential tool for supporting ecosystem service payments within a landscape-scale framework. EPA has approved the clustering or grouping of permits for wastewater, stormwater, and other related facilities. The most notable example of this clustering is that of the Tualatin Basin in Oregon. The local surface water utility's watershed jurisdiction includes a number of towns, four wastewater systems, and stormwater runoff from multiple locations. As described earlier in this chapter, the local water agency, Clean Water Services, bundled into a single permitting action the renewals of four wastewater treatment permits and the stormwater permit. Rather than investing $60 million in expensive refrigeration systems, the agency worked with the adjacent farming community to plant shade trees along thirty-five miles of riverbank to cool water temperatures to required standards for $6 million.

Federal Loan Programs: Several EPA loan programs provide similar opportunities to emphasize ecosystem services protection within a landscape scale.

EPA provides grant and loan programs under the Clean Water Act and the Safe Drinking Water Act with potential to support ecosystem services investments to protect water supplies, though these grants have seldom been used for these purposes.[36] These programs include the Clean Water State Revolving Fund (SRF, sec. 212), which offers loans for water quality improvements, most frequently for wastewater treatment infrastructure. However, these funds (over $1 billion, combined with another $4.7 billion in state monies) can be used to implement nonpoint source management plans and develop and implement estuary plans. Just 5 percent of projects target nonpoint source pollution mitigation.

Under the Safe Drinking Water Act, State Revolving Fund loans ($787 million in grants and $1.3 billion in loans in 2003) help fund public water system infrastructure. A third of these monies can be used for investment in water source protection that includes land acquisition. Two examples in which states have used the land protection provisions of these loans and grants include Ohio's Water Restoration Sponsorship Program and New Jersey's Green Acres Program. Ohio's program provides loan rate reductions for wastewater treatment projects if the recipient uses a portion of the savings to invest in watershed protection and restoration.[37] In New Jersey, the state revised its criteria to allocate CWA loan funds to give three times the weight to projects with a water supply protection benefit through land protections.[38]

VII. CONCLUSION

This thumbnail sketch of planning, mitigation, and payment policies that can potentially link ecosystem services investments with landscape-scale conservation illustrates the potential of a number of federal policies and programs to drive multibenefit ecosystem services investments on a landscape scale. Their uses to support landscape-scale conservation action to provide focused ecosystem services benefits has, however, been limited to date. Moreover, where activities, policies, and initiatives have focused on ecosystem services, generally these efforts have targeted a single benefit stream. They have not provided the foundations for generating integrated, multifunctional benefits nor have they generated the tools to support such integration. And few policy tools and practices actually require a landscape scale, including cross-jurisdictional focus, though such a focus is increasingly gaining traction. Nonetheless, all of the federal tools described here, with revised emphasis and policy guidance, have statutory underpinnings consistent with a landscape-scale focus that measures performance in terms of ecosystem goods and services.

NOTES

1. This chapter draws significantly from Lynn Scarlett & Jim Boyd, Ecosystem Services: Quantification, Policy Applications, and Current Federal Capabilities (Discussion Paper 11–13, March 2011).

2. For a discussion of the emergence of landscape-scale conservation, *see* Matthew McKinney, Lynn Scarlett, & Daniel Kemmis, Large Landscape Conservation (2010).

3. Gretchen C. Daily, *Introduction: What Are Ecosystem Services?, in* Nature's Services 1–10 (Gretchen C. Daily ed., 1997); Barton H. Thompson Jr., *Markets for Nature*, 25 Wm. & Mary Envtl. L. & Pol'y, 261 (2000).

4. Lynn Scarlett, Environmental Defense Fund, Clean, Green and Dollar Smart: Ecosystem Restoration in Cities and Countryside 33–34 (2010).

5. Trust for Public Lands, Protecting the Source (May 21, 2004).

6. *See* case studies at American Forests, http://www.AmericanForest.org (last visited at Dec. 12, 2011).

7. Scarlett, *supra* note 2, at 15–16.

8. Michael Gallis, Co-evolution 12 (2009).

9. *See* Army Corps of Eng'rs, Final Feasibility Report and Environmental Impact Statement on Hurricane Protection and Beach Erosion Control for Dare County Beaches, North Carolina, *available at* http://www.saw.usace.army.mil/Dare%20County/Dare%20County%20Bodie%20Is.%20FEASIBILITY%20REPORT.pdf (last visited Dec. 12, 2011).

10. Exec. Order No. 13352, 69 FR 52989 (August 26, 2004).

11. *See* a description of the Landscape Conservation Collaboration framework *available at* http://www.fws.gov/science/shc/lcc.html.

12. *See* U.S. Fish and Wildlife Service, Strategic Habitat Conservation: Landscape Conservation Cooperatives, *available at* http://www.blm.gov/wo/st/en/prog/more/climatechange/reas.html (last visited Sept. 14, 2011).

13. *See* Memorandum from the White House, Office of the Press Secretary, *available at* http://www.whitehouse.gov/the-press-office/presidential-memorandum-americas-great-outdoors (last visited April 16, 2010).

14. J.B. Ruhl, Steven E. Kraft & Christopher L. Land, The Law and Policy of Ecosystem Services (2007).

15. 42 U.S.C. § 4321 (2006).

16. CEQ National Environmental Policy Act Regulations, 40 C.F.R. §§ 1500–1508 (2011), *available at* http://ceq.hss.doe.gov/nepa/regs/ceq/toc_ceq.htm (last visited Dec. 12, 2011).

17. James Connaughton, Memorandum for the Heads of Federal Agencies, Cooperating Agencies in Implementing the Procedural Requirements of the National Environmental Policy Act. Provisions on cooperating agency status are at 40 CFR §§ 1501.6 and 1508.5.

18. Implementation of the National Environmental Policy Act (NEPA) of 1969, 43 CFR § 46 (2008)

19. The Principles and Guidelines are a document developed by the Water Resources Council under provisions of Sec. 103 of the Water Resource Planning Act (42 U.S.C. § 1962a-2), entitled "Economic and Environmental Principles and Guidelines for Water and Related Land Resources Implementation Studies," March 10, 1983. The Obama Administration is updating the guidelines, with details *available at* http://www.whitehouse.gov/administration/eop/ceq/initiatives/PandG.

20. Council on Environmental Quality, Updated Principles and Guidelines for Water and Land Related Resources Implementation Studies, *available at* http://www.whitehouse.gov/administration/eop/ceq/initiatives/PandG (last visited Dec. 12, 2011).

21. Ecosystem Marketplace, http://www.ecosystemmarketplace.com (last visited Dec. 12, 2011). *See also* Environmental Protection Agency, Mitigation Banking Factsheet, *available at* http://www.epa.gov/owow/wetlands/facts/fact16.html (last visited Dec. 12, 2011).

22. Jim Boyd & Lisa Wainger, Measuring Ecosystem Service Benefits: The Use of Landscape Analysis to Evaluate Environmental Trades and Compensation (Discussion Paper 02–63, April 2003).

23. *See* U.S. Fish and Wildlife Service, Endangered Species Program, *available at* http://www.fws.gov/endangered/landowners/conservation-banking.html (last visited Dec. 12, 2011).

24. U.S. Fish and Wildlife Service, Guidance for the Establishment, Use and Operation of Conservation Banks (April 30, 2003), *available at* http://www.ecosystemmarketplace.com/pages/dynamic/resources.law_policy.page.php?page_id=194&section=home&eod=1#close (last visited Dec. 12, 2011).

25. 16 U.S.C. §§ 791–828c as amended ch. 5, June 10, 1920; 41 Stat. 1063.

26. 15 C.F.R. § 990.53 (2011).

27. *See, e.g.*, U.S. Dep't of the Interior's October 2, 2008 regulatory revisions at 73 C.F.R. § 57259.

28. *See* U.S. Fish and Wildlife Service, Environmental Contaminants: Natural Resource Damage Assessment and Restoration Program for New England, *available at* http://www .fws.gov/newengland/Contaminants-NRDAR.htm (last visited Dec. 12, 2011); *see also* U.S. Fish and Wildlife Service, Natural Resource Damage Assessment Restoration Program: North Oil Spill, Rhode Island (October 2005), *available at* http://www.fws.gov/newengland/pdfs/ North%20Cape.pdf (last visited Dec. 12, 2011).

29. U.S. Dep't of the Interior, Natural Resources Damages Assessment and Restoration Office, personal communication from Frank DeLuise (2010) (on file with author).

30. Tadlock Cowan & Renee Johnson, Cong. Research Serv., RL 34557, Conservation Provisions of the 2008 Farm Bill (2008).

31. U.S. Dep't of Agric., Natural Resources Conservation Service, Conservation Beyond Boundaries: NRCS Landscape Initiatives, power point presentation, available from the NRCS, Office of the Chief, undated.

32. U.S. Dep't of Agric., Natural Resources Conservation Service, Assessment of the Effects of Conservation Practices on Cultivated Cropland in the Chesapeake Bay Region 12 (February 2011).

33. *See* Environmental Protection Agency, Water Quality Trading Handbook: Can Water Quality Trading Advance Your Watershed's Goals (Nov. 2004).

34. *Id.*

35. *See* CT Dep't of Energy & Environmental Protection, Nitrogen Control Program for Long Island Sound, *available at* http://www.ct.gov/dep/cwp/view.asp?a=2719&q=325572&depNav _GID=1635 (last visited Dec. 13, 2011).

36. Caryn Ernst, Trust for Public Lands, (2004).

37. *Id.*

38. *Id.*

IV
Finding the Right Tools
Going Forward

10 Wildlife Conservation, Climate Change, and Ecosystem Management

Robert B. Keiter, University of Utah

G rizzly bears have long roamed across Yellowstone National Park and beyond—a seminal fact that triggered a controversial early federal ecosystem management effort. Less than a quarter century ago, though protected under the Endangered Species Act, the Yellowstone grizzly population teetered on the edge of extinction, jeopardized by escalating development pressures around the park and an unresponsive federal bureaucracy seemingly indifferent to the mounting bear losses. Once a congressional investigation revealed the grizzly's worsening plight, the responsible agencies responded by proposing to coordinate their conservation efforts at an ecosystem scale and to revise their land management practices to reduce threats to the region's iconic bear population. Although the most grandiose of their plans, captured in a draft Greater Yellowstone Area Vision Document, fell prey to local political intermeddling, the concept of ecosystem management as a plausible—even essential—approach to wildlife management was validated. And the region's grizzly population began rebounding to the point where it is now a candidate for removal from the federal endangered species list.[1]

But the Yellowstone grizzly bear population is not yet secure, not in the face of climate changes that are inexorably altering the regional ecosystem and once again calling into question the adequacy of our conservation efforts. According to scientists, warming temperatures and mountain pine beetles are devastating the region's high elevation whitebark pine trees, which produce nuts that represent an important seasonal food source for the grizzly bear.[2] Two federal courts have agreed and ordered the U.S. Fish and Wildlife Service to reconsider whether the loss of this food source will affect the bear's long-term survival.[3] In fact, whitebark pine is dying at such an alarming rate over its range that it is now a candidate for listing under the Endangered Species Act.[4] The culprit, by most accounts, is global warming, which has unleashed a massive bark beetle infestation and intensified wildfires, both of which are taking a toll on the whitebark pine. As the grizzly loses this critical food source, it will be forced to roam further afield for nutrition, a journey that will likely result in more deadly encounters with humans. Future conservation efforts will plainly have to take account of these climate change impacts, providing yet another justification for an ecosystem-based management strategy.[5]

This chapter will address the relationship between wildlife conservation, climate change, and ecosystem management, suggesting that a coordinated and adaptive landscape-scale approach is the only viable strategy for ensuring that the grizzly bear and other species survive in a warming world. It begins by describing the relationship between wildlife conservation and ecosystem management, highlighting key ecosystem management principles designed to promote biodiversity conservation in a world where ecological change is often unpredictable. This includes a critical overview of the nature reserve system that we have created to promote wildlife conservation, explaining the system's ecological and managerial shortcomings. The chapter then introduces climate change, reviewing the impacts that rising temperatures will likely have on wildlife populations and potential adaptive responses that could ameliorate these impacts. The chapter concludes with an assessment of how the law underpinning ecosystem management might be employed to address the climate change challenge through landscape-scale, adaptive management conservation strategies. Given the dire consequences of failing to act in the face of climate change, adaptive ecosystem management merits careful consideration as a viable wildlife conservation strategy for addressing the uncertain and potentially drastic environmental changes that are afoot.

I. WILDLIFE CONSERVATION AND ECOSYSTEM MANAGEMENT

Wildlife conservation has evolved as scientists have come to better understand species and their needs. During the late nineteenth century, as the nation was filling up its western reaches, unregulated commercial hunting devastated the region's once plentiful bison and elk herds, driving bison to the brink of extinction and decimating local elk populations. Conservationists responded by advocating nature reserves—national parks, forests, and wildlife reserves— where the animals would be secure, and they transplanted animals from Yellowstone and elsewhere to repopulate bison and elk numbers in other locations. We also established and enforced limitations on hunting, including a permit system and license fees, to halt the indiscriminate killing. We were not so kind, however, to large predators like the wolf, which was labeled a 'beast of destruction' and hunted to extinction. Other species without commercial or utilitarian value were mostly ignored, while newly established state wildlife agencies focused on ensuring elk, deer, and other big game species were available for their hunting constituencies, whose fees supported the agencies and their conservation efforts.[6]

As the twentieth century progressed, the strictly utilitarian view toward wildlife began to shift, laying the foundation for modern wildlife conservation practices. Aldo Leopold, a universally respected biology professor at the University of Wisconsin, made key contributions to this shift: he identified the critical role habitat played in maintaining viable wildlife populations, and he introduced ethical notions into conservation policy, calling for adoption of an "ethic dealing with man's relation to land and to the animals and plants which grow upon it."[7] Leopold's view that "man is, in fact, only a member of a biotic team" helped legitimize the notion that all species mattered in nature's scheme and that the ecosystem was key to conservation.[8] One upshot was the emergence of biodiversity as a new focal point for conservation efforts, another was the reversal of past predator eradication policies, including resurrection of the wolf as an important ecological cog, and yet another was the realization that an effective conservation policy should focus at the ecosystem level. This latter point underscored the limitations of our nature reserve system, which scientists were coming to believe was inadequate alone to ensure the long term survival of many species.[9]

These realizations soon coalesced in the ecosystem management idea as a more effective conservation strategy to meet biodiversity, habitat, and other needs. One of the most prominent early efforts to move ecosystem management from an idea onto the ground occurred during the late 1980s in the Yellowstone region, when the principal federal land management agencies, facing considerable congressional pressure over the potential loss of the grizzly bear, joined together to produce a vision document that would establish Greater Yellowstone as a 'world-class model' for integrated and coordinated natural resource management. The draft document called for ecosystem management to administer "a landscape where natural processes are operating with little hindrance on a grand scale . . . a combination of ecological processes operating with little restraint and humans moderating their activities so that they become a reasonable part of, rather than encumbrance upon, those processes."[10] But faced with intense local political opposition to this federal foray into the new world of ecosystem management, the agencies retreated in the final document, dropping any reference to ecological management and emphasizing the separate missions of each agency. The high-profile nature of the initiative, however, gave ecosystem management a measure of legitimacy and considerable public exposure.[11]

Other ecosystem management initiatives soon ensued, focused primarily on the public lands and at-risk watersheds. Often born from conflict and driven by the powerful Endangered Species Act, these initiatives represented several major federal ecosystem management experiments, including the Northwest Forest Plan, Sierra Nevada Forest Plan Amendment Initiative, Interior Columbia Basin Ecosystem Management Plan, the Everglades Restoration Project, and the San Francisco Bay Delta Project. In the case of the Northwest Forest Plan, after a series of federal court injunctions halted logging on federal forests across a three-state region to protect the dwindling northern spotted owl population, the Clinton administration developed an expansive, science-based, multispecies ecosystem management plan designed to preserve and restore owls, salmon, and other at-risk species through a series of interconnected reserves, improved interagency coordination efforts, and use of adaptive management techniques. Although the details varied, the other large-scale federal ecosystem management initiatives each endorsed similar ecological conservation goals and employed similar management strategies. And pretty much the same can be said about the myriad smaller scale ecosystem management initiatives—the Quincy Library Group, Malpai Borderlands Group, Applegate Partnership, Canyon Country Partnership, and others—that arose during the 1990s. Some failed, some succeeded, and others have sputtered along over the intervening years, but all these efforts helped give meaning to the ecosystem management concept, demonstrating its strengths, weaknesses, and potential.[12]

Beyond the federal lands, the ecosystem management concept garnered less overt support, but still gained a foothold in some venues. With the exception of laws like the Endangered Species Act and the Clean Water Act, few federal environmental legal mandates extend to private lands, meaning that the legal support for any ecosystem management initiative on these lands must derive from state law, which is generally weak when it comes to biodiversity conservation, land use planning, and the like. But under the Endangered Species Act, private landowners whose property harbors federally protected species face the prospect of extensive federal regulatory constraints absent an approved habitat conservation plan.[13] The Clinton administration utilized these legal provisions to promote large-scale habitat conservation plans designed to address multiple species at an ecosystem-level, including a 'no surprises' policy designed to give property owners a level of security from future regulatory limitations once they developed such plans.[14] Under the Clean Water Act, the Environmental Protection Agency has used its regulatory authority to limit development in sensitive wetland areas, thus protecting critical wildlife habitat as well as important ecosystem services.[15] Moreover, the land trust movement has burst onto the scene, promoting conservation easements as an economically viable way to protect open spaces and vital habitat, and it has continued on an upward trajectory.[16] Even during the Bush administration, a number of collaborative private land stewardship projects were launched with a view toward promoting wildlife conservation and other ecological objectives.[17] And though many landowners object to any notion that their land is part of an ecological complex, these legal tools are available to promote conservation objectives on private lands.

A critical dimension to ecosystem management is the role that nature reserves play in providing sanctuary for wildlife, conserving biodiversity, and sustaining ecological processes. Indeed, one of our primary wildlife conservation strategies has involved creating and safeguarding national parks, wildlife reserves, and wilderness areas. These sanctuaries offer secure habitat in a relatively natural setting where fires, floods, and other natural disturbances shape the environment, and roads, buildings, and industrial development activities are excluded or minimized. Yet scientists now agree that our expansive nature reserve system—more than 250 million acres have been placed in some type of formally protected status—is inadequate to meet basic biodiversity conservation goals.[18] These protected lands do not represent a full suite of ecosystem types; most reserves are too small to ensure species survival over the long term; and they are not well connected enough to enable species to move back and forth. Even in an area the size of the Greater Yellowstone Ecosystem, as we have seen, the grizzly bear population remains at risk in the face of climate change and ongoing development pressures. The solution to this problem, according to conservation biologists, involves re-designing our nature reserve system at an ecosystem scale, enhancing its representativeness, and better knitting it together with protected corridors and other coordination strategies.[19]

These ecosystem management concepts and efforts have not escaped controversy, in large part because they elevate biodiversity conservation and environmental protection to a more prominent position on agency agendas and stretch conservation efforts beyond conventional boundary lines, even onto private property in some cases. Upon succeeding the Clinton administration, the Bush administration set about dismantling several ecosystem management initiatives and revising others, including the Northwest Forest Plan and the Sierra Nevada Forest Plan Amendment initiative. Bush administration officials also redirected the Bureau of Land Management's priorities away from conservation and toward accelerated energy development, a trend reflected in the agency's revised resource management plans, which have largely ignored ecological concerns. Further, they attempted to undermine the Endangered Species Act, refusing to list species and even rewriting agency scientists' professional conclusions. And while calling for a new '4 Cs' collaborative conservation approach, the Bush administration largely ignored the term 'ecosystem management' in its policies and plans.[20] Nonetheless, the ecosystem management concept has endured; it still appears explicitly in the Forest Service's new forest planning regulations and in the National Park Service's and U.S. Fish and Wildlife Service's management policies.[21] And the Obama administration has endorsed the concept, albeit under the rubric of landscape conservation.[22] Moreover, a few states have adopted an ecosystem approach for managing their lands and wildlife.[23]

Ecosystem management is not a concept that can be defined succinctly. In fact, the concept has generated a diverse assortment of definitions. Among those that have been advanced, the federal Interagency Ecosystem Management Task Force definition captures several key elements; "[t]he ecosystem approach is a method for sustaining or restoring natural systems and their functions and values. It is goal driven, and it is based on a collaboratively developed vision of desired future conditions that integrates ecological, economic, and social factors. It is applied within a geographic framework defined primarily by ecological boundaries."[24] Other definitions tend to emphasize one or more of these same elements, including the role of science, ecologically defined goals, large-scale planning, monitoring, and public participation.[25] On the ground, several additional factors have proven important to initiate and sustain ecosystem-based regional management efforts, namely strong and enforceable environmental laws, a gradual merging of agency missions, the ability to establish a common regional vision, and the absence of overpowering market forces.[26]

Alternatively, ecosystem management can be understood as a set of governing principles, for which there is general agreement. Simply put, these principles are: 1) to ensure healthy ecosystems and to address species extinction concerns, a primary goal of ecosystem management is to maintain and restore biodiversity and sustainable ecosystems; 2) because people are a part of nature and human values inform any natural resource policy, ecosystem management goals must be socially defined through public processes to incorporate ecological, economic, and social concerns; 3) because species and ecological processes transcend jurisdictional boundaries, ecosystem management requires coordination among federal agencies and collaboration with state, local, and tribal governments as well as opportunities for public involvement in planning and decision processes; 4) given the dynamic, nonequilibrium nature of ecosystems, ecosystem management requires management on broad spatial and temporal scales in order to accommodate ecological change and to address multiple rather than single resources; 5) because science plays an important role in understanding natural systems, ecosystem management is based on integrated, interdisciplinary, and current scientific information that can be used to address risk and uncertainty; and 6) because ecosystem management and the accompanying science are still experimental, ecosystem management requires an adaptive management approach that includes establishing baseline conditions, monitoring, reevaluation, and management adjustments to reflect new information and scientific knowledge as well as evolving human concerns.[27]

In sum, wildlife conservation and ecosystem management have become nearly synonymous. The science supporting the notion of managing wildlife at an ecosystem scale is unassailable, even as differences persist over how best to accomplish this objective on a dynamic and fragmented landscape with diverse owners and resource priorities. The fundamental principles underlying ecosystem management have gained widespread acceptance as an important conservation strategy, as reflected in the myriad federal and other ecosystem management initiatives that have taken hold. And federal law, as we shall see, provides a legal foundation for pursuing ecosystem management on the public lands, though these same laws have only limited application on state and private lands, where the concept still faces resistance. The emergence of ecosystem management thus represents an important advance in wildlife conservation designed to meet species' needs in an ever more crowded and industrialized world.

II. WILDLIFE CONSERVATION IN THE FACE OF CLIMATE CHANGE

Where the fear of losing species to escalating human development pressures drove most early ecosystem management initiatives, the specter of destabilizing climate change significantly strengthens the case for an ecologically-based adaptive conservation strategy. Scientists have become increasingly alarmed at the magnitude of the ecological dislocations that climate change portends, including potentially profound impacts on wildlife habitat and populations. Temperature increases linked to greenhouse gas emissions will inevitably alter terrestrial and aquatic ecosystems, forcing species to adapt to new environmental conditions or to relocate in order to survive, assuming they are able to disperse across the landscape. When the stresses associated with climate change are added to current development pressures, the impacts on wildlife and their habitat will only be magnified. To address these twin challenges, scientists concur that ecosystem-based management principles, including coordinated landscape-scale planning and adaptive management practices, represent one of the most viable strategies for safeguarding species over the long term.[28]

The likely impacts of climate change on the landscape and wildlife species are now well documented, though these impacts will vary from one location to another. Owing to escalating greenhouse gas emissions, the Intergovernmental Panel on Climate Change has predicted that global temperatures will rise, and that a temperature increase of 1.5 to 2.5 degrees Celsius could place 20–30 percent of the world's species at increased risk of extinction.[29] At a broad scale, such temperature increases portend rising sea levels, melting glaciers, prolonged droughts, fluctuations in freshwater flow patterns, more destructive wildfires, and other significant environmental impacts. As these changes occur, wildlife habitat will be affected as will the behavior patterns of individual species. Warming temperatures could alter some species habitats enough to force them to relocate, most likely to move northerly and to higher elevations. These dislocations could result in different species assemblages that will, in turn, trigger additional ecological changes as these new congregations adapt to one another and their new environment. A warmer atmosphere will also change the rate at which important seasonal events occur—a phenomenon that scientists know as 'phenology.' Earlier snowmelt and run-off in mountain streams, for example, could affect individual species' breeding, feeding, and other critical life cycle processes. The upshot is that various species will find themselves at greater risk if they are unable to adapt to these changes.[30]

Some examples of species-at-risk will illustrate the potential magnitude of the problem. In the case of the grizzly bear, besides facing the loss of whitebark pine nuts as an important food source, a warming climate could cause bears to exit their winter dens earlier when less food will be available at higher elevations, most likely prompting them to seek nourishment at lower elevations where they could come into conflict with humans or livestock.[31] For the mountain pika, which resides only at high elevations in western mountains, scientists have already documented an upslope shift in its habitat due to temperature-induced ecological changes, resulting in the loss of over one third of the pika population.[32] A similar fate could befall the wolverine, a charismatic but reclusive member of the weasel family that also dwells only at higher elevations in a few remote western mountain ranges.[33] Waterfowl and other migratory bird populations could face dried up watering holes and diminished food sources during their annual intercontinental journeys due to localized drought conditions.[34] Salmon and trout species that live in moving cold water streams could see their habitat significantly altered by reduced stream flows and warmer water temperatures.[35] Local drought conditions and invasive insect infestations, most scientists believe, will precipitate larger, more catastrophic wildfires in western forests, putting species like the Canadian lynx and pine marten (that depend on dense forest cover) in peril.[36] Add to these likely impacts the potentially devastating effect that sea-level rise will have on the nation's wildlife refuge system—nearly 30 percent of the refuges are situated in coastal locations—and the potential climate change impacts on wildlife are extremely serious, albeit still somewhat uncertain.[37]

Most scientists advocate an adaptive approach to address the risks that climate change poses for wildlife. Few of them believe that mitigation measures can be effectively deployed to safeguard wildlife from the coming changes.[38] It is too late, given energy consumption patterns, for any changes in our use of fossil fuels, either through conservation measures, new alternative energy sources, or otherwise, to forestall the ecological changes that are forecast for most terrestrial and riparian ecosystems. Besides, mitigation measures can present their own wildlife-related problems. Constructing wind turbine farms, for example, can imperil birds that might be killed or injured by the rotating blades, and these structures, along with the accompanying service roads and other infrastructure, can fragment wildlife habitat and migration corridors. This does not mean, however, that curtailing industrial activities on sensitive public lands would not reduce carbon emissions and benefit resident wildlife. But in the absence of a demonstrably effective mitigation strategy, an adaptive approach represents a logical risk management strategy for enabling wildlife to respond to the global and local impacts attributed to a warming climate.[39]

The adaptive approaches required to address climate-induced stresses on wildlife correspond closely to existing ecosystem management principles. Just as ecosystem management is designed to address and manage for ecological uncertainty in a natural world where change is endemic, the same basic concerns over the uncertain and disruptive impacts from climate change must be addressed for wildlife conservation purposes. And just as ecosystem management is designed to promote ecological integrity and resiliency, these same goals must be paramount to ensure a future for wildlife in the face of a warming climate. In brief, then, these ecosystem-based, adaptive management approaches include: 1) landscape-scale planning based on long range temporal considerations, including improved coordination efforts across jurisdictions and designated connective corridors to facilitate species dispersal; 2) the use of adaptive management techniques to address uncertainty, which entails establishing baseline conditions, science-based monitoring of ecological conditions, regular assessments to determine changes in habitat and species behavior, and readjustments in management approaches to address unanticipated developments; and 3) reconfiguring the existing nature reserve system to provide adequate sanctuaries for threatened or displaced species. These climate-focused adaptive approaches—in contrast to several versions of ecosystem management—could result in more (rather than less) active intervention in order to avoid potential extinctions.[40]

Landscape-scale planning is essential to promote and ensure ecological resiliency over the long term. Given the projected changes in temperature and moisture regimes, climate change will impact ecosystems across the landscape, generating widespread and even radical ecological changes that can only be addressed at the same geographic and temporal scales. In fact, where traditional ecosystem management strategies called for coordinated planning and decision making at an ecosystem scale, climate change may require expanding the scale of such efforts. As one ecosystem type gives way to another in the face of changes in temperature and moisture, an effective wildlife-based adaptive response will have to extend across these ecosystem types to enable species to adapt and move when necessary. Because most scientists doubt that species can consistently adapt to new surroundings on an in-situ basis, these planning efforts should also include connective corridors and dispersal routes to enable climate-displaced species to move to new habitats when their old ones can no longer sustain them.[41]

An important aspect of such an expanded, landscape-scale planning effort will be the need for improved coordination among landowners and agencies. This is a tall order when the landscape is already so fragmented by jurisdictional boundaries and different ownership goals, particularly in the case of private lands where most owners are interested in economic not ecological returns on their property. The problem could be addressed at the federal agency level, however, by new legislation or regulations that mandated preparation of an interagency coordination statement as part of the National Environmental Policy Act environmental assessment process, or by requiring interagency consultation whenever an agency action might adversely affect wildlife.[42] Outside the federal lands, the federally funded State Wildlife Action Plans offer a starting point for promoting state-federal coordination and for engaging private landowners in climate-based wildlife conservation strategies.[43] In any event, such coordinated planning efforts will also have to take into account economic concerns in order to secure local support for and cooperation with climate-related wildlife conservation efforts.

Adaptive management techniques are designed to enable both scientists and managers to assess the effectiveness of their wildlife conservation strategies and to adjust those strategies accordingly. Adaptive management traces its origins to recent understandings about the dynamic, nonequilibrium nature of ecosystems;[44] it is a science-driven management strategy designed to address uncertainty by monitoring the results of particular management actions on ecosystems, using this information to assess their effectiveness, and then adjusting them as necessary to accomplish the desired objectives.[45] In the case of wildlife, this means establishing baseline population and habitat data, identifying and pursuing a particular conservation strategy, monitoring the results achieved on the ground, reevaluating the strategy's effectiveness, and then making adjustments if it is not working as anticipated. In making adaptive management strategy decisions designed to sustain species levels, managers regularly draw upon scientists to identify likely wildlife responses, to assess the risks and uncertainties associated with climate change and other stressors, and to help identify alternate strategies as well as associated risks. For climate change, this is very much a site-driven process, because the precise effects that temperature and precipitation shifts will have on wildlife in specific locations remain quite uncertain. Moreover, a comprehensive adaptive management strategy should also enable managers to assess shifts in local values and concerns, providing them with valuable information as to which alternate conservation strategies can realistically be implemented on the ground.[46]

Our existing nature reserve system was largely designed and adopted during a time when we did not understand the complex nature of ecosystems nor even contemplate the dramatic effect that a warming climate could have on wildlife species. In fact, wildlife concerns were not ordinarily factored into our early national park designations, nor were they often part of the wilderness area designation process—both of which were generally driven by scenic, recreational, and political considerations rather than ecological concerns.[47] Though wildlife concerns have motivated the national wildlife refuge system design, most refuges are not large enough to meet the diverse ecological needs of the species that inhabit them. Only a few refuges sustain the full range of ecosystem processes that give resiliency to these systems, leaving resident native species vulnerable to ecological disturbances as well as human development pressures on the periphery.[48] When the profound changes associated with warming temperatures are taken into account, species that depend on these reserves are even more vulnerable. Moreover, the current nature reserve system is not representative of the diverse ecosystem types that exist across the nation, nor are the existing reserves well-connected one to another, except by political happenstance.[49]

Climate change virtually demands that we address these problems by redesigning the nature reserve system. To meet ecological integrity and resiliency goals in the face of uncertainty, the system must be large enough, diverse enough, redundant enough, and connected enough to provide sanctuary for all species. This will entail paying closer attention to science in the design and designation of nature reserves, locating reserves with an eye toward ensuring connectivity, and providing some level of redundancy in the system to accommodate unforeseen developments that could decimate local wildlife populations.[50] Several related strategies merit consideration: 1) expand existing parks, wilderness areas, and wildlife refuges or create buffer areas around them to enhance ecosystem resiliency; 2) designate protected movement corridors to connect reserves and enable wildlife to shift locations as their habitats change; 3) establish a new national restoration area designation to restore damaged ecosystems in order to provide new habitat options;[51] and 4) promote better planning and coordination for wildlife management purposes among the agencies responsible for overseeing the reserves and adjacent lands. While these strategies might be feasibly implemented on federal lands where managers share a common responsibility for providing wildlife habitat, the role of private lands in any future nature reserve system poses unique challenges given the rights that attach to property ownership and the limited scope of federal law that applies on these lands.[52]

One ongoing controversy that spills over from wildlife conservation to ecosystem management that is also relevant to climate change conservation strategy is the role that active rather than passive management should play in addressing the problem. In recent years, nature conservation policy in our national parks, wilderness areas, and wildlife refuges has been characterized by less, rather than more human intervention into ecological processes. The management policy goal is to restore historical conditions whenever possible, for example, by letting wildfires burn unless human life or property is at risk. As noted, the Greater Yellowstone Area vision document called for "a landscape where natural processes are operating with little hindrance on a grand scale."[53] Some scholars and others disagree, however, about the degree to which ecosystem management does—or should— allow for active management of ecosystems for utilitarian purposes.[54] Regardless, because climate change will accelerate the rate at which ecological change occurs on the landscape, it may well be impossible to even contemplate restoring historical conditions.[55] In the case of wildlife, if climate-induced changes raise the specter of extinction for a species, most scientists support using an 'assisted migration' strategy that would involve human intervention to help move a population from one location to another more suitable one. In short, more rather than less active management could be required to respond to the radical changes ahead.[56]

III. THE LAW OF ECOSYSTEM MANAGEMENT AND CLIMATE CHANGE

Federal law provides a patchwork legal foundation for ecosystem management. On the public lands, no single law expressly mandates an ecosystem management approach, but the sum total of an array of organic mandates, planning provisions, and environmental laws supports an ecological approach to natural resource management. That foundation rests principally upon the organic statutes governing the federal land management agencies, the Endangered Species Act, and the National Environmental Policy Act. In fact, after citing these laws, a landmark judicial opinion sustaining the Northwest Forest Plan put it this way, "[g]iven the condition of the forests, there is no way the agencies could comply with the environmental laws *without* planning on an ecosystem basis."[57] On private lands, the law relies primarily on acquisition and incentive strategies to encourage landowners to take nature into account. What follows briefly analyzes how the law can be interpreted and applied to promote an ecosystem approach to wildlife conservation in the face of climate change.[58]

Legal protection for wildlife, biodiversity, and ecosystem integrity can be derived from the organic mandates governing the principal federal land management agencies, which give each agency a statutory responsibility to conserve wildlife and to manage sustainably. For the Forest Service, this responsibility derives from the Multiple Use–Sustained Yield Act and the National Forest Management Act (NFMA), where wildlife is one of several multiple uses, but enjoys no special priority.[59] But the NFMA also contains a diversity provision that has been treated as a biodiversity conservation mandate, requiring forest managers to factor species protection into their planning and project decisions.[60] In addition, the Multiple Use–Sustained Yield Act instructs the agency not to impair the productivity of the land,[61] while the National Forest Management Act provides protection for soil, watersheds, and riparian areas from timber harvesting activities.[62] For the Bureau of Land Management (BLM), the Federal Land Policy and Management Act likewise lists wildlife as one among several multiple uses, and also includes the protection of ecological values and wildlife habitat as an express policy objective.[63] Unlike the NFMA, FLPMA does not contain a diversity provision, but it does prohibit "permanent impairment of the productivity of the land and the quality of the environment."[64] The National Park Service, under its organic act, is specifically directed to conserve wildlife as a primary management goal and must "leave [the national parks] unimpaired for the enjoyment of future generations."[65] And the U.S. Fish and Wildlife Service's organic laws give priority to wildlife conservation and restoration on its refuges "for the benefit of present and future generations of Americans," and establish a "biological integrity, diversity, and environmental health" management standard for the refuge system.[66] Moreover, state wildlife conservation laws aimed at managing wildlife populations provide a level of protection to big game and other species, though state law must recede if in conflict with federal law. Taken together, these laws ensure that species conservation, including habitat protection, is part of the management calculus on all federal lands, and even enjoys priority on park and refuge lands.

The federal land management agencies are each required to prepare land use plans that effectively zone their lands for resource management purposes, including designating wildlife habitat. In general, the relevant laws call for integrated and interdisciplinary planning, the use of science in the planning process, coordination with other federal agencies, the states, tribes, and local communities, and public involvement opportunities.[67] Because these plans are intended to accomplish the agencies' mission goals, most plans incorporate provisions designed to ensure wildlife habitat, promote biodiversity, and protect environmental values, while also meeting commodity production goals on national forest and BLM lands. Significantly, the FLPMA instructs the BLM to give priority to designating "areas of critical environmental concern" (ACECs) in order "to protect and prevent irreparable damage to important . . . fish and wildlife resources or other natural systems or processes."[68] The statutory coordination provisions enable the agencies to conceive their plans in ecosystem terms, taking account of how neighboring lands are managed and providing an opportunity to coordinate wildlife management goals.[69]

Drawing upon these planning provisions and related laws, the Forest Service has undertaken several ecosystem-wide planning efforts that transcend its conventional forest-based planning and expand the scale of its management focus in order to promote ecological integrity and biodiversity conservation goals, most notably in the case of the Northwest Forest Plan and the Sierra Nevada Forest Plan Amendments.[70] And the Department of the Interior, through a series of Secretarial Orders, has endorsed the concept of landscape-scale planning and created new Landscape Conservation Cooperatives to study and address climate change impacts.[71] The lesson is that legally mandated planning processes on the public lands provide an opportunity to assess and provide for ecosystem integrity and resiliency at a broad landscape scale in order to address wildlife conservation concerns related to global warming.

The Endangered Species Act (ESA) has played an important role in promoting ecosystem management, and it will undoubtedly have a similar impact for protecting wildlife in the face of climate change. The act serves as a safety net to guard against extinction, doing so through key provisions that can also be invoked to address climate change threats to individual species. These provisions include the Section 4(a) 'listing' requirements that must be met before a species can enjoy protection under the act as either threatened or endangered; the Section 4(b) critical habitat designation provision that denominates habitat regarded as vital to the species' recovery; the Section 7 jeopardy review process, which applies to all federal agencies and is designed to regulate any activities that could imperil protected species; the Section 9 'take' provision that protects individual animals and their habitat, even extending to private individuals and lands; the Section 10 habitat conservation planning process that has been expanded to include multispecies, ecosystem-level plans on private lands; and the Section 10(j) species reintroduction provision that can be used to restore and relocate extirpated species.[72] Indeed, the Endangered Species Act has already been invoked to protect the polar bear through listing from temperature-induced changes to its Arctic habitat where sea ice is melting at an alarming rate and reducing the bear's foraging opportunities.[73]

Knowledgeable legal analysts have parsed the Endangered Species Act and identified how it might be employed to address climate-induced impacts on wildlife as well as some potential problems.[74] In their view, the ESA listing criteria, namely the "threatened destruction, modification, or curtailment of [species'] habitat or range" and the "inadequacy of existing regulatory mechanisms" provisions, provide a clear basis for extending federal protection to species that are being adversely affected by climate-based changes.[75] The act's critical habitat designation provision, which by policy is currently limited to historically occupied territory, may need to be extended to lands that were never occupied by the listed species, including connective corridors, if warming trends are displacing a species from its historic habitat.[76] Under the Section 7 consultation process, one court has already ruled that the U.S. Fish and Wildlife Service must address the impact of climate-based changes to the species or its habitat in its jeopardy determination.[77] And the Section 9 take prohibition, which extends to habitat alterations that "actually kill or injure listed species,"[78] could be implicated if private land development activities exacerbate the effects of climate change by, for example, causing more carbon dioxide to be emitted into the atmosphere. Though exactly how these ESA protective provisions will be applied to address the impact of climate change on listed and not yet listed species largely remains to be seen, the law's commitment to protecting species from extinction—as well as the ecosystems that they depend upon—plainly extends to climate-related threats confronting wildlife.

   Wildlife managers, given the severely disruptive effect that climate change could have on species and their habitats, are already assessing the legality of relocating species that are unable to disperse from their home ranges due to climate disruption. A species listed under the Endangered Species Act represents the type of species most likely to be considered for translocation due to its present vulnerable status, a condition that may only worsen as climatic changes take hold on its present habitat. Under Section 10(j) of the act, the U.S. Fish and Wildlife Service is authorized to reintroduce listed species as an experimental population to "further the conservation of such species," if the reintroduction is outside its current range and the reintroduced population is kept wholly separate geographically from other populations of the species.[79] As a policy matter, Section 10(j) reintroductions have generally involved relocations only within a species' historic range, but climate change could force a reassessment of that policy given the likelihood that a species' historic range may be so fundamentally altered as to be uninhabitable. Depending on whether the reintroduced species is considered an 'essential' or 'nonessential' population, the legal protections available under the act will vary, which may argue for the more protective essential designation for climate displaced species given their already tenuous state. Beyond Section 10(j) assisted relocations, the individual federal land management agencies seem to have sufficient discretionary authority under their respective organic laws to undertake species relocations for non-ESA protected species on their own lands, though wildlife conservation policies that privilege 'native species' may limit their ability to move a species outside its historic range. In addition, state wildlife agencies, as well as private individuals, can claim some degree of legal authority to undertake similar species relocations, whether triggered by climate change or other concerns.[80]

The National Environmental Policy Act (NEPA) requires federal agencies to prepare an environmental analysis whenever contemplating a "major federal action significantly affecting the human environment."[81] Although NEPA is only a procedural law, the courts have read its Environmental Impact Statement (EIS) requirements strictly to ensure the agencies fully identify, assess, and consider potential environmental impacts, including cumulative effects, in their decision processes.[82] In fact, these NEPA EIS requirements have effectively compelled the land management agencies to adopt an ecosystem perspective for NEPA analysis purposes.[83] Because climate change constitutes a diffuse but potentially significant environmental impact, the agencies confront the question of how to address climate-based impacts in their NEPA analysis—a question that the Council on Environmental Quality has thus far deferred in the case of land use impacts. More specifically, for ecosystem management and climate change purposes, does NEPA require the agencies to frame an EIS at the ecosystem—or landscape—scale, the most pertinent level of analysis for understanding and responding to the impacts associated with global climate change? The answer would appear to be "yes" given NEPA's cumulative effects environmental analysis requirements, which are designed to expand the spatial and temporal dimensions of agency decisions, requiring full disclosure and examination of serial development proposals and trans-boundary environmental impacts. Under NEPA, the agencies also must consider a range of alternatives in their EIS analysis, which should compel them to compare the climate change implications of different courses of action, ensuring full disclosure and assessment of potential climate-related impacts on the ecosystem. Thus, through the NEPA cumulative effects and alternatives analysis requirements, the agencies have an opportunity—and perhaps even an obligation—to factor climate change impacts on wildlife into their planning and management decisions.[84]

The uncertainties associated with climate change and its impact on wildlife and habitat can only be addressed effectively through adaptive management protocols. As noted, adaptive management contemplates regular monitoring of ecological conditions and management adjustments when necessary to achieve desired conditions. By statute, regulation, or policy, each of the four principal federal land management agencies are obligated to inventory and monitor their resources,[85] requirements that the courts have proven willing to enforce. In the case of the Northwest Forest Plan, for example, a federal court enjoined the Forest Service from eliminating the plan's survey and manage requirements—a critical adaptive management component—until it undertook a thorough environmental analysis.[86] Adaptive management adjustments to a land use plan or project decision may require NEPA compliance if the changes portend significant environmental impacts. However, agencies might avoid preparing multiple EISs when implementing climate-related adaptive management decisions by tiering off an initial planning EIS that adequately addressed environmental conditions and potential changes at this larger scale.[87] In the case of climate change, then, threshold planning decisions that incorporate adaptive management techniques should enable the agencies to make efficient adjustments as the climatic impacts on wildlife become clearer over time.

For wildlife confronted with potentially destabilizing climate changes to habitat conditions, the current system of nature reserves may not offer adequate safe haven to allow them to adapt to these changes. Under existing law, Congress holds the key to creating new national parks, wildlife reserves, and wilderness areas, and to expanding existing ones to buffer wildlife and ecosystems against climatic disruptions. For purposes of biodiversity conservation, most scientists agree that our current nature reserve system is inadequate to ensure long term survival for species with expansive ranges or specific habitat requirements. Add climate change to the picture, and the case for expanding and connecting our nature reserve system is even stronger, and should perhaps include new national restoration areas, where strategically located degraded lands are nursed back to health to serve as nature refuges for climate-displaced species.[88] Mustering the political will in Congress to expand our nature reserves will not be easy, however, especially when scientists cannot say with certainty where additions will be necessary or even whether such additions will be sufficient given the uncertainties attached to climate change. If Congress cannot be persuaded to act, the President has authority to establish new national monuments strategically located to ameliorate climate change impacts on species, an executive action that falls well within the ambit of the Antiquities Act, which includes objects of scientific interest as the basis for new monument designations.[89] Alternatively, federal officials could utilize their land exchange authority to acquire property that might be of benefit to displaced wildlife, either as new habitat or as a connective corridor to reach more suitable habitat.[90]

The legal tools available to enlist private lands either for habitat or corridor purposes to meet climate-displaced wildlife needs are even more limited. Few private landowners are motivated solely by altruistic concern about the welfare of wildlife, and climate change is unlikely to change these sentiments. The federal government, besides exercising its eminent domain power, could purchase critical parcels, either in fee simple or as a conservation easement, or it could exchange other lands for them.[91] In the case of agricultural lands, the federal conservation reserve program, which essentially pays farmers for leaving lands fallow, has proven an effective strategy for protecting wildlife habitat on private lands.[92] Land trusts are already factoring wildlife-focused climate considerations into their conservation easement acquisitions, and such practices could be encouraged with appropriately targeted tax credits available to landowner participants. Absent the acquisition of ownership interests for wildlife conservation purposes, an alternative strategy is to involve private landowners in collaborative efforts that promote coordinated landscape-scale conservation initiatives designed to enable climate-threatened species to move across the landscape. Options for inducing landowners to endorse such efforts include tax incentives, technical assistance, and co-management arrangements.[93]

To be sure, the legal tools outlined above to advance ecosystem management as a viable wildlife conservation strategy in the face of climate change do not constitute the proverbial 'silver bullet' for solving the problem. Ecosystem management is built on a patchwork legal foundation that supports the use of landscape-scale planning and adaptive management strategies as a means to promote greater ecological integrity and to address the uncertainties posed by climate change. New laws could be helpful to surmount the existing legal and geographical fragmentation that persists and to improve adaptive management accountability. Although Congress has not yet adopted climate change legislation, the American Clean Energy and Security Act of 2009 climate bill that passed the House of Representatives did address wildlife conservation concerns, establishing a "National Climate Change and Wildlife Science Center" and requiring federal and state adaptation plans focused on ecosystem protection, secure corridors, monitoring programs, and interagency coordination mechanisms.[94] The present challenge, drawing upon this proposed framework, is to work creatively with the existing law to strengthen and expand ecosystem management strategies in order to secure a more wildlife-friendly landscape that enables species to adapt to the climate changes that are afoot.

IV. CONCLUSION

Though the Yellowstone's grizzly bear story we began with may epitomize the wildlife conservation challenges posed by climate change, it is the story of another Yellowstone area species—the pronghorn—that offers guidance for stitching the landscape together in a way that will enable wildlife to adapt to warming temperatures and related habitat changes. For centuries, pronghorn have seasonally migrated southward from Grand Teton National Park in northwestern Wyoming to sagebrush plains situated along the upper Green River in western Wyoming, covering 170 miles and representing the longest wildlife migration route in the lower 48 states. In recent years, however, unchecked subdivision activity, expansive new oil and gas development, and perilous highway crossings have disrupted the pronghorn's seasonal migration route, putting the herd at risk as it sought to navigate across an increasingly fragmented and dangerous landscape. Confronted with potential loss of this iconic natural spectacle, the area's federal land managers, state officials, and local landowners set about creating a Path of the Pronghorn migration corridor designed to ensure the antelope safe passage from their summer to winter habitat. The Park Service obtained a key parcel along the route through a land exchange with the state; the Forest Service amended its forest plan to designate a new wildlife migration corridor across national forest lands; the BLM essentially did the same by designating protective new ACECs along the route; the Wyoming Highway Department agreed to install wildlife overpasses across a busy state highway; and several strategically located private landowners entered into conservation easements that will allow the pronghorn to pass safely across their ranch lands.[95] The resulting secure migration corridor, though not related to climate change, illustrates nonetheless how the law can be creatively employed to promote wildlife conservation at a landscape scale through new designations and collaborative arrangements—the same type of strategies that will be essential to avoid dire consequences as the atmosphere continues to heat and displaced wildlife are forced to seek alternate habitat.

NOTES

1. On the Greater Yellowstone vision exercise, *see* Susan G. Clark, Ensuring Greater Yellowstone's Future 72–77, 123–26 (2008); Robert B. Keiter, Keeping Faith with Nature 67–68 (2003); Bruce E. Goldstein, *Can Ecosystem Management Turn an Administrative Patchwork into a Greater Yellowstone Ecosystem?*, 8 Northwest Envtl. J. 285 (1992); John Freemuth & R. McGreggor Cawley, Science, *Expertise and the Public: The Politics of Ecosystem Management in the Greater Yellowstone Ecosystem*, 40 Landscape & Urban Planning 211 (1998).

2. *See* U.S. Dep't of the Interior, Fish & Wildlife Service, Endangered and Threatened Wildlife and Plants; Final Rule Designating the Greater Yellowstone Area Population of Grizzly Bears as a Distinct Population Segment; Removing the Yellowstone Distinct Population Segment of Grizzly Bears From the Federal List of Endangered and Threatened Wildlife, 72 Fed. Reg. 14866, 14929–30 (March 29, 2007).

3. Greater Yellowstone Coalition v. Servheen, 665 F.3d 1015 (9th Cir. 2011), *aff'g in part* 672 F. Supp. 2d 1105 (D. Mont. 2009).

4. *See* Dep't of the Interior, Fish & Wildlife Service, Endangered and Threatened Wildlife and Plants; 12-Month Finding on a Petition to List *Pinus Albicaulis* as Endangered or Threatened With Critical Habitat, 76 Fed. Reg. 42631 (July 19, 2011).

5. *See* Stephen Saunders et al., Rocky Mountain Climate Org., Greater Yellowstone in Peril: The Threats of Climate Disruption (2011); Molly Cross & Chris Servheen, Climate Change Impacts on Wolverines and Grizzly Bears in the Northern U.S. Rockies: Strategies for Conservation (Workshop Summary Report, 2009); Louisa Wilcox, Natural Resources Def. Council, An Alternative Path to Grizzly Recovery in the Lower 48 States (May 2004).

6. *See* Peter Matthiessen, Wildlife in America (1959); Howard P. Brokaw, Wildlife and America (1978).

7. Aldo Leopold, A Sand County Almanac 238 (1966) (quote is from "The Land Ethic" essay).

8. *Id.* at 241.

9. On Leopold and his contribution to conservation policy and philosophy, *see* Curt Meine, Aldo Leopold (1988); Susan Flader, Thinking Like a Mountain (1974); Eric Freyfogle, *The Land Ethic and Pilgrim Leopold*, 61 U. Colo. L. Rev. 217 (1990).

10. Greater Yellowstone Coordinating Committee, Vision for the Future: A Framework for Coordination in the Greater Yellowstone Area (Draft) 3-1 (1990), http://fedgycc.org/documents/GYCC.Vision.pdf.

11. *See supra* note 1 and accompanying text.

12. On the Northwest Forest Plan, *see* Keiter, *supra* note 1, at 80–126; Kathie Durbin, Tree Huggers (1996); Stephen L. Yaffee, The Wisdom of the Spotted Owl (1994). On Clinton era ecosystem management initiatives, *see* Stephen L. Yaffee et al., Ecosystem Management in the United States (1995); Julia M. Wondolleck & Stephen L. Yaffee, Making Collaboration Work (2000).

13. 16 U.S.C. § 1539(a) (2006).

14. *See* Dep't of the Interior, Fish & Wildlife Service, Habitat Conservation Plans ("No Surprises") Rule, 63 Fed. Reg. 8859 (Feb. 23, 1998); Karin P. Sheldon, *Habitat Conservation Planning: Addressing the Achilles Heel of the Endangered Species Act*, 6 N.Y.U. Envtl. L.J. 279 (1998).

15. *See* Zygmunt J.B. Plater et al., Environmental Law and Policy 1233–45 (3rd ed., 2004).

16. *See* Nancy A. McLaughlin, *The Role of Land Trusts in Biodiversity Conservation on Private Lands*, 38 Idaho L. Rev. 453 (2002).

17. *See* U.S. Dep't of the Interior, Cooperative Conservation: Success through Partnerships (2004); Lynn Scarlett, *An Address to the Natural Resources under the Bush Administration Symposium*, 14 Duke Envtl. L. & Pol'y F. 281 (2004).

18. *See* Reed Noss & Allen Cooperrider, Saving Nature's Legacy (1994); Keiter, *supra* note 1, at 186–208; Robert B. Keiter, *Saving Special Places: Trends and Challenges for Protecting Public Lands*, in The Evolution of Natural Resources Law and Policy 266–67 (Lawrence J. MacDonnell & Sarah F. Bates eds., 2010).

19. *See* Michael Soule & John Terborgh, Continental Conservation (1999); Jonathan S. Adams, The Future of the Wild (2006).

20. *See* Robert B. Keiter, *Breaking Faith with Nature: The Bush Administration and Public Land Policy*, 27 J. Land, Resources & Envtl. L. 195 (2007).

21. *See* Dep't of Agric., Forest Service, Planning Rules, 36 C.F.R. §§ 219.1(b), 219.6, 219.8, 219.9 (2012); National Park Service, Management Policies 1.6, 4.1, 4.4.1 (2006); Fish & Wildlife Service Manual, 602 FW 3.3(C), *available at* http://www.fws.gov/policy/602fw3.html.

22. U.S. Dep't of the Interior et al., America's Great Outdoors: A Promise to Future Generations 56–65 (2011), http://www.doi.gov/americasgreatoutdoors.

23. *See, e.g.,* California's Natural Community Conservation Planning program, http://www.dfg.ca.gov/habcon/nccp/ (last visited on Dec. 7, 2011); *see also* John C. Nagle & J.B. Ruhl, The Law of Biodiversity and Ecosystem Management 87–108 (2002).

24. U.S. Environmental Protection Agency, Interagency Ecosystem Management Task Force, The Ecosystem Approach: Healthy Ecosystems and Sustainable Economies (1996).

25. *See* R. Edward Grumbine, *What is Ecosystem Management?*, 8 Cons. Bio. 27 (1994); Keiter, *supra* note 1, at 71–72.

26. Joseph L. Sax & Robert B. Keiter, *The Realities of Regional Resource Management: Glacier National Park and Its Neighbors Revisited*, 33 Ecology L.Q. 233 (2006).

27. *See* Keiter, *supra* note 1, at 72–73.

28. *See* Millennium Ecosystem Assessment, Ecosystems and Human Well-Being: Policy Responses 2–5 (2005); E. Jean Brennan, Defenders of Wildlife, Reducing the Impact of Global Warming on Wildlife 17–30 (2008).

29. Intergovernmental Panel on Climate Change Fourth Assessment Report, Climate Change 2007: Synthesis Report 54, 64 (2007), http://www.ipcc.ch/publications_and_data/publications_ipcc_fourth_assessment_report_synthesis_report.htm.

30. On the likely impact of climate change on wildlife and wildlife habitats, *see* Frederic H. Wagner, *Global Warming Effects on Climatically-Imposed Ecological Gradients in the West*, 27 J. Land, Resources & Envtl. L. 109 (2007); Stephen Saunders et al., Rocky Mountain Climate Org., Hotter and Drier: The West's Changed Climate (2008); Douglas B. Inkley, National Wildlife Fed., Investing in America's Natural Resources: The Urgent Need for Climate Change Legislation (2008).

31. *See* Fish and Wildlife Service, *supra* note 2, at 14932.

32. Liesl Erb et al., *On the Generality of a Climate-Mediated Shift in the Distribution of the American Pika (Ochotona princeps)*, 92 Ecology 173–35 (2011); Erik A. Beever et al., *Patterns of Apparent Extirpation Among Isolated Populations of Pikas in the Great Basin*, 84 J. Mammalogy 37–54 (2003).

33. *See* Molly Cross & Chris Servheen, Climate Change Impacts on Wolverines and Grizzly Bears in the Northern U.S. Rockies: Strategies for Conservation (Workshop Summary Report, 2010).

34. Lisa G. Sorenson et al., *Potential Effects of Global Warming on Waterfowl Populations Breeding in the Northern Great Plains*, 40 Climatic Change 343–69 (1998); W. Carter Johnson et al., *Vulnerability of Northern Prairie Wetlands to Climate Change*, 55 Bioscience 863–72 (2005).

35. Phillip W. Mote et al., *Declining Mountain Snowpack in Western North America*, 86 Bull. Amer. Meteorological Society 39–49 (2005); Stephen Saunders et al., Rocky Mountain Climate Org., Greater Yellowstone in Peril: The Threats of Climate Disruption 14–17, 28 (2011).

36. Donald McKenzie et al., *Climate Change, Wildfire, and Conservation*, 18 Cons. Bio. 1–13 (2004).

37. *See* Brad Griffith et al., *Climate Change Adaptation for the U.S. National Wildlife Refuge System*, 44 Envtl. Management 1043, 1044–46 (2009); *see also* J. Michael Scott et al., *National Wildlife Refuge System: Ecological Context and Integrity*, 44 Nat. Resources J. 1041 (2004).

38. *See* IPCC, Climate Change 2007: Synthesis Report, *supra* note 29, at 65; *see also* Robin Craig, *"Stationarity is Dead"—Long Live Transformation: Five Principles for Climate Change Adaptation Law*, 34 Harv. Envtl. L. Rev. 18–31 (2010); John D. Leshy, *Federal Lands in the Twenty-first Century*, 50 Nat. Resources J. 111 (2010).

39. For a brief comparison of mitigation and adaptive strategies for addressing climate change, *see* Craig, *supra* note 38, at 18–30; Leshy, *supra* note 38, at 116–35; *see also* IPCC, Climate Change 2007: Synthesis Report, *supra* note 29, at 55–62.

40. *See* National Park Service, Climate Change Response Strategy (2010); Climate Change Wildlife Action Plan Work Group, Voluntary Guidance for States to Incorporate Climate Change into State Wildlife Action Plans and Other Management Plans (2009); H. John Heinz III Center for Science, Economics and the Environment, Strategies for Managing the Effects of Climate Change on Wildlife and Ecosystems (2008); Douglas B. Inkley et al., Global Climate Change and Wildlife in North America (Wildlife Society 2004).

41. *See* Fish & Wildlife Service, Rising to the Urgent Challenge: Strategic Plan for Responding to Accelerating Climate Change (2010).

42. *See* Keiter, *supra* note 1, at 309.

43. *See* Association of Fish and Wildlife Agencies, State Wildlife Action Plans: Working Together to Prevent Wildlife from Becoming Endangered, *available at* http://www.wildlifeactionplan.org/pdfs/wildlife_action_plans_summary_report.pdf.

44. Daniel Botkin, Discordant Harmonies (1990); Fred P. Bosselman & A. Dan Tarlock eds., *Symposium on Ecology and the Law*, 69 Chi.-Kent L. Rev. 847–985 (1994).

45. For a description of the theory and practice of adaptive management, *see* J.B. Ruhl & Robert L. Fischman, *Adaptive Management in the Courts*, 95 Minn. L. Rev. 424, 427–43 (2010).

46. *See* Kai N. Lee, Compass and Gyroscope (1993); Gene Lessard, *An Adaptive Approach to Planning and Decision-Making*, 40 Landscape & Urban Planning 81–87 (1998).

47. *See* Alfred Runte, National Parks 11–32 (2d ed. 1987); Chad P. Dawson & John C. Hendee, Wilderness Management 6–12 (4th ed., 2009); Dyan Zaslowsky & T.H. Watkins, These American Lands 226–28 (1994).

48. *See* Noss & Cooperrider, *supra* note 18, at 172–74; Zaslowsky & Watkins, *supra* note 47, at 192–93; Scott et al., *supra* note 37, at 1050–55.

49. *See* Noss & Cooperrider, *supra* note 18, at 172–74; Patrick N. Halpin, *Global Climate Change and Natural-Area Protection: Management Responses and Research Directions*, 7 Ecological Applications 828–43 (1997).

50. *See* Robert L. Peters, Defenders of Wildlife, Beyond Cutting Emissions: Protecting Wildlife and Ecosystems in a Warming World (2008); Halpin, *supra* note 49, at 830–31; *see also* Reed F. Noss, *Some Principles of Conservation Biology, as They Apply to Environmental Law*, 69 Chi.-Kent L. Rev. 893, 900–904 (1994).

51. On the national restoration area concept, *see* Robert B. Keiter, *The National Park System: Visions for the Future*, 50 Nat. Resources J. 71, 96–99 (2010).

52. *See* Robert B. Keiter, *Ecosystems and the Law: Toward an Integrated Approach*, 8 Ecological Applications 332, 336–38 (1998); Symposium, *Biodiversity and Its Effects on Private Property*, 38 Idaho L. Rev. 291–520 (2002).

53. *See supra* note 10 and accompanying text.

54. *See, e.g.*, Bruce Pardy, *Ten Myths of Ecosystem Management*, 39 Envtl. L. Rep. 10917 (2009); Keiter, *supra* note 1, at 147–52.

55. *See* Emma Marris, *The End of the Wild*, 469 Nature 150 (2011).

56. For a more detailed discussion of this potential shift toward more active management, *see* Alejandro E. Camacho, *Assisted Migration: Redefining Nature and Natural Resource Law under Climate Change*, 27 Yale J. on Reg. 171, 225–28 (2010).

57. Seattle Audubon Society v. Lyons, 871 F. Supp. 1291, 1311 (W.D. Wash. 1994) (emphasis in original).

58. For an insightful analysis of the role federal law might play in addressing climate change impacts on the public lands, *see* Robert L. Glicksman, *Ecosystem Resilience to Disruptions Linked to Global Climate Change: An Adaptive Approach to Federal Land Management*, 87 Neb. L. Rev. 833 (2009).

59. 16 U.S.C. §§ 528, 1604(e)(1) (2006).

60. *See, e.g.*, Colorado Environmental Coalition v. Dombeck, 185 F.3d 1162 (10th Cir. 1999); Sierra Club v. Martin, 168 F.3d 1 (11th Cir. 1999); Seattle Audubon Society v. Evans, 771 F. Supp. 1081 (W.D. Wash. 1991), *aff'd*, 952 F.2d 297 (9th Cir. 1991).

61. 16 U.S.C. § 531 (2006).

62. 16 U.S.C. § 1604(g)(3)(E) (2006).

63. 43 U.S.C. §§ 1702(c), 1701(a)(8) (2006).

64. 43 U.S.C. § 1702(c).

65. 16 U.S.C. § 1 (2006).

66. 16 U.S.C. §§ 668dd(a)(2), (a)(3)(C) (2006). *See* Symposium, *Managing Biological Integrity, Diversity, and Environmental Health in the National Wildlife Refuges*, 44 Nat. Resources J. 931–1238 (2004).

67. *See* 16 U.S.C. § 1a–7(b) (national parks); 16 U.S.C. § 668dd(e)(1)(A) (providing for "comprehensive conservation planning" on national wildlife refuges); 16 U.S.C. § 1604 (national forests); 43 U.S.C. § 1712 (BLM lands).

68. 43 U.S.C. §§ 1702(b), 1711(a).

69. Coordination provisions are found at 16 U.S.C. § 1604(a) (national forests); 16 U.S.C. § 668dd(e)(3) (national wildlife refuges); 43 U.S.C. § 1712(c)(9) (BLM lands).

70. These regional planning efforts are examined in Keiter, *supra* note 1, at 81–113, 274–99.

71. Secretary of the Interior, Order No. 3289 (Sept. 14, 2009); *see also* Secretary of the Interior, Order No. 3226 (Jan. 19, 2001). On Landscape Conservation Cooperatives, *see* Office of the Science Advisor, Fish & Wildlife Service, LCC Information Bulletin #1: Form and Function (Jan. 2010).

72. The cited ESA provisions are codified at 16 U.S.C. §§ 1533(a), (b), § 1536(a)(2), § 1538(a) (1)(B), § 1539(a), § 1539(j).

73. Dep't of the Interior, Fish & Wildlife Service, Endangered and Threatened Wildlife and Plants; Determination of Threatened Status for the Polar Bear (*Ursus maritimus*) Throughout Its Range, 73 Fed. Reg. 28212 (May 15, 2008). However, the Fish & Wildlife Service also adopted a special Section 4(d) rule for the polar bear exempting activities that produce greenhouse gas emissions from consultation under Section 7, concluding that the required causal connection between the action and affect on the species cannot be made with reasonable scientific certainty at this time. 50 C.F.R. § 17.40(q) (2011); 73 Fed. Reg. 76249, 76265 (Dec. 16, 2008). *See* Maggie Kuhn, *Climate Change and the Polar Bear: Is the Endangered Species Act Up to the Task*, 27 Alaska L. Rev. 125 (2010).

74. John Kostyack & Dan Rohlf, *Conserving Endangered Species in an Era of Global Warming*, 36 Envtl. Law Rptr. 10203 (2008).

75. *Cf.* Greater Yellowstone Coalition v. Servheen, 665 F.3d 1015 (D. Mont. 2009) (reversing the Fish & Wildlife Service's Greater Yellowstone Area grizzly bear delisting decision because the agency failed to adequately assess "other natural or man-made factors affecting [the bear's] continued existence," namely loss of the whitebark pine nut food source to climate change and bark beetles).

76. *Id.* at *6.

77. Natural Resource Def. Council v. Kempthorne, 506 F. Supp. 2d 322, 367-70 (E.D. Cal. 2007); *but cf.* Alliance for the Wild Rockies v. Lyder, 728 F. Supp. 2d 1126 (D. Mont. 2010) (rejecting claim that Fish & Wildlife Service failed to consider best available evidence on Canada lynx habitat and climate change).

78. 16 U.S.C. § 1538(a)(1)(B); 50 C.F.R. § 17.3 (2012); *see also* 16 U.S.C. § 1532(19).

79. 16 U.S.C. § 1539(j).

80. For a detailed analysis of the legal and policy implications of assisted relocation, *see* Camacho, *supra* note 56.

81. 42 U.S.C. § 4332(2)(C).

82. *See, e.g.*, Robertson v. Methow Valley Citizens Council, 490 U.S. 332 (1989); Thomas v. Peterson, 753 F.2d 754 (9th Cir. 1985); Conner v. Burford, 848 F.2d 1441 (9th Cir. 1988).

83. *See* Robert B. Keiter et al., *Legal Perspectives on Ecosystem Management: Legitimizing a New Federal Land Management Policy*, in Ecological Stewardship 20–21 (W.T. Sexton et al. eds., vol. 3, 1999).

84. For a perceptive analysis of the role of NEPA on the public lands in the face of climate change, *see* Mark Stephen Squillace, Guidance for Public Land Managers in Assessing Climate Change under NEPA (University of Colorado Law Legal Studies Research Paper No. 11-13, 2011), *available at* http://ssrn.com/abstract=1912811.

85. 16 U.S.C. § 1604(g)(3)(C) (national forests); 16 U.S.C. § 668dd(a)(4)(N), § 668dd(e)(1) (E) (national wildlife refuges); 43 U.S.C. § 1711(a), § 1712(c)(4) (BLM lands); National Park Service, Management Policies 4.1.2, 4.2.1 (2006) (national parks).

86. Northwest Ecosystem Alliance v. Rey, 380 F. Supp. 2d 1175 (W.D. Wash. 2005).

87. On the tiering of NEPA documents, *see* 40 C.F.R. 1502.20, 1508.28 (2010); *see also* Squillace, *supra* note 84.

88. *See* Keiter, The National Park System, *supra* note 51, at 96–99.

89. 16 U.S.C. § 431.

90. 43 U.S.C. § 1716.

91. For an overview of federal conservation-oriented acquisition programs, James R. Rasband & Megan E. Garrett, *A New Era in Public Land Policy? The Shift Toward Reacquisition of Land and Natural Resources*, 53 Rocky Mtn. Min. L. Inst. 11-1, 11-21 to 11-26 (2007).

92. *See* Barton H. Thompson, Jr., *Providing Biodiversity Through Policy Diversity*, 38 Idaho L. Rev. 355, 366 (2002).

93. *See generally*, Symposium, *Biodiversity and Its Effects on Private Property*, 38 Idaho L. Rev. 291–520 (2002).

94. H.R. 2454 §§ 471–82 (111th Cong., 1st Sess., 2009).

95. For a more detailed description and analysis of the Path of the Pronghorn migration corridor initiative, *see* David L. Cherney, *Securing the Free Movement of Wildlife: Lessons from the American West's Longest Land Mammal Migration*, 41 Envtl. Law 599 (2010).

11 From Principles to Practice

Developing a Vision and Policy
Framework to Make Ecosystem
Management a Reality

Sara O'Brien & Sara Vickerman, Defenders of Wildlife

For decades, natural resource managers have struggled to reconcile the political realities of jurisdictional boundaries, land ownership, and agency mandates with the ecological realities of boundary-less ecosystems and ecological processes. As managers are increasingly confronted with issues that cross boundaries—like wildland fire, invasive species, climate change, and land use change—the need to better coordinate management across ownerships and spatial scales has become even more pronounced. But the idea of working across legal and political boundaries carries with it a great deal of important social, economical, and political baggage. Coordinating fire management or land use decisions across land ownerships brings up sensitive questions about authority and control, the different mandates of different public agencies, and the rights and responsibilities of private property owners. How can policies best promote coordination at ecologically-relevant scales while respecting the individual mandates of distinct agencies and the rights and independence of private landowners?

The concept of ecosystem management is, in part, a response to these challenges and concerns. Ecosystem management has been defined and implemented in a variety of ways, but as the concept has evolved, it almost always includes at least four salient features:

- A focus on managing resources and natural systems at a scale that is eco-logically meaningful, rather than being confined to artificial political boundaries;
- Attention to multiple resources, such as water quality or quantity, carbon storage, and biodiversity, rather than a siloed, single-resource approach;
- A collaborative approach that seeks to involve both public and private part-ners in planning and decision making; and
- A commitment to managing natural systems in a way that balances sustain-able human uses with long-term ecological health.[1]

Interest in the ecosystem management approach has ebbed and flowed in the roughly two decades since its conceptual beginnings. As early as 1995, an article in the *New York Times* pronounced ecosystem management to be "dead on arrival" at Capitol Hill[2], and today the term has been largely supplanted by closely related concepts such as landscape-scale conservation, the all-lands approach, collaborative natural resource management, and ecosystem services. But these four key concepts—collaboration, ecological scale, attention to multiple resources, and balancing human and natural values—have persisted. The increas-ing number of conservation success stories that follow these basic approaches suggests that, regardless of the terms used to describe these concepts, the funda-mentals of ecosystem management remain meaningful and useful.

We argue that broader implementation of ecosystem management princi-ples has been inhibited by at least two factors, both of which would seem to follow logically from its primary origins in the academic community and among the professional staff of the federal land management agencies.

First, implementation has been hampered by the absence of a clearly artic-ulated and cohesive vision for what the goals and outcomes of the ecosystem management process ought to be. Ecosystem management outlines a useful and logical process for achieving an implied goal—natural landscapes that accom-modate both sustainable human uses and healthy natural communities—but the lack of a clearer, more explicit vision of the future has almost certainly checked the enthusiasm of policy makers and the public. Difficulty in describ-ing what ecosystem management could be expected to achieve if it were broadly implemented may well have been one of the reasons the *Times* found such limited support for the concept on Capitol Hill.

Second, ecosystem management has not been well supported by existing environmental and land management policies, nor have its advocates met with much success in developing policies that support collaborative, landscape-scale conservation. Instead, at best, practitioners have made use of policies like the Endangered Species Act or the National Environmental Policy Act to help motivate stakeholder and agency participation in collaborative processes; at worst, policies have constrained agencies' abilities to cross jurisdictional lines and meaningfully engage the public.[3]

Both of these shortcomings are resolvable, but neither has so far appeared as a significant focus of the ecosystem management community. Nonetheless, progress is being made on both fronts, albeit often by different players and under different guises. Many in the conservation community are coalescing around a vision of creating an interconnected, resilient network of conservation lands and waters. While this goal does not currently enjoy broad popular support, it is based around ideas that have proven to be both successful and popular with the public. We argue that the concept of a national network of conservation lands has the potential to provide the ecosystem management approach with the guiding vision that it currently lacks.

Changes are also afoot in the world of environmental policy. The emergence of innovative policy approaches to conservation on both public and private lands has created new opportunities for ecosystem management to work across borders. At the same time, political pressure on command-and-control approaches to environmental regulation is creating a sense that the foundational environmental policies of the 1970s are in danger of losing their popular support. For better or worse, some of the policies that are taken for granted today are likely to change over the next several years. The principles of ecosystem management can help guide those changes in a direction that is beneficial for both natural and human communities.

In this chapter, we present a potential vision and policy future for ecosystem management. We describe the 'network of lands' concept now circulating throughout the conservation community and its rationale and essential elements. We briefly identify the main sources of conflict between existing environmental policy and the ecosystem management approach, and we describe policy changes and emerging approaches that have some potential to help resolve these conflicts.

I. CONSERVATION VISION

The United States is often described as having a fragmented or piecemeal approach to environmental policy. In a way, this situation reflects a deeper level of discord, a lack of national consensus about what we expect and desire for—or from—our natural ecosystems.

Nonetheless, national polls have reflected a surprising level of public agreement on the need to address many environmental issues. In 2009, a bipartisan national poll showed that despite a very difficult economic situation, three in five voters said they would support a small increase in taxes to increase public investments in conservation at the local, state, or federal levels.[4] Results of recent state and local ballot initiatives tend to support these numbers. In November 2008, for example, 62 of the 87 proposed state and local conservation finance ballot measures passed.[5] In the 2010 elections, following two years of high unemployment and flagging state economies, Iowa passed a measure to fund wildlife habitat conservation out of state sales tax revenues, and Oregon made permanent a previous constitutional commitment of funding for conservation projects, with the support of almost 70 percent of the voters and majorities in every county across the state's usually deep urban-rural divide. Spending on conservation similarly reflects a high level of enthusiasm. By the turn of the century, funding for land conservation from federal, tribal, state, local, and private sources had reached more than $9 billion per year.[6] As of 2005, private land trusts had protected more than 37 million acres of land from development through acquisitions, easements, and other tools.[7]

Despite these investments, however, many indicators continue to show a decline in environmental health.[8] Traditional piecemeal approaches to land conservation and natural resource management have failed to provide Americans with healthy, abundant natural ecosystems. Environmental regulations, set-asides of public and private lands for conservation purposes, and local efforts to improve resource management and land use planning have not been sufficient to halt or reverse the loss of native ecosystems and species.

We argue that providing for the long-term health of our ecosystems and the values they provide will require developing a clear vision of conservation success to which the American public can relate. The growing popularity among the conservation community of the concept of a network of conservation lands and waters that support native species and natural ecosystems provides a promising start.

The idea of a national conservation system has been with us for a long time. Many of our great thinkers in the conservation world have pictured the establishment of a system of lands that will represent a cohesive national approach to resource conservation.[9] Realistically, however, relatively few important decisions about land use and conservation are or will be made at the national level or by the federal government. Centralized, and especially command-and-control, approaches to environmental protection have led to significant gains but have also resulted in push-back from business interests and citizens concerned about land values and government intrusiveness. Given the current legal and social environment, most decisions about land use will continue be made at smaller scales, by individuals, local communities, and county and state governments. But past experience has clearly shown that uncoordinated conservation actions among private and public landowners are not sufficient to meet conservation goals, especially in the face of rapid, uncontrolled land development and degradation.

In the long run, the best way to conserve biodiversity and the human values it sustains is to change the way people think about the natural world and manage our natural resources. In the short run, the surest way to ensure that species, habitats, and ecosystem processes are maintained over time is to devote some portion of the landscape to those specific purposes. Because this task is beyond the capacity of any single agency or organization, public or private, maintaining healthy ecosystems, species, and resources will require the creation of a virtual network of conservation lands, one that stretches across land ownerships and builds on the strengths of a variety of land ownerships and land uses.

As a result, the idea of developing a national network of conservation lands is perhaps best seen as a shared goal that can be implemented in different ways and to different extents at different scales. The network would function not as a monolithic system created by expanding the federal land base but, as with other kinds of networks, as a loosely confederated and related set of hubs and connections with a common purpose but no single authority in charge. Decisions made at local, state, and regional levels about conservation and land development could be seen as independent but nested, with priorities and goals identified at each level and used to guide decision making.

Past investments in land conservation have created an impressive system of conservation lands that can form the backbone of this network. This system includes not only national parks and refuges, but also local greenspaces and farm-scale habitat protection projects, and many things in between. These lands and the services they provide are critical to the nation's physical, economic, and social health, but they have not been sufficient to protect native biodiversity, clean air and water, open spaces for recreation, carbon storage, and other ecological values. Lands that are managed primarily for human use can and must also play a key role. When stewardship of private lands is thoughtful, responsible, and sustainable, they have much to contribute, regardless of their primary management objectives.

Working toward this vision will require an array of policy tools and approaches, including public lands management, incentives for private land-owners, regulations, and market-based approaches; and monitoring and evaluation to support adaptive changes in management strategies over time. This network of lands does not have to be limited—and likely cannot be limited—to traditional forms such as preserves, parks, and refuges, and some of the lands that are valuable for conservation can still generate significant financial benefits for their owners. However, management must focus on sustaining ecological values, not as a collateral or subsidiary benefit, but as a primary goal for the land. Because so many of our landscapes have already been heavily modified by human actions, this network of conservation lands will not be a collection of pristine habitats. But our goals for the conservation network can be aspirational, and these goals can help articulate the vision that most people find lacking in traditional explanations of ecosystem management.

To be successful in a time of rapid human-caused climate change, continuing habitat loss and land conversion, and other major environmental challenges, we argue that a network must meet the following criteria.

First, a national conservation network should be *comprehensive, adequate, and representative,* according to established principles of conservation planning. These principles are well accepted and well documented in the literature as the cornerstone of the conservation of species and ecological systems, although slightly different terms are sometimes used.[10] Significantly, the government of Australia has identified these three elements as "nationally agreed criteria" for establishing a conservation network there.[11] These three criteria describe the importance of developing a reserve system that includes the full range of eco-logical communities; is large enough to maintain species diversity, support inter-action among species and systems, and accommodate disturbance events; and is representative and redundant enough to protect genetic and species diversity indefinitely.[12]

J. Michael Scott and others have shown that the current system of conserva-tion lands in the United States falls far short of these criteria.[13] It has long been observed that our reserves have typically been established opportunistically, away from competing uses. As a result, high-elevation ecosystems with steep slopes, colder climates, and less productive soils are relatively well protected, while lower, more productive ecosystems such as grasslands and valley bottomlands have been almost completely converted to human uses.[14] Many significant and unique ecosystems have all but disappeared. According to one estimate, nearly one-third of North American ecoregions have few or no large blocks of natural habitat remaining.[15]

Because human-induced climate change is expected to have profound effects on species and ecosystems, developing a comprehensive, adequate, and representative system of conservation lands is more important than ever. Formal consideration of future climate conditions, and how the changing climate might affect the outcomes of today's investments, is now considered a critical step in conservation planning. Recently, a new approach to representativeness has emerged as a result of concerns that current vegetation communities no longer provide a reasonable basis for identifying a conservation portfolio that will support biodiversity under future climates. Instead, some ecologists have proposed using geophysical variables such as soil type, geology class, and elevation to determine whether a given set of conservation areas is expected to support future diversity.[16] In developing a conservation network, the rapidly-evolving field of climate change adaptation science should be used to identify areas that are likely to support important ecological values given projections of future climate. Because the existing mix of habitats and species in some areas may change over time, maintaining ecological processes and the geophysical drivers of biological diversity may be among the best ways to protect the resilience of natural systems.

Second, such a network must also be *interconnected* at multiple scales, with significant roles played both by public and private lands set aside for conservation and by those lands managed primarily for human use. In our existing system, conservation lands are often isolated from one another, separated by anthropogenic barriers such as developed areas, intensively managed agricultural lands, blocks of clear-cut forests, and roads. This ad hoc arrangement of land uses presents obvious barriers to wildlife, as very few species have the ability to move long distances through highly altered landscapes to migrate, search out new habitat or mates, or shift their range in response to changing climate conditions. Furthermore, isolated natural areas, especially in smaller patches, are highly susceptible to invasion by nonnative species or destruction by fire and other natural disturbance events, and they often cannot be managed to accommodate ecological processes such as fire and flood.

With increased concern about habitat fragmentation and climate change, the idea of improving connectivity among conservation lands has received a great deal of attention in conservation circles in recent years.[17] Interconnected landscapes allow wildlife to repopulate an area after a disturbance; facilitate the exchange of genes and individuals among otherwise isolated populations and thereby reduce the likelihood of inbreeding or a localized extinction event; and provide space for species with an inherent need to move across large areas to meet habitat and resource needs.[18]

Nonetheless, connectivity cannot be seen as a simple solution for limiting the effects of either habitat loss or climate change. Research suggests that keeping protected areas well connected is a poor substitute for protecting the overall area of habitat available, that "the effects of habitat loss far outweigh the effects of habitat fragmentation . . . [and] details of how habitats are arranged cannot usually mitigate the risks of habitat loss."[19] Furthermore, poorly designed wildlife corridors can create negative effects, for example, by leading individuals away from one habitat area but failing to move sufficient numbers to a new area to create a viable population.[20] Connectivity may also work against efforts to protect species that are vulnerable to climate change, by leaving them open to invasion by nonnative or 'new native' species or to fire; reducing connectivity of selected unique or diverse areas has even been proposed as a possible climate change adaptation strategy.[21]

In light of these challenges, the most effective approaches to landscape connectivity will likely be those that focus on:

- Increasing the overall quantity and quality of land managed for conservation[22]
- Minimizing barriers to species movement and protecting connectivity where it already exists[23]
- Increasing the permeability of the "matrix" lands, areas in agricultural, extractive, residential, or other uses that separate and can isolate conservation lands[24]

Each of these approaches highlights the importance of private lands in a conservation network. Conservation on private lands can take many forms and can be encouraged through a variety of policy tools. Improving the management of these lands can greatly increase the permeability of the landscape to at least some species. Research indicates that the type of land cover in matrix areas surrounding patches of habitat is a better predictor of species occurrence than the size or arrangement of habitat patches, suggesting that improving the quality of land management in nonprotected areas could provide better conservation outcomes than altering the size or distribution of existing protected areas to improve landscape connectivity.[25]

Finally, we argue that a national conservation network should be designed to be as *resilient* as possible to rapid climate change, increased climate variability, and other landscape-scale threats. In the context of climate change, promoting resilience may be a matter of conservation planning—identifying conservation sites that are most likely to retain desired values in future climates—or one of managing conservation lands in ways that increase their ecological health and capacity to maintain values in the face of disturbances.[26]

In his seminal paper on ecological resilience, C.S. Holling took a somewhat broader view, contrasting the term with stability and using it to describe the persistence of a system and its capacity to "absorb change and disturbance and still maintain the same relationships" among parts of the system.[27] Holling saw resilience as the direct result of instability—"large fluctuations"—in a system, and noted the tendency for efforts to impose stability on natural systems, as through fire suppression or regulation of water flow, to decrease the long-term viability of those systems.[28]

The resiliency of conservation lands has often been seen primarily as a function of patch size; the larger a conservation area, the more likely it is to be resilient in the face of disturbances and other changes.[29] In conservation planning, attention to resilience was originally motivated by a concern that conservation areas should be sufficiently large, connected, functional, and protected to meet conservation goals in the face of large- and small-scale disturbances and other changes.

Like many ecological principles, however, the concept of resiliency has new complexity and significance in the era of rapid, human-caused climate change. In many systems, the size and distribution of conservation areas can no longer be expected to ensure ecological resilience in the face of massive changes to climate and disturbance regimes. Much more work is needed to define what these goals mean in terms of managing natural ecosystems. We do know, however, that under conditions of rapid, human-caused climate change, it is critical that we find ways to manage and protect ecosystems that help them, as Holling says, better "absorb disturbance and reorganize while undergoing change so as to still retain essentially the same function, structure, identity, and feedbacks."[30]

There is now a sense of urgency behind efforts to improve the conservation network and ensure that it can support the full diversity of species, habitats, and ecological values. Increased awareness of climate change is one major driver of this urgency. As we look to our existing conservation network to both sequester carbon dioxide naturally and to support human and wildlife populations under rapidly changing conditions, the shortcomings of that system become increasingly apparent. Developing an interconnected network of conservation lands and waters will help both natural and human communities adapt to climate change by allowing plant and animal species more opportunities to move or adjust to new conditions and help to buffer human populations from droughts, floods, fires, and other threatening features of increasingly volatile disturbance regimes.

II. EMERGING POLICIES FOR ECOSYSTEM MANAGEMENT

The widespread application of ecosystem management principles and the implementation of the conservation network vision have been limited in part by our existing policy environment. The United States has never had a comprehensive policy aimed at protecting biodiversity, the functioning of natural ecosystems, or the services they provide. Instead, we have taken a patchwork approach, depending on federal land protection, the Endangered Species Act, and resource-specific environmental policies such as the Clean Air and Clean Water Acts to achieve ecosystem protection by proxy. This system has not been sufficient to halt or reverse the loss of native ecosystems, species, or ecosystem services.

Recently, however, the emergence of innovative policy approaches is creating new opportunities for conservation that better reflect an ecosystem management approach. At the same time, many existing policies and programs are being examined in the light of new challenges, perspectives, and priorities. Some of the policies that form the foundation of the conservation world—the Endangered Species Act, the Clean Water Act, and the Farm Bill conservation programs, for example—are being used in new ways, to meet new goals. In many cases these emerging approaches better reflect the principles of ecosystem management, as consensus builds around the importance of collaborative, landscape-scale, and cross-sectoral initiatives.

A diversity of new and existing policy tools will be needed to make the conservation network vision a reality. New approaches, such as market-based tools and payments for ecosystem services, hold impressive promise, but they should not be expected to replace other types of policies. For example, the ecosystem markets that are most successful today are those that are driven by regulatory requirements; voluntary ecosystem markets lag far behind. Thus, these emerging policy tools should be combined with improvements to regulatory approaches, landowner incentives programs, and other more traditional approaches to form the policy basis of a national conservation framework. What kinds of policy innovations are emerging, and what others might be needed, to help implement the principles of ecosystem management and fulfill the promise of that approach?

A. *Looking Beyond the Boundaries on Federal Lands*

Managing species, resources, and ecosystems at ecologically meaningful scales creates significant challenges for the existing body of federal lands and federal environmental policies. Federal lands form most of the largest blocks of natural habitat in the United States and are often seen as the backbone of the nation's conservation system, but they have proven to be insufficient for achieving broader biodiversity conservation goals in part because their management often does not 'make it to scale.' Most federal land units are not large enough to allow them to individually manage species and processes at ecologically relevant scales, and their boundaries were often not drawn along ecologically meaningful features. Federal lands are also not representative of the full diversity of the nation's ecosystems.[31] In contrast, some 60 percent of land in the United States is privately owned, and about one-third of federally listed species occur only on nonfederal lands.[32]

With considerable political resistance to a major expansion of federal lands, the opportunities for crafting the federal land base into a stand-alone conservation network would be limited. Nonetheless, the large areas of habitat and level of protection provided by these policies present a significant opportunity for landscape-scale management through better coordination among land-management agencies and between public and private entities. Unfortunately, the large land-management agencies have often avoided collaborative relationships with other federal agencies, state land-management agencies, or private or industrial landowners.[33]

In his book on collaboration among public agencies, Craig W. Thomas points out that top-down mandates have most often been seen as the only way to achieve better interagency collaboration, but they are rarely effective.[34] Indeed, each of the federal land management agencies already has a policy mandate to use the ecosystem management approach.[35] Using a series of interagency efforts in California as case studies, Thomas finds that individuals tend to engage in collaborative efforts only when they believe they have something to gain by doing so. Often, he argues, partners collaborate when it helps them resolve problematic uncertainties—uncertainty about whether a given project will spark litigation, for example, or about whether public lands will continue to provide livelihoods for local communities.[36] The policy challenge for ecosystem management, then, is to find ways to create meaningful incentives for federal agency staff to plan and manage conservation efforts that cross political boundaries.

Federal environmental policies have played a key role in creating incentives for increased interagency collaboration over the last few decades. Thomas found that the Endangered Species and National Environmental Policy Acts created significant incentives for cross-agency collaboration in California. As a result of the Endangered Species Act, many professional staff recognized a mandate for biodiversity conservation that required them to look beyond their own boundaries and participate in larger-scale efforts; at the same time, "as environmental activists successfully sued local, state, and federal agencies under the ESA and other environmental laws . . . agency managers began to see benefits to cooperative planning and management."[37] Of course, lawsuits may also have the opposite effect in some situations, discouraging agencies from engaging in more significant interactions with the public than is legally required.

Similar dynamics have also motivated many public-private collaborative partnerships. Starting in the mid-1990s, a number of collaborative groups have emerged that aim to find new solutions to the messy set of issues surrounding the intersection of private lands and uses with public forests and rangelands. Early examples included the Quivira Coalition, the Quincy Library Group, and the Greater Yellowstone Coordinating Committee, but countless similar efforts have emerged around the country. Many of these collaborative efforts are also motivated by the "blunt hammer" of the Endangered Species Act.[38] Conflicts over endangered species and extractive uses of public lands set the stage for 'nature versus jobs' battles that proved to have few real winners. At the same time, decades of fires suppression in national forests began to shift fire regimes in some dry forest types, increasing the risk that large, stand-replacing fires would cross over into private lands and threaten rural communities. These trends created a strong sense that neighbors of federal lands had a large and growing stake in their management. As antagonistic approaches over these issues increasingly led to stalemates and economic damage to rural communities, the conditions were ripe in some areas for a new approach.

Funding for planning and management also plays a key role in creating incentives for collaboration at larger scales. Practitioners of ecosystem management in the 1990s quickly found that "ecosystem management does not fit into the appropriations structures of most governments."[39] Working outside of federal boundaries created a need for some creative thinking around how to fund work that crosses agencies, units, sectors, projects, and budget years. There is also an increasing need to find more effective ways to combine multiple funding sources—public and private—for large, landscape-scale projects. A number of recent initiatives intended to encourage collaboration among federal agencies and between private and public entities may provide some ideas:

- The Fish and Wildlife Service-funded Habitat Joint Ventures, originally created to improve cross-boundary collaboration in conserving waterfowl and their habitats, has proven to be an effective model for promoting cooperative conservation efforts on a landscape scale. Although the joint ventures do not in themselves provide significant funding for on-the-ground conservation, their expanded focus on all birds and increasingly sophisticated conservation planning have made them a powerful, decentralized driver for expansion of the network of conservation lands.
- With department-level interagency initiatives, such as the Great Lakes Restoration Initiative and America's Great Outdoors, funds may be routed through a single federal agency with a specific mandate to provide funding to other agencies' projects, or a budget line may be created at the secretarial level to fund work and distributed among agencies and other stakeholders. The more recent Landscape Conservation Cooperatives are intended to employ a similar model.
- In dealing with the threat of wildland fire, Congress has tried a combined approach by creating a mandate for two departments to develop a collaborative management plan and also providing a pool of funding that both agencies can use to implement the plan. The Federal Land Assistance, Management, and Enhancement (FLAME) Act required the Department of the Interior and the Department of Agriculture to work together to create a strategy for collaboratively addressing wildland fire issues on both departments' lands and also created a joint pool of funding for emergency fire activities.[40]
- The Forest Service's 'all-lands approach' aims to extend the capacity of that agency to work outside national forest boundaries. For example, the service's Collaborative Forest Landscape Restoration Program seeks to encourage interagency collaboration by funding landscape-scale collaborative management efforts. Previously, a 1998 amendment to a Senate appropriations bill (the Wyden amendment) gave the Forest Service the authority to spend existing appropriations on projects on nonfederal lands that benefit wildlife and other resources on national forestlands, paving the way for projects that span that divide.

B. Using Single-Resource Policies to Drive Integrated Planning

While policies like the Endangered Species Act, National Environmental Quality Act, and Clean Air and Clean Water Acts form an important foundation for environmental health in the United States, they have generally proven to be insufficient to protect ecological values in the face of a rapidly expanding human footprint. Many also inherently conflict with the ecosystem management approach, because they generally embody a narrow focus on a single resource, create few incentives for managing across political boundaries, favor a punitive over a collaborative approach, and generally fail to provide much guidance in how to balance ecological health with human uses. These policies play a vital role in preventing ecological disasters and we do not argue for dismantling them. Rather, finding ways to allow federal environmental policies to better accommodate an ecosystem management approach could vastly improve both their ecological results and social acceptability.

For example, after nearly forty years on the books the Endangered Species Act has proven to be relatively successful at halting species extinctions.[41] It serves as an important final safety net for species on the brink of extinction, and it has played an important role in encouraging federal agencies and other stakeholders to pursue more collaborative and cross-boundary approaches to management. It has, however, been less successful in terms of helping species recover to sustainable population levels and protecting the ecosystems upon which imperiled species depend, the other two stated goals of the act.[42] Traditional implementation of the act, with its linear, process-oriented approach of listing a species, identifying its critical habitat, developing and implementing a recovery plan, and delisting the species when recovered, fits poorly with the ecosystem management approach.

Alternative compliance mechanisms developed for the implementation of the Endangered Species Act on private lands may indicate a potential way forward. Habitat conservation plans, which allow an incidental take of listed species on private lands that will not threaten the survival of the species, are increasingly being used to cover multiple species and multiple stakeholders, creating promising opportunities to bring together diverse stakeholders for the kind of large-scale strategic conservation planning that typifies an ecosystem management approach.[43]

In Pima County, Arizona, for example, a multispecies regional habitat conservation plan, the Sonoran Desert Conservation Plan, has developed into a county-wide process that combines science-based conservation planning, a comprehensive land-use plan, and a plan for Endangered Species Act compliance.[44] This process provided a venue for diverse stakeholders to engage in meaningful, productive conversations about balancing economic concerns with endangered species and other ecological priorities and to use the outcomes of those conversations in an integrated conservation and land use planning process.

Alternative compliance mechanisms, including regional habitat conservation plans, programmatic safe harbor agreements, and candidate conservation programs with assurances can also be combined with market-based approaches described below to create incentives for landscape-scale conservation and collaborative management.

C. Using Market-Based Approaches to Balance Economic and Ecological Values

The limitations of regulatory approaches to conservation on private lands have led in recent years to an increasing interest in quantifying the value of the ecosystem services[45] that those lands provide and in creating markets for some of these services that have generally functioned as external costs in more traditional markets. Many of our most intractable environmental problems are arguably the result, in Kenneth Arrow's words, of "the failure of markets to exist."[46] The idea of making payments for ecosystem services—either through markets or incentive programs or other policy means—is rapidly emerging as a viable response to that failure.

Market-based approaches include voluntary, pre-compliance, and compliance markets.[47] Compliance markets, those driven by regulations, have so far seen the most implementation. For example, wetland and conservation banks are used increasingly to address wetland and endangered species protection needs, and water quality trading has emerged as another way for regulated parties to meet federal requirements. Examples of voluntary markets include potential pre-compliance markets for resources that are likely to be regulated in the future. For example, where there is interest in avoiding future listing under the Endangered Species Act by establishing credit trading programs within the range of the species.

Mitigation banking offers an opportunity to implement existing land-use regulations in a way that creates more flexibility in compliance while resulting in a net gain in ecological function. For example, if land development can be steered toward sites with low ecological value and developers are required or encouraged to conduct any mitigation projects on areas of high ecological value, it will be possible to move beyond the zero-sum (or, too often, net loss) transactions that have typified wetland mitigation banking so far. Similarly, if mitigation projects and other conservation actions, such as landowner incentives and conservation easements purchased by local governments, can be directed toward the agreed-upon priority areas, the overall conservation benefit of all of these programs will be much enhanced.

Taking advantage of these opportunities will require a process for clearly identifying areas that are of high value and high priority for conservation and development, respectively. Research by the Environmental Law Institute and the Nature Conservancy suggests that "a more comprehensive approach to mitigation informed and guided by" state-level conservation plans can provide much more effective results than the current approach.[48] They argue that State Wildlife Action Plans and other federally-required conservation plans should be used to strategically direct mitigation efforts, although county and city comprehensive plans and other local-level documents may also prove useful for identifying priority areas at a finer scale.

Properly structured and implemented, market-based approaches and payments for ecosystem services have the potential to make conservation investments more efficient and effective, tap new funding sources for conservation, direct development away from the most ecologically sensitive areas, provide alternative revenue to rural landowners to supplement traditional commodity production, and facilitate the use of more natural infrastructure to address conservation challenges. In doing so, they may also create new options for implementing an ecosystem management approach by providing innovative ways to fund conservation on private lands and help balance human economic and social needs with ecological health.

However, the development of ecosystem services markets has been hampered by a number of policy limitations. Although regulations often drive markets, they were not written with that purpose in mind. Most environmental laws and regulations do not explicitly authorize payments for services or market-based approaches. They were generally developed to address a single resource issue, making the development of markets that address multiple resource values a particular challenge. Furthermore, many important ecological challenges, including non-point-source pollution, climate change, and biodiversity loss, are virtually unregulated in the United States, which tends to limit demand and inhibit market development. These structural barriers also tend to result in high administrative and transaction costs for market-based programs.

D. Payments for Ecosystem Services on Private Lands

Better understanding of the importance of private lands has led many in government and in the conservation community to shift their attention to the use of landowner incentives and other voluntary programs to achieve conservation goals. Where ecosystem services are minimally regulated and markets are impractical or are not sufficiently developed, strategic use of landowner incentive programs can help bridge the gap. The term 'payments for ecosystem services' is often used to describe the combined set of market-based approaches and incentive programs in which governments pay landowners to implement conservation practices.

Incentives are appropriately seen as an important supplement to other conservation tools, such as regulation and land acquisition.[49] However, existing incentive programs face many of the same challenges as other types of policy tools. They are often focused on a single species or resource, and they tend to be applied on an opportunistic, site-by-site basis, without attention to the broader, landscape-scale impacts. Many incentive programs lack adequate funding and technical support, and outcomes are not always monitored in a way that allows for meaningful evaluation.[50] Many new incentive tools are now emerging, and old incentives such as those tied to the Farm Bill are being reworked and repurposed, but, in general, the development of conservation incentive programs for private landowners has received relatively insufficient time, effort, and funding.

The idea of governments making 'green payments' to landowners for their provision of ecosystem services is gaining significant traction. The National Resources Conservation Service, which invests $5 billion annually in conservation programs, is looking at new ways to measure results and to ensure that its expenditures result in clear progress toward strategically-defined conservation goals. State and local ballot initiatives are also being used to steer more funding toward conservation priorities by dedicating small portions of tax or lottery proceeds.

Federal agencies could help push incentive programs toward a more integrated approach by supporting state grant programs that could be set aside to provide matching funds for local or regional entities to use in conservation planning and implementation, or by directing some portion of funding toward projects that help implement comprehensive conservation plans created at the local, state, ecoregional, or regional scale.

Projects such as the Clean Water Services program in northwestern Oregon, which combines state and federal funding with utility ratepayer fees to invest in habitat restoration instead of hard infrastructure to provide water cooling capacity, have led some communities to consider developing Ecosystem Services Districts, a new institution that would be designed to deliver a wide range of ecosystem services to the public. This institution would be responsible for coordinating and funding local conservation efforts through a combination of local, state, and federal funding sources. As with the Clean Water Services example, state and local sources of funding can be matched with federal investments in natural flood control projects, habitat restoration for water quality, and other ecosystem-based approaches. Priority could be given to providing funding for projects that align with a local conservation plan, ensuring that the provision of clean water, carbon storage, biodiversity, and other ecosystem services is funded equitably by the many beneficiaries of those services and that conservation work at multiple scales and across multiple resources is coordinated and efficient.

Proponents of payments for ecosystem services—both market-based and publicly funded—have proposed and discussed several federal-level policy options that could help improve the effectiveness of existing programs and encourage innovation in this arena. These include directing relevant federal agencies to:

- Encourage and provide incentives for state and local governments to use natural infrastructure, where feasible, in place of hard engineering projects.
- Consider meeting mitigation requirements by purchasing tangible conservation outcomes, such as those measured in credits for habitat, water quality, carbon, etc.
- Jointly fund the development of consistent, multiscale, integrated measurements for ecosystem services that can be used in markets and payment programs.
- Support the development of integrated, landscape-scale plans by multiple agencies and stakeholders to guide strategic investments in ecosystem services.
- Provide direct financial support, loan guarantees, or revolving loan funds to support landowners offering ecosystem services for sale in new, fragile markets.
- Use federal funds for pilot projects as proof of concept for more coherent, efficient, and effective public and private investment strategies.

III. CONCLUSION

The ecosystem management approach, while innovative, practical, and well-grounded in science, creates significant challenges for those who would see it more broadly implemented. While widely accepted within the land and resource management communities that created it, the concept has failed to garner popular support, even while its foundational principles have grown in application and popularity.

We have argued here that the lack of an overarching vision and supportive policy framework has hindered implementation of the ecosystem management approach, but other factors are certainly at work. A lack of public understanding or appreciation of the values nature provides likely plays a role; being able to put a dollar value on these services may make these benefits more tangible to many people. A tendency toward reductionist thinking may be another element. Our culturally-driven impulse to intellectually divide natural systems into small parts leads us to ignore connections and the larger whole. As Margaret Wheatley, a researcher in organizational behavior and management, describes it, "[w]hen we study the individual parts or try to understand the system through discrete quantities, we get lost. Deep inside the details, we cannot see the whole."[51] Finally, American beliefs about private property and preference for prioritizing property rights over other rights and values can make parts of the ecosystem management approach seem intrusive or threatening.

Despite these challenges, ecosystem management provides one of the most compelling, comprehensive, holistic, and pragmatic approaches to land and resource management yet to emerge. It evolved in response to the clear failure of previous approaches to negotiate some of the difficult trade-offs described here. While much progress has been made toward infusing the principles of ecosystem management into conservation and land management practices, a great deal more effort will be needed to develop ecosystem management into an approach with broad public and policy support.

ACKNOWLEDGMENTS
The authors thank Dennis Figg, Bruce Taylor, Noah Matson, Peter Nelson, and Tim Male for their review of the chapter and thoughtful input and the Doris Duke Charitable Foundation for generous support of this project.

NOTES

1. *See, e.g.*, Ecosystem Management (Mark S. Boyce & Alan Haney eds., 1999); R. Edward Grumbine, *What is ecosystem management?*, 8 Conserv. Biology 27 (1994); Steven L. Yaffee, et al., Ecosystem Management in the United States (1995).

2. Yaffee et al., *supra* note 1, at 40.

3. *Id.* at 33–34.

4. David Metz & Lori Weigel, Public Opinion Strategies & Fairbank, Maslin, Maullin, Metz & Assoc., Opinion Research & Analysis, Key Findings from National Voter Survey on Conservation 1 (Sept. 25, 2009).

5. The Trust for Public Land. Conservation Finance: 2010 Results, at 1 (2010), *available at* http://www.tpl.org/assets/files/publications/conservation-finance/conservation -finance-2010.pdf (last visited Dec. 13, 2011).

6. National Council for Science and Environment, Wildlife Habitat Policy Research Program: Completing a Wildlife Habitat System for the Nation, at 2 (Oct. 2010), *available at* http://ncseonline.org/CMS400Example/uploadedFiles/03_NEW_SITE/3_Solutions/ WHPRP/WHPRP%20Statement.pdf (last visited Dec. 13, 2011).

7. Land Trust Alliance, 2005 National Land Trust Census Report, *available at* http://www .landtrustalliance.org/about-us/land-trust-census/census (last visited Dec. 13, 2011).

8. J. John Heinz Center for Science, Economics and Environment, The State of the Nation's Ecosystems 2008: Measuring the Lands, Waters, and Living Resources of the United States (2008).

9. *E.g.*, Mark L. Shaffer, J. Michael Scott & Frank Casey, *Noah's options: Initial cost estimates of a national system of habitat conservation areas in the United States*, 52 BioScience 439 (2002).

10. *E.g.*, James A. Fitzsimons & Hugh A. Robertson, *Freshwater Reserves in Australia: Directions and Challenges for the Development of a Comprehensive, Adequate and Representative System of Protected Areas*, 552 Hydrobiologia 87 (2005).

11. Australian and New Zealand Environment and Conservation Council, Comprehensive, Adequate and Representative Reserve System for Forests in Australia 1 (1997), *available at* http://www.daff.gov.au/__data/assets/pdf_file/0011/49493/nat_nac.pdf (last visited Dec. 13, 2011).

12. *Id.* at 5–6. *See also* Craig R. Groves, Drafting a Conservation Blueprint 30–33 (2003); Mark L. Shaffer & Bruce A. Stein, *Safeguarding Our Precious Heritage, in* Precious Heritage 301–22 (Bruce A. Stein, Lynn S. Kutner & Jonathan S. Adams eds., 2000).

13. Michael Scott et al., *Nature Reserves: Do They Capture The Full Range of America's Biological Diversity?* 11 Ecological Applications 999, 1004 (2001).

14. *Id.*

15. Taylor H. Ricketts et al., Terrestrial Ecoregions of North America 67 (1999).

16. Mark Anderson & Charles E. Ferree, *Conserving the Stage: Climate Change and the Geophysical Underpinnings of Species Diversity*, 5 PLoS One e11554 (2010).

17. N.E. Heller & E.S. Zavaleta, *Biodiversity Management in the Face of Climate Change: A Review of 22 Years of Recommendations*, 142 Biological Conserv. 14, 18 (2009).

18. Daniel Simberloff, James A. Farr, James Cox & David W. Mehlman, *Movement Corridors: Conservation Bargains or Poor Investments?*, 6 Conserv. Biology 493, 495–96 (1992).

19. Lenore Fahrig, *Relative Effects of Habitat Loss and Fragmentation on Population Extinction*, 61 J. Wildlife Mgmt. 603, 603 (1997).

20. Michael E. Soulé, *Theory and Strategy, in* Landscape Linkages and Biodiversity 91, 92–93 (Wendy E. Hudson ed., 1991).

21. Suzanne M. Prober & Michael Dunlop, *Climate Change: A Cause for New Biodiversity Conservation Objectives But Let's Not Throw The Baby Out With The Bathwater*, 12 Ecological Mgmt. & Restoration 2, 3 (2011).

22. Jenny A Hodgson, Chris D. Thomas, Brendan A. White & Atte Moilanen, *Climate Change, Connectivity and Conservation Decision Making: Back to Basics*, 46 J. Applied Ecology 964, 967 (2009); Laura R. Prugh, Karen E. Hodges, Anthony R.E. Sinclair & Justin S. Brashares, *Effect of Habitat Area and Isolation on Fragmented Animal Populations*, 105 Proc. Nat'l Acad. Sci. 20770, 20773 (2008).

23. Reed F. Noss, *Landscape Connectivity: Different Functions at Different Scales*, *in* Landscape Linkages and Biodiversity 27, 36 (Wendy E. Hudson ed., 1991).

24. Prugh et al., *supra* note 22, 20773; Jerry F. Franklin & David B. Lindenmayer, *Importance of Matrix Habitats in Maintaining Biological Diversity*, 106 Proc. Nat'l Acad. Sci. 349, 349 (2009).

25. Prugh et al., *supra* note 22, at 20773–20774; Franklin & Lindenmayer, *supra* note 24, at 349–50.

26. Jonathan R. Mawdsley, Robin O'Malley & Dennis S. Ojima, *A Review of Climate-Change Adaptation Strategies for Wildlife Management and Biodiversity Conservation*, 23 Conserv. Biology 1080, 1082–83 (2009).

27. C.S. Holling, *Resilience and Stability of Ecological Systems*, 4 Ann. Rev. of Ecology & Systematics 1, 14 (1973).

28. *Id*. at 15.

29. *E.g.*, Shaffer & Stein, *supra* note 12, at 310.

30. C.S. Holling, *Resilience and Stability of Ecological Systems*, 4 Ann. Rev. of Ecology & Systematics 1, 14 (1973).

31. Scott et al., *supra* note 13, at 1000.

32. Ruben N. Lubowski et al., U.S. Dep't of Agric., Economic Research Service, Major Uses of Land in the United States, 2002, at 35 (2006). U.S. Gov't Accountability Office, GAO/RCED-95-16, Endangered Species Act: Information on Species Protection on Nonfederal Lands, at 5 (1994), *available at* http://www.gao.gov/products/RCED-95-16.

33. Craig W. Thomas, Bureaucratic Landscapes 11–14 (2003).

34. *Id*. at 21, 23.

35. John B. Loomis, Integrated Public Lands Management 530–32 (2nd ed. 2002).

36. Thomas, *supra* note 32, at 52, 276.

37. *Id*. at 20.

38. *Id*. at 17, 20; Yaffee et al., *supra* note 1, at 7.

39. Yaffee et al. *supra* note 1, at 32.

40. The Federal Land Assistance, Management and Enhancement Act of 2009 Report to Congress (2011), *available at* http://www.forestsandrangelands.gov/strategy/documents/reports/2_ReportToCongress03172011.pdf (last visited Dec. 13, 2011).

41. J. Michael Scott, Dale D. Goble, Leona K. Svancara & Anna Pidgorna, *By the Numbers*, *in* The Endangered Species Act at Thirty 31 (Dale D. Goble, J. Michael Scott & Frank W. Davis eds., 2006).

42. Kieran Suckling & Martin Taylor, *Critical Habitat and Recovery, in* The Endangered Species Act at Thirty 75 (Dale D. Goble, J. Michael Scott & Frank W. Davis eds., 2006).

43. Stanford Environmental Law Society, The Endangered Species Act 134–39 (2001).

44. Sonoran Desert Conservation Plan, http://www.pima.gov/cmo/sdcp/index.html (last visited Dec. 13, 2011).

45. Here, we use that term as it was defined in one of the first pieces of legislation on eco-system services, as "the benefits that human communities enjoy as a result of natural processes and biological diversity." SB 513, 75th Leg. Assemb., Reg. Sess. (Or. 2009), *available at* http://www.leg.state.or.us/09reg/measures/sb0500.dir/sb0513.en.html.

46. Kenneth J. Arrow, *The Organization of Economic Activity: Issues Pertinent to the Choice of Market versus Non-market Allocations, in* Analysis and Evaluation of Public Expenditures: The PPP System 58 (1969).

47. Oregon Sustainability Board, Senate Bill 513 Ecosystem Services and Markets: Report from the Oregon Sustainability Board to the 2011 Legislative Assembly (2011).

48. Jessica B. Wilkinson et al., Environmental Law Society, The Next Generation of Mitigation: Linking Current and Future Mitigation Programs with State Wildlife Action Plans and Other State and Regional Plans, (2009), *available at* http://www.elistore.org/Data/products/d19_08.pdf.

49. Frank Casey, Sara Vickerman, Cheryl Hummon & Bruce Taylor, Defenders of Wildlife, Incentives for Biodiversity Conservation: An Ecological and Economic Assessment 8–9 (2006).

50. *Id.* at 10.

51. Margaret J. Wheatley, Leadership and the New Science 125 (2nd ed. 1999).

12 Valuation and Payment for Ecosystem Services as Tools to Improve Ecosystem Management

Deborah McGrath, Sewanee: The University of the South &
Travis Greenwalt, Cardno ENTRIX

Ecosystems are biological communities comprised of living organisms interacting with each other and the nonliving physical environment. Natural ecosystem processes resulting from these interactions provide and support countless goods and services enjoyed by human society. Ecosystem (or environmental) services are simply the benefits that people derive from ecosystems. At any given time, an ecosystem provides a 'flow' of diverse services, depending upon the type of ecosystem and its condition, as well as how it is managed.[1] The value of some ecosystem benefits, such as food, fiber, and fuel, is captured in financial markets and thus easily recognized. Other services are less apparent and thus taken for granted because they are of a 'public good' nature, accruing directly to humans without passing through a market exchange.[2] This is particularly true of the natural processes underlying and sustaining the provision of environmental goods and services.

Consider a square meter of Danish pasture soil in which researchers identified more than 50,000 worms, 48,000 small insects, and 10 million nematodes. A soil ecosystem such as this provides a range of critical services: the recycling of dead organic matter and breakdown and disposal of wastes; the renewal of soil fertility, physical support, and the supply of nutrients and water for plants that, in turn, buffer and moderate the hydrological cycle and provide food for animals and humans.[3] Our agricultural systems are utterly dependent upon healthy soil ecosystems and yet most of these environmental services are unrecognized and thus undervalued by society. Similarly, few consumers are aware of the many ecosystem services that contribute to the production of clean water. Consumers pay utilities or municipalities for the services of capturing, treating, and delivering water, but not for producing the water. Healthy watersheds or aquifers provide clean water through ecosystem services such as flow regulation, filtration, flood control, and protection against runoff, erosion, sedimentation, and evaporation. While the service of water purification might be partially substituted for by technology, a forested watershed provides this and a variety of other benefits, such as carbon storage, fisheries maintenance, recreation, and aesthetic fulfillment, usually at a fraction of the cost of building a water treatment facility.[4]

Although economists and environmental scientists have discussed the concept of ecosystem services for decades, private and public-sector decisions about land and resource use often ignore the value of less tangible ecosystem services because they are underpriced—or not priced at all—in markets. As a result, policies and practices that protect and enhance the flow of multiple benefits from ecosystems lag behind the degradation of these services, leading to inefficient and unsustainable resource use. This is due, in part, to 'immature markets' that fail to value environmental goods or services that are not formally bought or sold[5] and a lack of information on the ecological value of these flows that can be readily used to inform decision making. The development of common metrics for communicating the value of ecosystem services has been critical to more fully assessing the benefits and costs to society of natural resource use and management decisions. However, even when more information is available, there are trade-offs associated with managing for different ecosystem services, and often there is less incentive to manage for nonmarket environmental values.

In this chapter, we discuss two critical components for improved ecosystem management: effective communication about the value to society of environmental service flows and financial incentives to manage for nonmarket environmental values. Ecosystem service (ES) indicators condense complex information about environmental flows and their ecological status into information that can be easily communicated.[6] Economists can then utilize this information to develop monetary values for these services provided by nature. Ecosystem service valuation employs different economic approaches or methodologies in evaluating environmental services for which no market prices exist. In this manner, the valuation of ecosystem services requires a multidiscipline approach resulting in findings that can be useful for land managers, policy makers and others. The combination of environmental indicators and economic analysis methodologies can be highly effective at communicating the state of, and value derived from, ecosystem service flows. One application of the findings from this type of research is in developing incentives for protecting, conserving, or enhancing ecosystem services. Payment for Ecosystem Service (PES) programs offer a mechanism for translating nonmarket utility (benefits) derived from ecosystems into appropriate financial incentives for local actors to provide environmental services.[7] Using examples, we demonstrate how ES indicators, economic valuation, and PES programs contribute to ecosystem management that sustains and enhances the flow of environmental services.

I. ASSESSING THE MAGNITUDE AND STATUS OF ECOSYSTEM SERVICES

In 2005, the Millennium Ecosystem Assessment (MEA), a four-year study conducted by 1,300 experts from around the world, reported that 60 percent of Earth's ecosystem services had been degraded and that this destructive trend would likely accelerate through the middle of the century.[8] The MEA report outlined the vast array of benefits flowing from ecosystems to humans, dividing them into four functional categories: (i) provisioning services that directly provide products or goods (e.g., fish and timber), (ii) regulating services that provide benefits from the regulation of ecosystem processes (e.g., air quality, erosion, and pest regulation), (iii) cultural services that provide nonmaterial benefits (spiritual, aesthetic, recreation) and (iv) supporting services that sustain the provision of other services (table 1).[9]

Table 1. Examples of Environmental Benefits Provided By Provisioning, Regulating, Cultural, and Supporting Ecosystem Services.

Provisioning Services Products obtained from ecosystems	Regulating Services Benefits obtained from regulation of ecosystem processes	Cultural Services Nonmaterial benefits obtained from ecosystems
Food	Climate regulation	Spiritual & religious
Fresh water	Disease regulation	Recreation & ecotourism
Biological raw materials	Pest regulation	Aesthetic
Biochemicals & medicines	Pollination	Inspirational
Genetic resources	Water regulation	Educational
Biomass fuel	Water purification	Sense of place
	Air quality regulation	Cultural heritage

Table 1. Continued

Supporting Services **Services necessary for the production of all other ecosystem services**
Habitat provision & stabilization, Soil formation, Nutrient cycling, Primary production, Oxygen production, Waste assimilation, detoxification, & purification

Source: Adapted from Ecosystems and Human Well-Being: A Framework for Assessment[10], Measuring Natures Benefits: A Preliminary Roadmap for Improving Ecosystem Service Indicators.[11]

In 1997, the annual value of all ecosystem services on Earth (calculated by each biome) was estimated to be between 16 to 54 trillion U.S. dollars, with an average annual value of \$33 trillion, or 1.8 times global gross national product.[12] This and subsequent attempts to estimate the value of nature underscore the importance of all ecosystem services to sustaining and enriching human livelihoods and economies. Nonetheless, society generally fails to appreciate the importance of these services until we suffer the impacts of their loss. Indeed, there are numerous recent examples of situations in which ecosystem service functions have been impaired or destroyed, resulting in emergency measures to correct the problem:

- Salem, Oregon, had to declare a water emergency in 1996 after the North Santiam River flood, and was forced to install a \$1 million pretreatment facility to stabilize the sand filtration system.[13]
- Soil erosion, as a result of the Hayman fires in Colorado, caused the Denver Water Utility to undertake an emergency program to remove ash and soil from mountain reservoirs and unclog pipes at a projected cost of \$31 million.[14]
- Wetland and marsh destruction, both before and after Hurricane Katrina, has left Gulf communities more susceptible to storm surges, and reduced fish and bird habitat in the area. To restore these environmental services, islands are being 'rebuilt' by state agencies under the Coastal Wetlands Planning, Protection, and Restoration Act (CWPPRA) at a funding level of \$80 million annually, which does not begin to approach the level needed to restore the ecosystem to sustainable levels.[15]

As these examples demonstrate, restoring ecosystem services can be a very costly endeavor, and degradation of ecosystem function can have catastrophic consequences. However, while investment in ecosystem services clearly provides a wide range of economic and environmental benefits, effectively measuring these benefits is a difficult task, in part because different disciplines, philosophical views, and schools of thought conceive of the value of ecosystems differently.[16] Very often metrics assessing ecosystem function and service-provision are not easily translated across disciplines or into a dollar value. Moreover, for policy makers and economists, ecosystem service values are most usefully examined at the margin to determine how much a change in the quantity or quality of a service affects human welfare.[17]

II. ECOSYSTEM SERVICE INDICATORS

Over the last decade, the measurement of ecosystem function and service flow has become more standardized through the use of indicators intended to relay a complex message in a simplified and useful manner. Ecological indicators characterize current ecosystem status and help track or predict significant change or stress to ecosystem processes and function.[18] With baseline data, indicators can also be used to quantify how environmental flows are maintained or altered over time in response to human impact and or changes in ecosystem function. Indicators are also monitored to assess the effectiveness of measures taken to increase and enhance environmental service flows. Strong ecosystem service indicators provide a means of measuring marginal change and communicating that information to entities which can respond by altering practices and implementing policies to maintain the flow of environmental benefits.

Ecosystem service indicators are derived from many fields, including ecology, hydrology, economics, and public health, to name a few. Examples of indicators that measure ecosystem services directly include forest or soil carbon sequestration (metric tons of carbon per hectare per year), species richness (number of species per hectare), and stream flow (cubic meters per day). Indicators that serve as indices for broadly defined services, such as water purification, might include a reduction in surface water nutrient loading (kg nutrient/day) and an increase in dissolved oxygen (mg DO/L). Commonly, river sediment loads are used as a measure of erosion control and number and types of fish species per area as an indicator of fisheries' health.[19]

Methods used to measure indicators vary across scales and disciplines. For example, forest carbon sequestration can be measured as annual tree growth in inventoried plots, or through the use of aerial or satellite imagery to identify an increase in forested land cover. Economists convey the value of carbon sequestration as 'credits' or tradable certificates that can be purchased by entities desiring to offset their carbon emissions (metric tons of CO_2). Provisioning services are often characterized using economic data, such as crop production (metric tons per year), employment in forest sector (number of people), total fish catch (metric tons) and value of pharmaceuticals developed from natural systems (currency), to name a few. The value of regulatory ecosystem services is often measured indirectly, using a variety of indices. For example, the value of pollination is inferred from the monetary value of the pollinated crop and disease regulation can be measured as a reduction in new cases of a disease, such as malaria, per year.[20]

A list of forty commonly used indicators for provisioning, regulating, and cultural services are assessed in "Measuring Nature's Benefits."[21] This assessment identified several constraints to more effective use of these indicators by policy makers. For example, few strong indicators have been developed for less tangible environmental benefits, such as cultural and regulating services.[22] In some cases, data collection methods are not standardized, or long-term data sets are not available, making it difficult to compare ecosystem service flows over time. In contrast, provisioning services have numerous widely-used indicators because humans have a long history of measuring, trading, and communicating information about these services.[23] Some regulating services, such as water purification and waste assimilation, have become integrated into economic markets because they serve as substitutes for engineering projects, and in such cases, more robust indicators exist, which are also supported by better data. Ecosystem services subject to government regulation (e.g., air and water quality) also have strong indicators to monitor their status.

Another challenge is to identify indicators that convey the true value to society of 'avoided change' because successful maintenance of a regulating service does not necessarily produce a difference in the magnitude of ecosystem service delivery. In addition, a decline in environmental service delivery is not always linear because of the resilient nature of many ecosystems. Therefore, a system may suffer a great deal of degradation before service delivery is compromised, or reach a critical threshold at which service provision collapses.[24] Lagging indicators provide information about impaired ecosystem function and lost service provision after the negative consequences have occurred, but are of little use in preventing ecosystem degradation. For example, by the time a loss of species diversity is measured, the ecosystem itself has been at least partially degraded, and other environmental services have likely been lost.

Thus, while great strides have been made in characterizing the condition and magnitude of ecosystem services through the interdisciplinary use of indicators, continued efforts are needed to improve indicator strength, as well as the availability and quality of data they represent.[25] This is especially important for the economic valuation of ecosystem services. Quantifying the initial condition (or stock) of environmental services is a necessary first step to understanding the magnitude of the service provided, while tracking service flows helps predict or identify changes in the 'value' of these benefits to consumers.[26]

III. ECONOMIC VALUATION OF ECOSYSTEM SERVICES

Ecosystem valuation from an economic perspective allows society to assess the overall contribution of ecosystems to social and economic well-being.[27] Economic valuation can be used to measure lost utility resulting from ecosystem service loss, and, conversely, to demonstrate the benefits of environmental investments. It can also enhance understanding of the role of incentives in managing ecosystem services, aid in evaluating the outcomes of different management alternatives, and assist in determining conservation and restoration priorities.

Perhaps the most common school of thought for economic valuation of ecosystem services is the utilitarian approach which is based on the premise that people derive benefit (utility) from the use of ecosystem services (directly or indirectly), and are willing to exchange something for maintaining these services. Some of these services, such as the provision of commercially harvested plants and animals, are priced and traded in traditional marketplaces. However, due to the complexity associated with measuring all benefits obtained from ecosystems, economists must use both market exchange data, as well as observations of nonmarket uses from ecosystem services. For example, people also place value on ecosystem services that they are not currently using. So called nonuse, or existence, values often involve deeply held historical, national, ethnic, religious, and spiritual values that people assign to ecosystems.

Each type of benefit provided by ecosystem services can be quantified through an understanding of the nature and magnitude of the service or benefit; how conditions would change if the ecosystem was degraded or destroyed; who makes use of the service, and in what way and for what purpose, as well as alternative uses and trade-offs that exist between different services flowing from an ecosystem. Nonmarket methodologies that economists use to quantify the economic value of ecosystem services are generally divided between revealed preference methods and stated preference methods.[28] Revealed preference methods include productivity, travel costs, replacement costs, human capital, and hedonic prices and use observable data or market prices as a basis to understand value provided by ecosystem services.[29] In contrast, stated preference methods, such as contingent valuation and choice modeling, generally employ data collected from questionnaires or surveys.[30] Below we briefly describe each of these methods, along with relevant real-world examples of how these methodologies have been used in informing regulations or advising natural resource developments.

Productivity, or a change in productivity, is commonly used to estimate the value of ecosystem services, especially in water transactions and water leases throughout the Western United States. An evaluation of the contribution of irrigation water to net revenue from agricultural production can be used to determine the value of water, and is often referred to as the income capitalization method. A typical example of this is the comparison of net returns from irrigated agriculture to dry land agriculture or grazing. The margin, or change, in net return is a direct result of irrigation, attributable to the value of having access to water. For example, if a farmer earns net returns of $350 per acre for hay irrigated with 3 acre feet (AF) of water, but only $150 per acre without irrigation, the value of irrigation water can be computed as ($350-$150) / 3 AF or approximately $67 per AF annually. It should be noted that the value of water (AF) may vary, depending upon on a host of factors such as water availability, priority date of the water right, alternative uses of water in the region, and market prices for goods produced with the water.[31]

Travel Cost is used to evaluate the value of recreation benefits from ecosystem services and relies on observed behavior instead of market data. This method generally employs surveys to collect monetary and time costs of travel to a specific destination. The measurable benefit to the region can then be calculated and attributed specifically to access for recreational pursuits. In this manner, users reveal their preferences for the ecosystem service of recreation and a demand curve can be derived from observed data and actual travel costs.[32]

In 2005 the Bureau of Reclamation (BOR) was considering the acquisition of water rights in the Yakima River Basin in order to restore natural stream flows for fish habitat. Part of the BOR decision-making process included an analysis of the economic value of improved fishing in the Upper Yakima River Basin. Local anglers were surveyed to determine how much a recreational fishing trip would have to cost before they would stop taking such trips altogether. The survey results indicated that anglers would pay, on average, between $376 and $393 for food, transportation, and other expenses before they would stop fishing, which was $220 more than what they were actually spending on fishing trips at the time. The anglers were also asked if their fishing trip frequency would increase with improved stock rates. From the survey results, it was calculated that each 5 percent increase in stock rates would account for an additional 0.25 to 0.5 trips per year, up to an increase in stock rate of 30 percent. By comparing angler benefits under current and hypothetical conditions, it was possible to estimate the value to recreationists of improving fish stock rates. Moreover, the study found that a 10 percent increase in the fish stock rate in the area of interest would likely result in an increased economic benefit of over $500,000 using a 3 percent discount rate over 20 year time period.[33]

Replacement cost is a revealed preference approach that estimates the value of an ecosystem service using the cost of replacing a lost good or service with the next best alternative. The Charles River Natural Valley Storage Project is frequently cited as an example of this method.[34] In lieu of building a $100 million dam to store water and control flooding around Boston, Massachusetts, the U.S. Army Corps of Engineers spent $10 million in land and preservation easement purchases for 7,000 acres of existing wetlands to accomplish the same goal. Not only did the project save taxpayers nearly $90 million, it added benefit to adjacent land uses. Based upon replacement cost, ecosystem services provided by the 7,000 acres of wetlands in the Charles River Natural Valley Storage Project were valued at $100 million or more.[35]

Ecosystem service valuation work done on behalf of the Carson Water Subconservancy District in Nevada provides an interesting example of the replacement cost approach used to compare the value of a natural versus 'developed' floodplain. In this analysis, the replacement cost data was found nearby, in the Truckee River, where past development had resulted in the construction of manmade flood control structures in the 1950s. Significant efforts are currently underway to restore the watershed back to its natural state so that the services of the natural floodplain can be utilized once again. Restoration and rehabilitation costs on a per acre basis were formulated and compared to similar sites on the Carson River. Based on that comparison, it was estimated that the cost to replace the ecosystem services provided by natural floodplains of the Carson watershed were somewhere between $564 and $1,679 per acre on an annual basis.[36] These valuation results could provide justification for developing and implementing a new PES program to provide financial incentives to landowners who maintain their land in the floodplain for ecological benefits such as flood control.

Human capital represents another revealed preference method that may use the cost of illness to assess the value of ecosystem services by tracing impacts of environmental change on morbidity and mortality. This method is used when impaired health is the primary issue associated with management or regulation actions. This is a difficult measurement to make, because response functions linking environmental conditions to health are often lacking and the value of life cannot be estimated. Nonetheless, the use of this method is illustrated by an interesting case from the state of Washington in which a cost-benefit analysis for a proposed prohibition rule on grass seed field burning was conducted by the state's Department of Ecology. Inland grass seed growers burned their fields after harvest to clear the crop residue and promote growth for the following year. The smoke created from the field burning was blamed for a variety of respiratory health problems suffered by residents in eastern Washington, as well as for aesthetic and nuisance issues. The costs evaluated in the analysis included (i) reduced net return (productivity) for grass seed growers as a result of not being able to burn the fields, (ii) displaced workers, and (iii) the costs of administering the regulation program. The state found potential costs to be somewhere between $4 and $6 million annually. Conversely, their estimate of health care costs attributable to grass smoke was between $4 and $10 million (derived by comparing the health impacts in counties exposed to smoke from seed field burning against per capita health care costs for unexposed counties). Other ecosystem benefits in unburned fields (not used in the analysis) were identified as increased visibility, enhanced recreation and tourism, and higher desire to live in Eastern Washington. Ultimately the state of Washington used the benefits received from avoided health or human capital costs to justify the prohibition of grass seed field burning.[37]

Hedonic price is used to extract the effect of environmental factors on the price of goods that include those factors, such as the impact of ecosystem service amenities (e.g., parks and wetlands) on real estate prices. In a hedonic price model, specific attributes of a natural resource asset are isolated and evaluated for the marginal impact that they have on the economic value of the asset itself. An example of the use of a hedonic analysis comes from a previously mentioned case study involving the Charles River Natural Valley Flood Storage Project.[38] A hedonic price model of real estate in the region revealed that properties abutting the preserved wetlands had property values 1.5 percent higher than properties not abutting the wetlands.[39]

Contingent valuation is a stated preference method that asks respondents directly about their 'willingness to pay' for a specified service. One of the first contingent valuation studies to measure values that citizens have for public trust resources occurred in Los Angeles, where the State Water Resources Control Board was faced with a decision regarding the allocation of water to Los Angeles from sources flowing into Mono Lake. The reduced water flows to the lake were affecting food supplies for nesting and migratory birds. In a mail survey, California households were asked whether they would pay more on their water bill for higher cost replacement water supplies, so that natural flows could once again go into Mono Lake. Survey respondents were informed that higher flows to the lake were needed to maintain food supplies for nesting and migratory birds. Survey responses indicated that the average willingness to pay per household was $13 per month, or $156 per year. When multiplied by the number of households in the state, total benefits exceeded the costs of replacing the water supply by a factor of 50. The results of this study shaped the future debate over water from Mono Lake as well as how water was allocated. Ultimately the California Water Resources Control Board did reduce the Los Angeles water rights by half to allow for more flows into Mono Lake. Instead of debating "fish or people" it was recognized that people care about fish and birds, but also about inexpensive water supplies.[40]

Choice modeling is another stated preference method that asks respondents to choose their preferred option from a set of alternatives with particular attributes. A choice modeling survey was implemented to elicit information about whether visitors to Blue Ridge Parkway in Virginia and Tennessee prefer more hiking trails, overlook areas, roadside landscape management, or some combination of these services. In addition, the survey included a monetary attribute to estimate the value that respondents would be willing to pay to maintain the current quality of the scenic views along the Blue Ridge Parkway. The survey results indicated that 60 percent of all respondents would be willing to pay a fee of between $5 and $125 to ensure their park experience. Also, a majority of respondents were willing to pay in order to enhance their scenic experiences on the Blue Ridge Parkway. Specifically, if given a choice between the status quo experience, with no fee, or one that included an improvement in roadside and overlook scenic quality with supplemental $50 annual fee, 69 percent of the time people would choose to pay the fee to improve scenic quality.[41]

The benefits transfer method is one of the most commonly used, but least specific, approaches to economic valuation of ecosystem services. Using this approach, results from one context are transferred or applied to another context. This method provides a comparison of relevant factors that can be useful in considering the potential value of an ecosystem service. However, results from these analyses can vary wildly, and may be inaccurate due to the different factors, even when contexts seem similar.

In Kern County, California, this method was used to estimate the total annual benefit of environmental services flowing from this region. A wide range of land covers was identified in the region and economic values from nonmarket valuation studies in other areas were applied to the various land covers. There were thirteen land covers evaluated, including agriculture, forest-conifer, desert shrub, desert woodland, fresh wetland, hardwood oak, herbaceous, mixed conifer, riparian forest, shrubs, urban and barren, urban green, and open fresh water. After applying benefits from other studies related to the services supplied by each of these land covers, the total benefit of ecosystem goods and services was estimated at $2.5 billion on an annual basis.[42]

These examples demonstrate that economic valuation results in benefits measured in terms of 'dollars,' providing a common metric in evaluating the diverse array of services that are provided by ecosystems. Economic valuation of ecosystem services can aid in both setting regulation and creating incentives for land management practices that enhance or preserve environmental benefits. PES programs have been used as a mechanism for creating financial incentives to manage for these ecosystem service benefits.

IV. PAYMENT FOR ECOSYSTEM SERVICES (PES) PROGRAMS

The methodologies used for valuation of ecosystem services can aid in making land management decisions for public lands, and can also be used to justify investment in ecosystems using public monies. Payment for ecosystem service (PES) programs have taken these valuation concepts and applied them to many different ecosystem services to provide incentives for local actors to manage ecosystems for one or more environmental benefits. Specifically, PES programs are voluntary transactions where an ecosystem service provider is paid by, or on behalf of, ecosystem service beneficiaries for management practices that are expected to result in continued or improved service provisions.[43] Ecosystem service markets can be regulatory, where policies and regulations establish the demand and supply for the services, or voluntary, in which land owners and businesses are encouraged to engage in conservation activities, resource restoration, or mitigation of adverse environmental impacts.[44] PES programs can be financed publicly through government tax credits or subsidies, privately by third-party entities such as nongovernmental organizations, or through user fees by the beneficiaries themselves. Thus, one of the most important aspects of establishing a PES program is clearly defining, valuing, and measuring the ecosystem service(s) being provided to beneficiaries.[45] Once a payment mechanism has been identified, the initial condition and/or magnitude of the service provided can be established by measuring baseline ecosystem service indicators. Thereafter, the monitoring of ecosystem indicators helps verify that what the service users are paying for is being provided.

PES programs are often used in watersheds to link payment for hydrological services to consumers, generating funds for conservation, restoration, or land acquisition projects. Two of the most widely cited PES programs in the United States are found on opposite coasts, separated by a century, and involved a much different approach to elicit the same response: watershed protection for the procurement of water. The Great Seattle Fire of 1889 destroyed the entire downtown business district and exposed the inadequacy of Seattle's water supply system. Shortly after the fire, voters approved bonds to finance construction of the Cedar River system, and city leaders formulated a plan to acquire all land surrounding the entire 100,000-acre watershed and thus control all activities on those lands. Their plan was finally realized a hundred years later in the 1990s, securing Seattle's drinking water supply for future generations.[46]

New York City was found to be out of compliance with the EPA standards for clean water and faced cleaning up their water supply by constructing and operating a new filtration plant, which required investing $6 billion in capital and $300 million in annual operating costs. Instead, the city worked with landowners in the Catskill and Delaware watersheds, investing $1.5 billion in incentives for landowners to protect the watershed and upgrade infrastructure in the watersheds.[47] These examples show both collaboration with landowners and the use of land acquisition policies; however, the objective of both projects was to protect and secure future water supplies using PES programs. Unfortunately, as often is the case, these PES programs were spurred by a catastrophic event that revealed the insecurity of the ecosystem services provided by watersheds. Numerous examples exist where actions to protect ecosystem services with a PES program have occurred only after similar events expose the vulnerability of the ecosystem service in question.

Today, many agencies, municipalities, and conservation organizations are proactively taking measures to ensure that ecosystem services are protected into the future. One example is Santa Fe, New Mexico, which is taking progressive measures to protect their water supply against future threats through watershed restoration. The main concern is protecting the watershed against erosion and sedimentation from uncontrolled forest fires, which would drastically reduce the quality, quantity, and longevity of municipal water supplies, as exemplified by the 2000 Cerro Grande forest fire around the neighboring town of Los Alamos.[48] The combined cost of reducing fuel loads in the Santa Fe watershed by removing small diameter trees (for which there is low market demand) and controlled burning ranges from $1,000 to $2,000 per acre. The cost to retain the restored forest condition over 20 years would be $4.3 million. In contrast, the avoided cost associated with a 7,000-acre fire in the watershed was estimated to be $22 million, with the likelihood of 1 in 5 that such a fire would occur during any given year.[49] Thus, the Santa Fe Municipal Watershed Plan 2009–2029 uses PES to develop a local sustainable source of funding that accounts for the true costs of watershed maintenance and water procurement. After a 5-year period of public outreach and education about the value of services supplied by a healthy watershed, water users will pay a separate fee on their utility bill for watershed maintenance, estimated to cost anywhere from $3.13 to $9.40 per year, depending upon household water consumption. In addition, the plan would monitor indicators, such as water temperature and chemistry, stream flow, and reservoir level, to quantify the impact of watershed maintenance on water quality and quantity and assure users that they are receiving the services they are paying for.[50]

Another example of a community addressing the protection of ecosystem services on a proactive basis is Sonoma County, California. In 1990, the public voted to create the Agricultural Preservation and Open Space District to preserve and protect the working farms and ranches, scenic hillsides, and natural areas of Sonoma County. The district does this through providing funding to landowners for conservation easements and acquiring ecologically sensitive lands that are then turned over to the park systems to maintain for open space and public access. The Agricultural Preservation and Open Space District is funded by a quarter cent sales tax, and to date has permanently preserved more than 85,000 acres.[51] The district is currently developing a framework whereby ecosystem service values are used as a guide in planning and leveraging current investments into land preservation.[52]

PES programs aimed at protection and conservation of ecosystem services are not restricted to the local level. The federal government has many programs that provide incentives to landowners to manage for different environmental services, such as the Wildlife Habitat Incentive Program (WHIP) through the Natural Resource Conservation Service (NRCS); the Conservation Reserve Program (CRP) through the United States Department of Agriculture (USDA), and the Environmental Quality Incentives Program (EQIP), also through NRCS. These are just a few of the options available to landowners that involve payment for certain environmental measures or conservation practices.

V. SUMMARY

As the examples highlighted in this chapter demonstrate, evaluation of ecosystem service indicators and the use of economic valuation methods serve numerous purposes, including setting priorities for public land management and educating the public about how investment into conservation and protection of ecosystem services can lead to economic and social utility. The first step in the process of placing a value on ecosystem services is to have an understanding of the ecological status and magnitude of the service(s) provided. Meaningful ecosystem service indicators communicate such information to landowners, policy makers, and other stakeholders in terms that are intuitive, easily comparable, and supported by strong data.

The various approaches to economic valuation require different levels of precision and effort in the analysis. Furthermore, some methodologies may be more appropriate for targeted services than for others. As new tools for economic valuation continue to emerge, society is increasingly recognizing the importance of ecosystem services to livelihoods, cultures, and economies. As a result, economic valuation methodologies are more commonly used in determining strategies for land use management. Of course, there are limitations in the use of valuation of ecosystem services. Economic valuation studies are sensitive to both the methodology implemented and the assumptions used in the analysis. For example, market data is generally readily available for products that are exchanged (mainly provisioning services), but focusing solely on these services when a suite of ecosystem services is provided will likely omit important nonmarket and nonuse values.[53] Conversely, the valuation of nonmarket or nonuse values often overstates the true 'willingness to pay' value of an ecosystem service.

PES programs have proven to be a useful application of economic nonmarket values provided by nature as well as environmental indicators. A properly designed PES program will provide appropriate financial incentives for managing natural resources in ways that maintain, enhance, or preserve ecosystem support services. Furthermore, bundling and stacking multiple environmental services together can provide additional incentive for landowners to manage for and enhance the flows of ecosystem benefits.[54]

NOTES

1. Millennium Ecosystem Assessment, Ecosystems and Human Well-Being: A Framework for Assessment 30 (2003) [hereinafter Ecosystems].

2. Robert Costanza et al., *The value of the world's ecosystem services and natural capital*, 387 Nature 253–60 (1997).

3. James Salzman, *Designing Payments for Ecosystem Services*, 48 PERC Policy Series 2–3 (2010).

4. Ecosystems, *supra* note 1, at 30; Travis Greenwalt & Deborah McGrath, *Protecting the City's Water: Designing a Payment for Ecosystem Services Program*, 24 Nat'l Res. & Env't 9 (2009).

5. Salzman, *supra* note 3, at 7, 30.

6. Ecosystems, *supra* note 1, at 30; Christian Layke, *Measuring Nature's Benefits: A Preliminary Roadmap for Improving Ecosystem Service Indicators*, 35 (World Res. Inst., Working Paper, September 2009), *available at* http://www.wri.org/project/ecosystem-service-indicators.

7. Stefani Engel, Stefano Pagiola & Sven Wunder, *Designing payments for environmental services in theory and practice: An overview of the issues*, 65 Ecological Economics 663, 664–65.

8. Millennium Ecosystem Assessment, Ecosystems & Human Well-Being: Synthesis 6–11 (2005).

9. *Id.* at 38–48.

10. Ecosystems, *supra* note 1, at 6.

11. Layke, *supra* note 6.

12. Costanza et al, *supra* note 2, at 253.

13. U.S. Gov't Accountability Office, GAO-98-220, Resources, Community and Economic Development: Oregon Watershed: Many Activities Contribute to Increased Turbidity During Large Storms 33 (1998).

14. Greenwalt & McGrath, *supra* note 4, at 10.

15. U.S.G.S. National Wetlands Research Center, About CWPPRA, Coastal Wetlands Planning, Protection, and Restoration Act, http://lacoast.gov/new/About/Default.aspx (last visited Oct. 9, 2011).

16. Ecosystems, *supra* note 1, at 30.

17. Costanza et al., *supra* note 2.

18. Office of Research and Development, U.S. Envt'l. Prot. Agency EPA/620/R-99/005, Evaluation Guidelines for Ecological Indicators (Laura E. Jackson, Janis C. Kurtz & William S. Fisher eds., 2000).

19. Janet Ranganathan et al., World Resources Institute, Ecosystem Services: A Guide for Decision Makers 34–36 (2008), *available at* http://www.wri.org/publication/ecosystem -services-a-guide-for-decision-makers.

20. Layke, *supra* note 6, at 30.

21. *Id.* at 14–17.

22. *Id.*

23. *Id.* at 18.

24. *Id.* at 21.

25. *Id.* at 25–26.

26. *Id.* at 5.

27. Ecosystems, *supra* note 1, at 30.

28. Unai Pascual & Roldan Muradian et al., *The Economics of Valuing Ecosystem Services and Biodiversity*, March 2010 version *available at* http://www.teebweb.org/LinkClick.aspx?fileticket =JUukugYJHTg%3D&tabid=1018&language=en-US; Dennis M. King & Marisa J. Mazzotta,

Dollar-based Ecosystem Valuation: Contingent Valuation Method, Section 6, Ecosystem Valuation, http://www.ecosystemvaluation.org/contingent_valuation.htm (last visited Oct. 12, 2011). James Salzman, *supra* note 3.

29. *Id.*

30. *Id.*

31. Brandon Scarborough & Hertha Lund, Saving Our Streams 27 (2007).

32. Unai & Muradian et al., *supra* note 28; King & Mazzotta, *supra* note 28.

33. Northwest Economic Associates, Economic Benefits of Improved Instream Flow in the Upper Yakima Basin: Taneum Creek Report, prepared for Bureau of Reclamation, Upper Columbia Area Office, August 11, 2005.

34. Jim Morrison, *How Much Is Clean Water Worth?*, National Wildlife, Feb. 1, 2005, at 43, *available at* http://www.nwf.org/News-and-Magazines/National-Wildlife/News-and-Views/Archives/2005/How-Much-Is-Clean-Water-Worth.aspx.

35. *Id.*

36. ENTRIX, Inc. & ENVIRON, Carson Water Subconservancy District, Floodplain Ecosystem Services Valuation for Carson Valley (2010).

37. Washington State Dep't of Ecology, Estimates of Costs of the Rule to Certify Alternatives to Grass Field Burning (May 21, 1998), *available at* http://www.ecy.wa.gov/programs/air/pdfs/cba.pdf.

38. *See* Morrison, *supra* note 34.

39. *Id.*

40. King & Mazzotta, *supra* note 28.

41. Leah Greden Mathews et al., *Using Economics to Inform National Park Management Decisions: A Case Study on the Blue Ridge Parkway, in* Crossing Boundaries in Park Management, Proceedings of the 11th Conference on Research and Resource Management in Parks and on Public Lands 326–31 (David Harmon ed., 2011), *available at* http://www.georgewright.org/56mathew.pdf.

42. Jay E. Noel, Eivis Qenani-Petrela & Thomas Mastrin, A Benefit Transfer Estimation of Agro-Ecosystems Services, Western Economics Forum (2009).

43. Salzman, *supra* note 3, at 6–7.

44. Gina L. LaRocco & Robert L. Deal, U.S. Dep't of Agric., PNW-GTR-842, Giving Credit Where Credit is Due: Increasing Landowner Compensation for Ecosystem Services 32 (2011).

45. Greenwalt & McGrath, *supra* note 4, at 9–13.

46. Sandra L. Postel & Barton H. Thompson, Jr., *Watershed Protection: Capturing the Benefits of Nature's Water Supply Services* 29 Nat'l Res. Forum 98 (2005).

47. *Id.*

48. Greenwalt & McGrath, *supra* note 4, at 10.

49. Santa Fe Watershed Association, Santa Fe Municipal Watershed 20-Year Protection Plan 2010–2029 (Feb. 18, 2009), http://www.santafewatershed.org/index.php?option=com_content&task=view&id=55&Itemid=74 (last visited Oct. 12, 2011).

50. *Id.*

51. Sonoma County Agricultural Preservation and Open Space District, Protected Lands, http://www.sonomaopenspace.org/Content/10134/view_by_area.html.

52. The District, Sonoma County Agricultural Preservation and Open Space, http://www.sonomaopenspace.org/Content/10142/district.html (last visited Oct. 12, 2011).

53. Ranganathan et al., *supra* note 19.

54. LaRocco & Deal, *supra* note 44.

Index